OCR

Mathematics

for GCSE B

Foundation Silver/Higher Initial
Foundation Gold/Higher Bronze

Series Editor:
Brian Seager
Howard Baxter
Mike Handbury
John Jeskins
Jean Matthews
Colin White

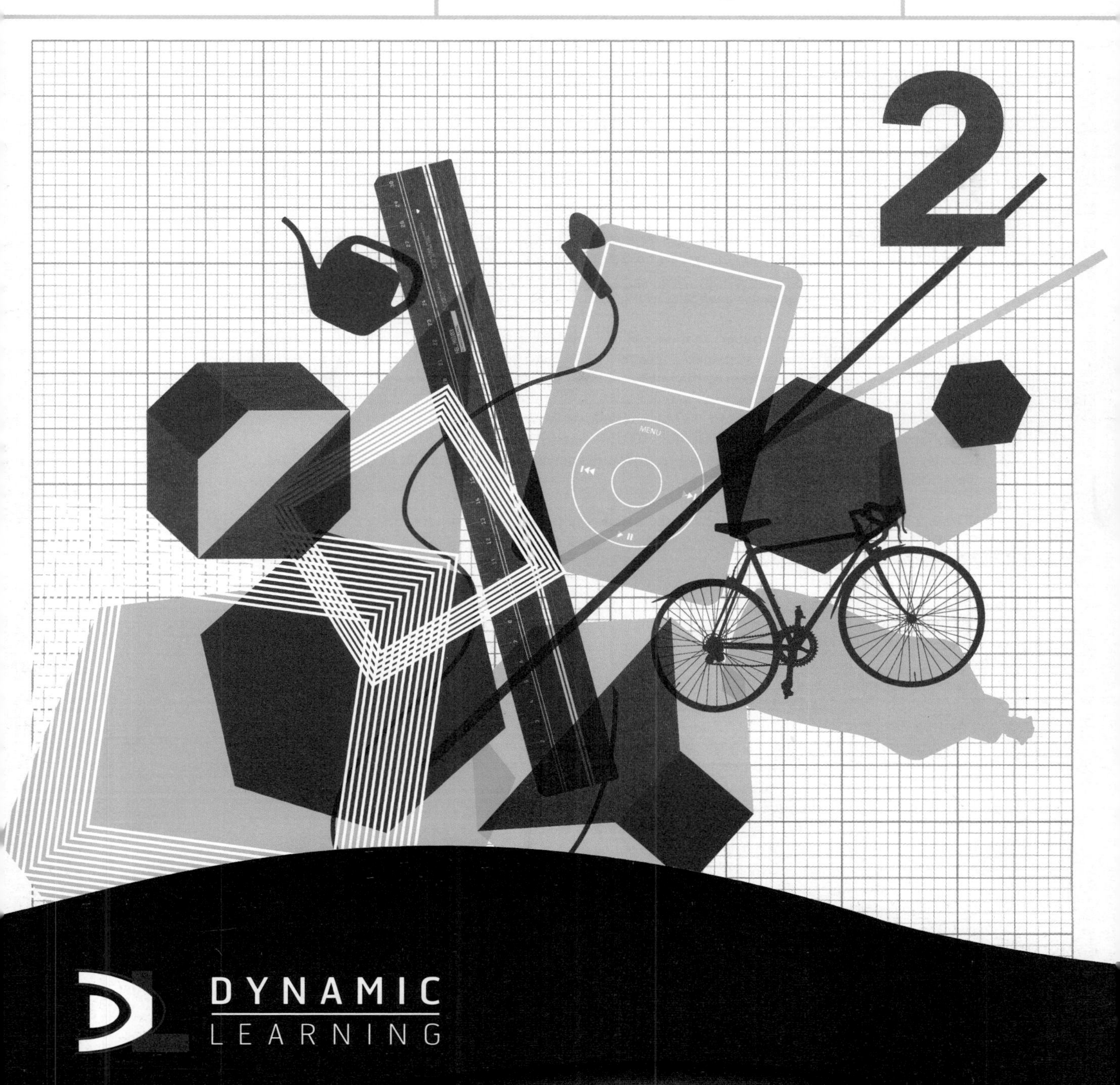

DYNAMIC
LEARNING

The Publishers would like to thank the following for permission to reproduce copyright material:

Photo credits p93 © Dennis Hallinan/Alamy; p94 © David L. Moore/Alamy; p101© F1online digitale Bildagentur Gmbtt/Alamy; p140 © www.purestockX.com; p269 © Horizon International Images Limited/Alamy

Orders: please contact Bookpoint Ltd, 130 Milton Park, Abingdon, Oxon OX14 4SB. Telephone: (44) 01235 827720. Fax: (44) 01235 400454. Lines are open 9.00 – 5.00, Monday to Saturday, with a 24-hour message answering service. Visit our website at www.hoddereducation.co.uk

First published in 2010 by
Hodder Education
An Hachette UK Company
338 Euston Road
London NW1 3BH

Impression number 5 4 3 2 1
Year 2015 2014 2013 2012 2011 2010

Cover illustration by Crush Design
Typeset in Avenir 35 Light 10/12pt by Pantek Arts Ltd, Maidstone, Kent
Printed in Italy

A catalogue record for this title is available from the British Library

ISBN: 978 1444 11851 3

Contents

Contents

About this book

This book covers part of the specification for GCSE Mathematics. It has been written especially for students following OCR's 2010 Specification GCSE Mathematics B (J567). This book covers two stages, Foundation Silver / Higher Initial and Foundation Gold / Higher Bronze.

There are three important elements in the new specifications: **problem solving**; elements of **functional mathematics** and **quality of written communication**. All of these are covered in this book.

- Each chapter is presented in a way which will help you to understand the mathematics, with straightforward explanations and worked examples covering every type of problem. These are models of **written communication** for you to follow.
- At the start of each chapter there are two lists, one lists topics that you should already know about before you begin and the other lists the topics you will be learning about in the chapter.
- 'Activities' offer another way to approach the core content, and give opportunities for you to develop your skills.
- 'Challenges' are more searching and designed to make you think mathematically and enhance your **problem solving** skills.
- There are plenty of exercises to work through to practise your skills. Many of the questions require **problem solving**. These are indicated by ▶ Many exercises include questions which involve elements of **functional** mathematics.
- Some questions are designed to be done without a calculator so that you can practise for the non-calculator paper. These are indicated by
- Look out for the 'Exam tips' – these give advice on how to improve examination performance, direct from the experienced examiners who have written this book.
- At the end of each chapter there is a short summary of what you have learned.

Other components in the series

- **A Homework Book**

  This contains parallel exercises to those in this book to give you more practice.
- **Dynamic Learning online**

  This offers interactive online assessment and access to audio-visual worked examples to help you to understand fully key concepts. Access to online versions of the textbooks and homework books is also available.

Top ten tips

Here are some general tips from the examiners who wrote this book to help you to do well in the examinations.

Practise

1. taking time to work through each question carefully.
2. answering questions without a calculator.
3. answering questions which require explanations.
4. answering unstructured questions.
5. accurate drawing and construction.
6. answering questions which need a calculator and using it efficiently.
7. checking answers, especially for reasonable size and degree of accuracy.
8. making your work concise and well laid-out.
9. checking that you have answered the question.
10. rounding numbers, but only at the appropriate stage.

CHAPTER 1

Using a calculator efficiently

YOU WILL LEARN ABOUT	YOU SHOULD ALREADY KNOW
○ The order of operations ○ Using special function keys on your calculator ○ Interpreting and rounding the answer	○ How to use a calculator to do simple calculations ○ How to find the mean of a set of numbers ○ How to round decimal numbers using decimal places

Order of operations

If you are asked to work out 2 + 3 × 4, without seeing it written down, what answer will you give?

Some people will give the answer 20 because they work from 'left to right' that is, 2 + 3 = 5 then multiply the 5 by 4. However, 20 is the wrong answer according to the rules of arithmetic.

If you enter 2 + 3 × 4 = on your calculator, you should get the correct answer of 14 because nearly all calculators follow the rules of arithmetic and do the multiplication first, that is 3 × 4 = 12, and then add 2.

Here is the correct order of operations when carrying out a calculation.

- Work out brackets first
- Then powers (such as squares)
- Then multiplication and division
- Lastly addition and subtraction

Example 1	Solution
Find the value of $3x^2$ when $x = 6$.	The value $= 3 \times 6^2$ $= 3 \times 36$ Work out the power before doing the multiplication. $= 108$

Example 2	Solution	
Without a calculator, work out $(6 - 2)^2 + 5 \times 3$	$(6 - 2)^2 + 5 \times 3$	
	$= 4^2 + 5 \times 3$	Work out the brackets first.
	$= 16 + 5 \times 3$	Then the powers.
	$= 16 + 15$	Then the multiplication.
	$= 31$	And lastly the addition.

When doing these calculations for yourself, you may do some of the steps mentally and not write them all down, but you need to make sure you keep the correct order of operations.

Activity 1

Using only + and × and the given numbers, try to make a sum to give the totals required.

a) Use 3, 5 and 6 to get **(i)** 33. **(ii)** 23.
b) Use 17, 11 and 5 to get **(i)** 72. **(ii)** 96.

Activity 2

Use a calculator to work out these.

a) $13.75 + 2.25 \times 3$
b) $14.6 + 3 \times 9.4$

Check that you get 20.5 and 42.8 as the answers.

Exercise 1.1

Work out these without a calculator.

1 $2 \times 5 + 4$

2 $2 \times (5 + 4)$

3 $2 + 5 \times 4$

4 $2 + 5^2$

5 5×4^2

6 $13 - 2 \times 5$

7 $(13 - 2) \times 5$

8 2×3^3

9 $(2 \times 5)^2$

10 $(5 - 3) \times (8 - 3)$

11 $6 + \frac{8}{2}$

12 $\frac{6 + 8}{2}$

13 $\frac{12}{4 \times 3}$

14 $\frac{12}{4} \times 3$

15 $\frac{6+4}{5} - 2$

16 $\frac{20}{5 + 3}$

17 $\frac{20}{5} + 3$

18 $3 \times 5 - \frac{6 \times 4}{8}$

19 $(20 - 2 \times 6)^2$

20 $12 - 6 \times 3$

21 $4 \times 3 - 5 \times 4$

22 3×2^3

Use a calculator to work out these.

23 $23.4 + 4 \times 5$

24 $19 - 4 \times 2.9$

25 $19.8 - 3.2 \times 4$

26 $25.5 - 5 \times 4.9$

27 $13.75 - 2.25 \times 3$

28 $37.8 + 12 \times 1.9$

29 $14.6 + 4 \times 1.9$

30 $100 + 2.2 \times 100$

31 $9.8 + 9.8 \times 9.8$

32 $1000 - 25 \times 25$

Changing the order of operations

If you want to change the normal order of doing things you need to give your calculator different instructions.

For instance, to get the calculator to do the addition first in working out $2 + 3 \times 4$, you can press the [=] button in the middle of the calculation.

[2] [+] [3] [=] [×] [4] [=]

Another way is to use brackets.

[(] [2] [+] [3] [)] [×] [4] [=]

You can use brackets to make more complicated calculations straightforward.

Example 3

Use a calculator to work out this.

$$\frac{9.7 + 4.6}{3 \times 2}$$

Solution

$\frac{9.7 + 4.6}{3 \times 2}$ You need to do the addition and the multiplication before you do the division. You tell your calculator to do this by using brackets.

These are the keys to press.

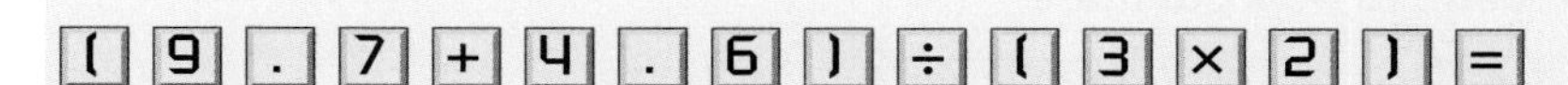

Try the calculation on your calculator and check that you get the answer 2.383 (to 3 decimal places).

Exam tip

The fraction line means you must divide all of the numerator by all of the denominator, so don't forget to use brackets.

Exercise 1.2

Use a calculator to work out these.

1 $^{-}4.73 + 2.96 - 1.71 + 3.62$

2 $^{-}14.7 + 6.92 - 1.41 - 2.83$

3 $(^{-}4.6 \times 7.2) + (3.1 \times {}^{-}4.3)$

4 $(^{-}1.2 \times {}^{-}2.4) - (9.2 \times {}^{-}3.6)$

5 $\frac{^{-}4.7 + 2.6}{^{-}5.7}$

6 $\frac{^{-}4.72}{^{-}1.4} \times \frac{8.61}{^{-}7.21}$

7 $\frac{7.92 \times 1.71}{^{-}4.2 + 3.6}$

8 $\frac{3.14 - 8.16}{^{-}8.25 \times 3.18}$

9 $\frac{5.2 + 10.3}{3.1}$

10 $\frac{127 - 31}{25}$

11 $\frac{9.3 + 12.3}{8.2 - 3.4}$

12 $6.2 + \frac{7.2}{2.4}$

13 $2.8 \times (5.2 - 3.6)$

14 $\frac{5.3}{2.6 \times 1.7}$

15 $\frac{5.3}{2.6 + 1.7}$

Special function keys

Some of the more common operations you will be asked to do use powers (the ^ button) and square roots (the √ button).

Example 4

Use a calculator to evaluate 4.2^3.

Solution

$4.2^3 = 74.088$
$= 74.1$ (to 1 decimal place)

This can be done by keying 4.2 × 4.2 × 4.2, but using the ^ button is quicker.

These are the keys to press.

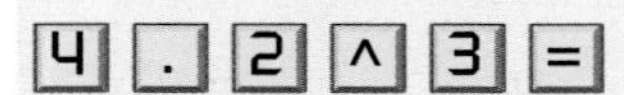

Your calculator may have a cube button. Look for a button marked x^3.

Example 5

Use a calculator to work out these.

a) $\dfrac{14.73 + 2.96}{15.25 - 7.14}$ **b)** $\sqrt{17.8^2 + 4.3^2}$

Solution

a) $\dfrac{14.73 + 2.96}{15.25 - 7.14} = 2.1812...$
$= 2.18$ (to 2 d.p.)

There are various ways to do this. One is to press these keys.

1 4 . 7 3 + 2 . 9 6 =

÷ (1 5 . 2 5 − 7 . 1 4) =

b) $\sqrt{17.8^2 + 4.3^2} = 18.312...$
$= 18.3$ (to 1 decimal place)

One way to do this is to press these keys.

√ (1 7 . 8 x^2 + 4 . 3 x^2) =

Another way of working is to press these keys.

1 7 . 8 x^2 + 4 . 3 x^2 =

And then press these keys.

√ Ans =

The Ans key uses the last result your calculator worked out.

Accuracy

If you work out $\sqrt{6.8}$ on your calculator the display will show 2.607 680 9… . (The number of decimal places will be different on some calculators.)

You will probably not need to give so many decimal places in your answer. Usually rounding your answer to 2 or 3 decimal places will be accurate enough. The question may tell you how accurately to give your answer.

$2.63^2 = 6.92$ or 6.917 is probably accurate enough.

$\sqrt{6.8} = 2.61$ or 2.608 is probably accurate enough.

Only round your final answer. Do not use rounded numbers in a calculation.

Exam tip

If you round a number, always state the number of decimal places (d.p.), you have rounded to. For example, 2.61 correct to 2 d.p.

If you have rounded a number for an answer to part of a question and need to use that number in another part of the question, always use the unrounded number.

It is a good idea to keep an answer in your calculator until you know that you don't need to use it again.

Exercise 1.3

1 Work out these.

a) 21^3
b) 9^6
c) 3.8^4
d) 0.4^4
e) 7.31^5

2 Work out these.

a) $7^2 + 14^2$
b) $43.73^3 + 17.1^2$

3 Work out these.

a) $(6.2 + 1.7)^2$
b) $(9.2 + 15.3)^2$

4 Work out these square roots. Give your answers correct to 2 decimal places.

a) $\sqrt{17.32}$
b) $\sqrt{29.8}$
c) $\sqrt{88}$
d) $\sqrt{567}$
e) $\sqrt{2348}$

5 Work out these.

a) $\sqrt{1.2^2 + 0.5^2}$

b) $\sqrt{15.7 - 3.8}$

c) $\dfrac{2.6^2}{1.7 + 0.82}$

d) $\dfrac{6.2 \times 3.8}{22.7 - 13.8}$

e) $\dfrac{5.3}{\sqrt{6.2 + 2.7}}$

f) $\dfrac{5 + \sqrt{25 + 12}}{6}$

g) $2.7^2 + 3.6^2 - 2 \times 2.7 \times 3.6 \times 0.146$

h) $\dfrac{6.2}{2.6} + \dfrac{5.4}{3.9}$

i) $\dfrac{2.6 + 4.25}{7.8 \times 3.6^2}$

6 The area of a square is $480\,\text{cm}^2$. Find the length of a side of the square. Give your answer to 1 decimal place.

7 The side of a cube is 12 cm. Calculate the volume of the cube.

Calculating with time and other measures

You need to be able to use your calculator to answer questions involving measures. This is straightforward with measures such as length and weight because they use a decimal system, provided that you take care with the units.

With time, however, you need to remember that there are 60 seconds in a minute and 60 minutes in an hour – time is not a decimal system. You may find it easier to add the minutes and seconds separately.

Example 6

The weights of eight bags of potatoes are as follows.

5 kg 800 g	5 kg 200 g	6 kg 500 g	4 kg 900 g
4 kg 750 g	6 kg 100 g	5 kg 600 g	5 kg

Find the mean weight.

Solution

Convert the weights to decimals as you enter them into your calculator.

$$\frac{5.8 + 5.2 + 6.5 + 4.9 + 4.75 + 6.1 + 5.6 + 5}{8}$$

$$= \frac{43.85}{8}$$

= 5.481 kg (to 3 d.p.) or 5 kg 481 g (to the nearest gram)

Example 7

The times it took four children to run a relay race were as follows.

9 minutes 10 seconds	9 minutes 43 seconds
9 minutes 49 seconds	9 minutes 53 seconds

Find the total time the race took.

Solution

The total time = (9 + 9 + 9 + 9) minutes + (10 + 43 + 49 + 53) seconds
= 36 minutes + 155 seconds
= 36 minutes + 2 minutes + 35 seconds
= 38 minutes 35 seconds

Example 8

a) Convert 8 hours 36 minutes into hours.

b) Write 8.4 hours in hours and minutes.

Solution

a) 8 hours 36 minutes is $8\frac{36}{60}$ hours = 8.6 hours

b) 8.4 hours = 8 hours and 0.4 × 60 minutes = 8 hours 24 minutes.

Activity 3

Some calculators have a [° ' "] button. You can use this for working with time on your calculator.
When you do a calculation and the answer is a time in hours but your calculator shows a decimal, you change it to hours and minutes by pressing [° ' "] and then the [=] button.
To change a time in hours and minutes back to a decimal time, use the [SHIFT] key.
The [SHIFT] key is usually the top left button on your calculator, but it might be called something else.
To enter a time of 8 hours 32 minutes on your calculator, press this sequence of keys.

The display should look like this.

8°32°0

You may wish to experiment with this button and learn how to use it to enter and convert times.

Exercise 1.4

1 Find the total weight of eight sacks of gravel with these weights.

25 kg 750 g 26 kg 800 g
30 kg 7 g 29 kg 85 g
29 kg 430 g 28 kg 106 g
26 kg 200 g 25 kg 400 g

2 These are the volumes of petrol put into a moped on six visits to the garage in March.

2 litres 350 ml 3 litres 100 ml
2 litres 930 ml 5 litres 255 ml
3 litres 700 ml 750 ml

Find the total volume of petrol bought in March.

3 Here are the times that the first seven patients spent with a doctor one morning between 9 am and 10 am.

3 minutes 20 seconds
5 minutes 35 seconds
1 minute 3 seconds
4 minutes exactly
2 minutes 50 seconds
4 minutes 40 seconds
6 minutes 25 seconds

a) What was the total time?
b) How much time did the doctor have left before 10 am?

4 These were the times that Chris was stuck in traffic on six days last week.

38 minutes 20 seconds
41 minutes 10 seconds
40 minutes 40 seconds
39 minutes 39 seconds
43 minutes 32 seconds
42 minutes 18 seconds

How long was Chris stuck in traffic altogether?

5 Find the mean weight of four suitcases with these weights.

36 kg 600 g 29 kg 840 g
43 kg 130 g 44 kg 436 g

6 The sign in a lift states 'Maximum weight 800 kg'.
There are five people in the lift with these weights.

78 kg 74 kg 100 g
70 kg 350 g 69 kg 850 g
72 kg 360 g

How many kilograms short of the maximum weight is their total weight?

7 The total weight of seven girls is 374 kg and the total weight of five boys is 314 kg.
Calculate the mean weight of all the children.

8 Tony weighs his three cats.
The weights are 5.3 kg, 6.2 kg and 6.1 kg.
Find the mean weight.

9 Six artists estimated the length of a canvas.
Here are their estimates.

1 m 25 cm	1 m 40 cm
1 m 12 cm	1 m 30 cm
1 m	1 m 10 cm

Find the mean of their estimates.

10 Jodi keeps a record of how long it takes her to drive from her home to work each morning for a week.
Here are the times.

45 minutes	36 minutes
34 minutes	42 minutes
32 minutes	

Find the mean time for her journey.

11 Convert these times to hours.
- **a)** 4 hours 15 minutes
- **b)** 7 hours 30 minutes
- **c)** 1 hour 50 minutes
- **d)** 3 hours 20 minutes
- **e)** 45 minutes

12 Write these times in minutes and seconds.
- **a)** 4.7 minutes
- **b)** 3.75 minutes
- **c)** 5.25 minutes
- **d)** 25.3 minutes
- **e)** 0.4 minutes

Working to a sensible degree of accuracy

Measurements and calculations should always be expressed to a suitable degree of accuracy. For example, it would be silly to say that a car journey took 4 hours 46 minutes and 13 seconds, but sensible to say that it took four and three-quarter hours, or about five hours. In the same way, saying that the distance the car travelled was 93 kilometres 484 metres and 78 centimetres would be giving the measurement to an unnecessary degree of accuracy. It would more sensibly be stated as 93 km.

As a general rule the answer you give after a calculation should not be given to a greater degree of accuracy than any of the values used in the calculation.

Example 9

Ben measured the length and width of a table as 1.8 m and 1.3 m.

He calculated the area as $1.8 \times 1.3 = 2.34\,\text{m}^2$.

How should he have given the answer?

Solution

Ben's answer has two places of decimals (2 d.p.) so it is more accurate than the measurements he took.

His answer should be $2.3\,\text{m}^2$.

Exercise 1.5

1 Write down sensible values for each of these measurements.

a) 3 minutes 24.8 seconds to boil an egg.
b) 2 weeks, 5 days, 3 hours and 13 minutes to paint a house.
c) A book weighing 2.853 kg.
d) The height of a door is 2 metres 12 centimetres and 54 millimetres.

2 Work out these and give the answers to a reasonable degree of accuracy.

a) Find the length of the side of a square field with area 33 m^2.
b) The length of a field is given as 92 m, correct to the nearest metre, and the width is given as 58.36 m, correct to the nearest centimetre. Calculate the area of the field.
c) Three friends win £48.32. How much will each receive?
d) A book has 228 pages and is 18 mm thick. How thick is Chapter 1 which has 35 pages?
e) It takes 12 hours to fly between two cities on an aeroplane travelling at 554 km/h. How far apart are the cities?
f) The total weight of 13 people in a lift is 879 kg. What is their average weight?
g) The length of a strip of card is 2.36 cm, correct to 2 decimal places, and the width is 0.041 cm, correct to 3 decimal places. Calculate the area of the card.
h Last year a delivery driver drove 23 876 miles. Her van travels an average of 27 miles to the gallon. Diesel costs 92p per litre and 1 gallon equals 4.55 litres. Calculate the cost of the fuel used.

Key Ideas

- The correct order of operations is brackets, powers, multiplication and division then addition and subtraction.
- You can use the brackets keys [( ] [ )] on your calculator to make calculations more straightforward.
- You should know how to use the power key [^] and the square root key [√] on your calculator.
- When working with time, remember there are 60 seconds in a minute and 60 minutes in an hour.
- When working with time on your calculator, think carefully about the units. For example, when working in hours, 6.25 relates to 6.25 hours and not 25 past 6.
- The accuracy of the result of a calculation depends on the accuracy of the measurements used.

CHAPTER 2 Brackets and factors

YOU WILL LEARN ABOUT	YOU SHOULD ALREADY KNOW
o Expanding brackets o Finding common factors	o How to calculate with negative numbers o How to calculate squares of numbers o How to substitute numbers for letters

Brackets

What is $2 \times 3 + 4$? Is it 14? Is it 10?

The rule is 'do the multiplication first', so the answer is 10.

If you want the answer to be 14, you need to add 3 and 4 first.

You use brackets to show this $2 \times (3 + 4)$.

Notice that this is equal to $2 \times 3 + 2 \times 4$.

Activity 1

You may have used brackets when finding the area of a shape and/or when learning about multiplication.

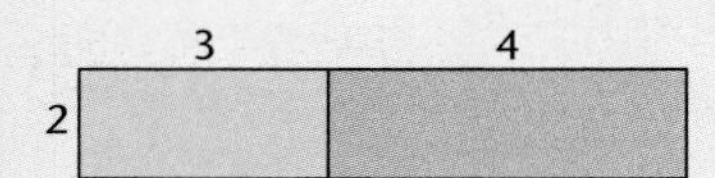

For example

The area is $2 \times (3 + 4) = 2 \times 3 + 2 \times 4$

$= 6 + 8$

$= 14$.

In the same way, write down the area of each of the following shapes using brackets and then find the area of the whole shape.

a)

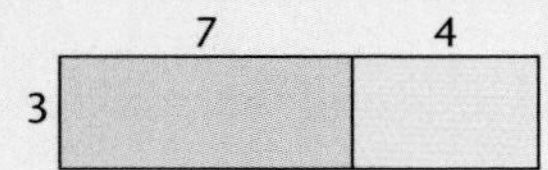

b)

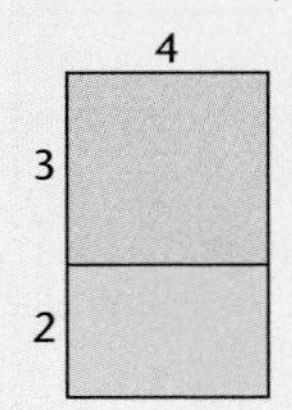

c)

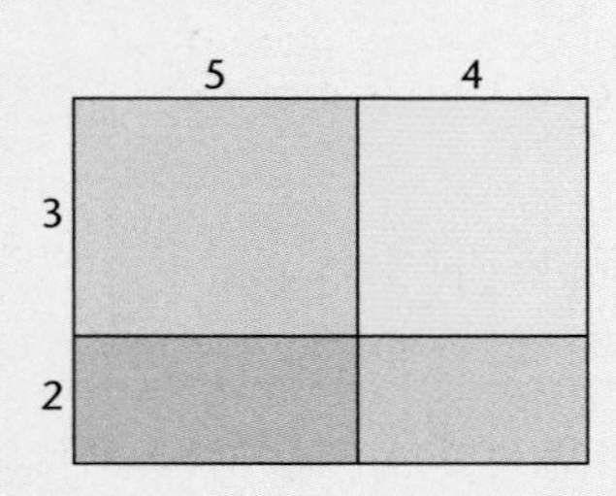

d)

b *c*

a

Activity 2

Draw a diagram to show each of these expressions.

a) $4(6 + 2)$ **b)** $a(c + d)$ **c)** $y(y + 1)$

> **Exam tip**
> Remember to multiply each of the terms inside the brackets by the number or letter outside the brackets.

It is the same in algebra.

$a(b + c)$ means 'add b and c and then multiply by a' and this is the same as 'multiply b by a, multiply c by a and then add the results'.

So $a(b + c) = ab + ac$.

This is called **expanding the brackets**.

Example 1

Expand these brackets.

a) $5(2x + 3)$

b) $4(2x - 1)$

Solution

a) $5(2x + 3) = 5 \times 2x + 5 \times 3$
$= 10x + 15$

b) $4(2x - 1) = 4 \times 2x + 4 \times {}^{-}1$
$= 8x - 4$

> **Exam tip**
> Remember about multiplying with negative numbers. For example,
> ${}^{-}3 \times x = {}^{-}3x$
> ${}^{-}3 \times {}^{-}x = 3x$.

Exercise 2.1

Expand these brackets.

1 a) $2(a + b)$ b) $3(x + 2)$
  c) $4(2x + 1)$

2 a) $3(x + y)$ b) $2(p + 3)$
  c) $4(3x - 2)$

3 a) $3(3x - 1)$ b) $2(3x + 2)$
  c) $2(2x - 3)$

4 a) $7(3y + z)$ b) $4(3 - 8a)$
  c) $2(4g - 3)$

5 a) $10(2a + 3b)$ b) $3(2c + 7d)$
  c) $5(3e - 8f)$

6 a) $2(1 - x)$ b) $5(p - q)$
  c) $a(a + 2)$

7 a) $b(b + 3)$ b) $x(x + 3)$
  c) $y(y - 1)$

8 a) $y(3 - 2y)$ b) $x(2 - x)$
  c) ${}^{-}y(2 + y)$

9 a) ${}^{-}z(z - 2)$ b) $3c(c + 4)$
  c) ${}^{-}2x(5x - 3)$

10 a) $2(3i + 4j - 5k)$
  b) $4(5m - 3n + 2p)$
  c) $6(2r - 3s - 4t)$

Challenge 1

Expand these.

a) $6(5a + 4b - 3c - 2d)$ b) $4(3w - 5x + 7y - 9z)$

c) $9(4p - 7q - 8r + 3s)$ d) $12(7e - 9f + 12g - 16h)$

Factorising algebraic expressions

Factors are numbers or letters which will divide into an expression.

The factors of 6 are 1, 2, 3 and 6.
The factors of p^2 are 1, p and p^2.

Remember that multiplying or dividing by 1 leaves a number unchanged, so 1 is not a useful factor and it is often ignored. For example $2y + 3$ would not be written as $1(2y + 3)$.

Factor trees can help. Here are two to remind you.

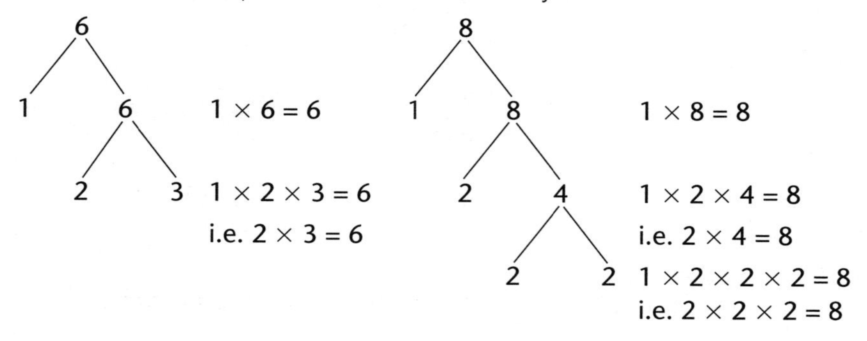

To factorise an expression, look for common factors. For example, the common factors of $2a^2$ and $6a$ are 2, a and $2a$.

Example 2

Factorise these fully.

a) $4p + 6$ **b)** $2a^2 - 3a$ **c)** $8x^2 - 12x$

Solution

a) The only common factor is 2.
You divide each term by the common factor, 2.

$4p \div 2 = 2p$ and $6 \div 2 = 3$
$4p + 6 = 2(2p + 3)$

b) The only common factor is a.
You divide each term by the common factor, a.

$2a^2 \div a = 2a$ and $3a \div a = 3$
$2a^2 - 3a = a(2a - 3)$

c) Think about the numbers and the letters separately and then combine them. The highest common factor of 8 and 12 is 4 and the common factor of x^2 and x is x. Therefore, the common factor of $8x^2$ and $12x$ is $4 \times x = 4x$.

$4x(\quad)$ You write this factor outside the brackets.

You then divide each term by the common factor, $4x$.

$8x^2 \div 4x = 2x$ and $12x \div 4x = 3$
$8x^2 - 12x = 4x(2x - 3)$

Exam tip

Make sure that you have found all the common factors. Check that the expression in the brackets will not factorise further.

Exercise 2.2

For questions **1** to **3**, copy and complete the factorisations.

1 a) $12a + 3 = 3(\square + 1)$
b) $6b - 4 = \square(3b - 2)$
c) $y^2 + 2y = y(\square + \square)$

2 a) $9a + 18 = 9(a + \square)$
b) $5y - 30 = 5(y - \square)$
c) $4x + 16 = \square(x + 4)$

3 a) $2b + 6b^2 = 2b(\square + \square)$
b) $8a^2 + 20a = \square(2a + 5)$

For questions **4** to **10**, factorise each of the expressions fully.

4 a) $2x + 6$ b) $10x + 15$
c) $8x - 12$

5 a) $4x - 20$ b) $8 + 12x$
c) $15 - 10x$

6 a) $9 - 12x$ b) $14 + 7x$
c) $15z - 3$

7 a) $3x^2 + 5x$ b) $5x^2 + 20x$
c) $6x^2 - 8x$

8 a) $5a^2 + 10b$ b) $3p + 4p^2$
c) $12q^2 - 18q$

9 a) $24 + 36a^2$ b) $3p^2 - p$
c) $x^2 + 7x$

10 a) $10x^2 - 100x$ b) $5 - 15x^2$
c) $21y^2 - 7y$

Challenge 2

Factorise these.

a) $24x + 32y$ b) $15ab - 20ac$ c) $30f^2 - 18fg$ d) $42ab + 35a^2$

Key Ideas

- Each term inside the brackets must be multiplied by the term outside the brackets.
- You factorise an expression by finding all the common factors.

CHAPTER 3

Translations

YOU WILL LEARN ABOUT	YOU SHOULD ALREADY KNOW
o Translating shapes and describing translations	o How to plot points in all four quadrants o What *congruent* means

When a shape is translated, all its points move the same distance in the same direction. The image and the original shape are **congruent**.
For example,

this shape could translate to this .

Activity 1

This is a pattern made by translations.

Copy it and add four more shapes.

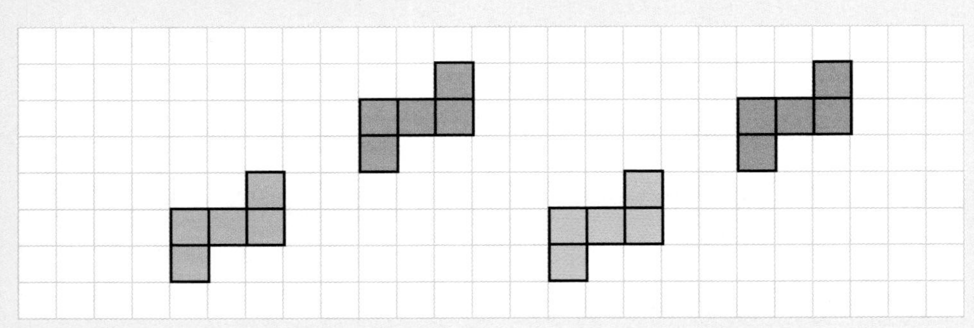

Activity 2

Draw some more patterns made with translations.

Activity 3

Triangle B is a translation of triangle A.

a) How can you tell it is a translation?

b) How far across has it moved?

c) How far down has it moved?

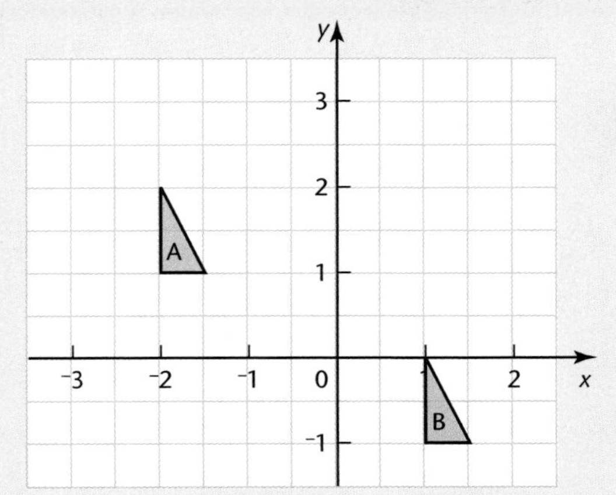

Exam tip

Take care with the counting. Choose a point on both the original shape and image and count the units from one to the other.

The distance and direction that a shape moves in a translation can be written as a **column vector**.

The *top* number tells you how far the shape moves *across*, or in the *x*-direction.
The *bottom* number tells you how far the shape moves *up or down*, or in the *y*-direction.
A *positive* top number is a move to the *right*. A *negative* top number is a move to the *left*.
A *positive* bottom number is a move *up*. A *negative* bottom number is a move *down*.

A translation of 3 to the right and 2 down is written as $\begin{pmatrix} 3 \\ -2 \end{pmatrix}$.

Example 1

Translate the triangle by $\begin{pmatrix} -3 \\ 4 \end{pmatrix}$.

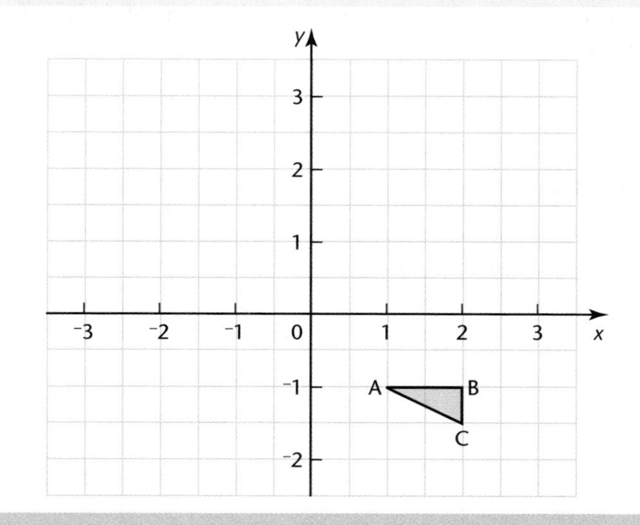

Solution

$\begin{pmatrix} -3 \\ 4 \end{pmatrix}$ means move 3 units left and 4 units up.

Point A moves from (1, −1) to (−2, 3).

Point B moves from (2, −1) to (−1, 3).

Point C moves from (2, −1.5) to (−1, 2.5).

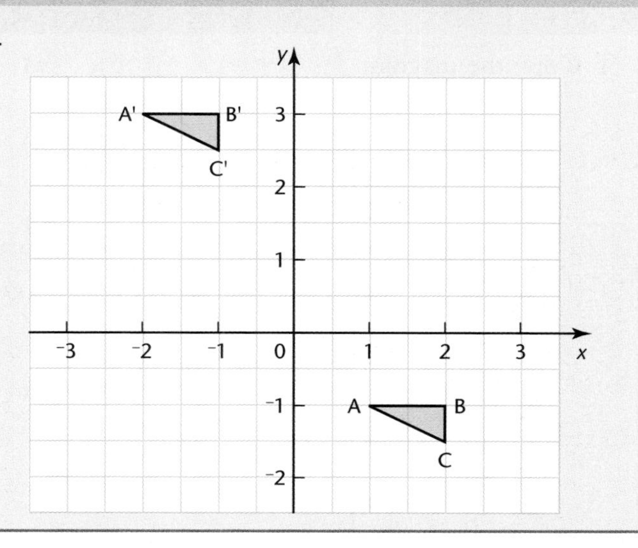

Example 2

Describe fully the single transformation that maps shape A on to shape B.

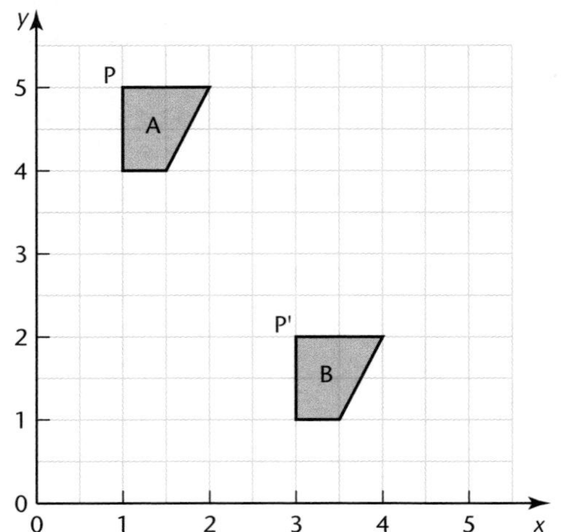

Solution

It is clearly a translation as the shape stays the same way up and the same size.

To find the movement, choose one point on the original shape and the image and count the units moved.

For example, P moves from (1, 5) to (3, 2). This is a movement of 2 to the right and 3 down.

The transformation is a translation of $\begin{pmatrix} 2 \\ -3 \end{pmatrix}$.

Exam tip

You must state that the transformation is a translation and give the column vector.

Exam tip

Try not to confuse the words *transformation* and *translation*.

Transformation is the general name for all changes made to shapes.

Translation is the particular transformation where all points of an object move the same distance in the same direction.

Exercise 3.1

1 Copy the diagram.

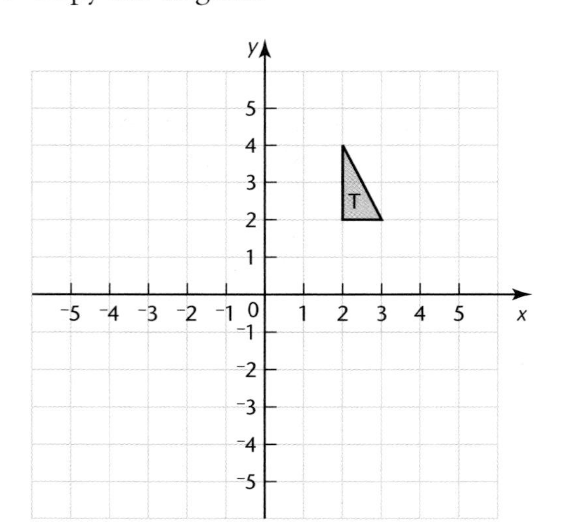

a) Translate triangle T 2 units to the right and 1 up. Label the image A.

b) Translate triangle T 1 unit to the right and 3 down. Label the image B.

c) Translate triangle T by $\begin{pmatrix} -4 \\ 1 \end{pmatrix}$. Label it C.

d) Translate triangle T by $\begin{pmatrix} -6 \\ -6 \end{pmatrix}$. Label it D.

e) Translate triangle T by $\begin{pmatrix} -2 \\ -5 \end{pmatrix}$. Label it E.

2 Copy the diagram.

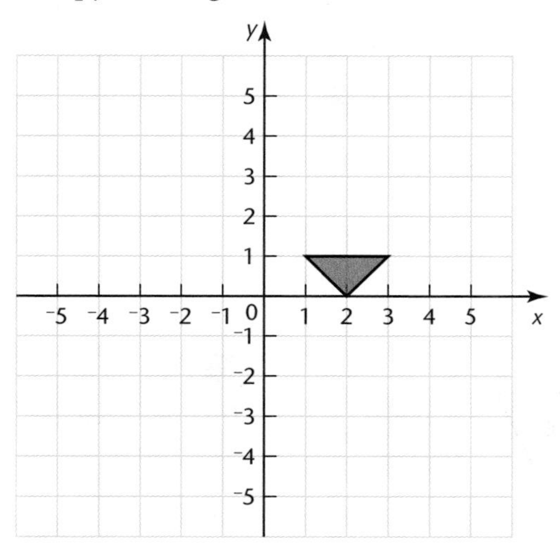

a) Translate the red triangle 2 units to the right and 4 up. Label the image A.

b) Translate the red triangle 1 unit to the right and 5 down. Label the image B.

c) Translate the red triangle by $\begin{pmatrix} -4 \\ -3 \end{pmatrix}$. Label the image C.

d) Translate the red triangle by $\begin{pmatrix} -6 \\ 3 \end{pmatrix}$. Label the image D.

e) Translate the red triangle by $\begin{pmatrix} -3 \\ -4 \end{pmatrix}$. Label the image E.

3 Look at the diagram.

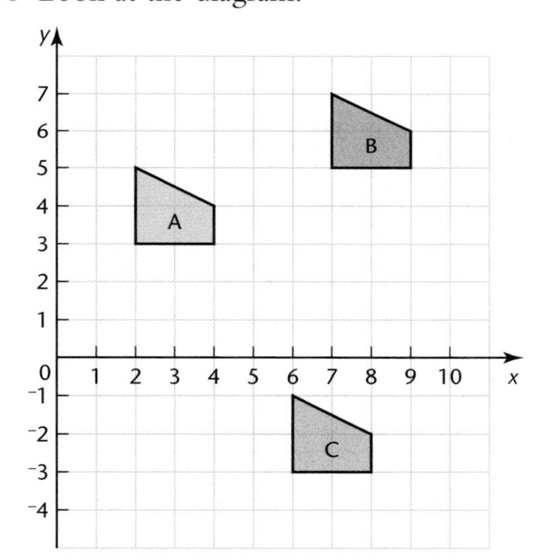

Describe the translation that maps

a) shape A on to shape B.

b) shape A on to shape C.

c) shape C on to shape B.

d) shape C on to shape A.

4 Look at the diagram.

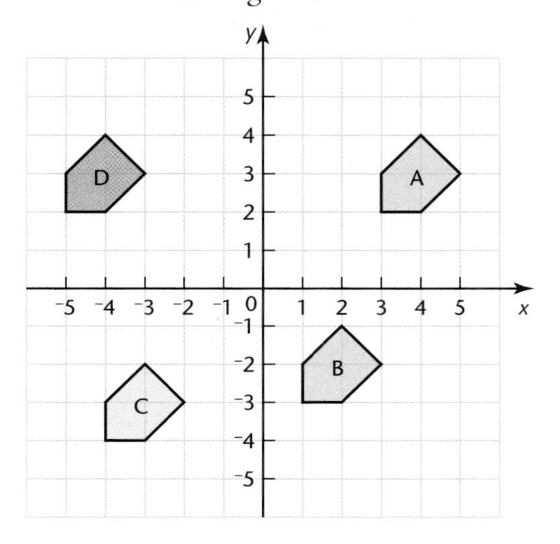

Describe the translation that maps

a) shape A on to shape B.

b) shape A on to shape C.

c) shape A on to shape D.

d) shape D on to shape C.

Challenge 1

- Copy the diagram.
- Translate shape A by $\begin{pmatrix} -3 \\ 6 \end{pmatrix}$. Label the image B.
- Translate shape B by $\begin{pmatrix} 7 \\ -8 \end{pmatrix}$. Label the image C.
- Translate shape C by $\begin{pmatrix} 1 \\ 5 \end{pmatrix}$. Label the image D.
- Translate shape D by $\begin{pmatrix} -5 \\ -3 \end{pmatrix}$. What do you notice?

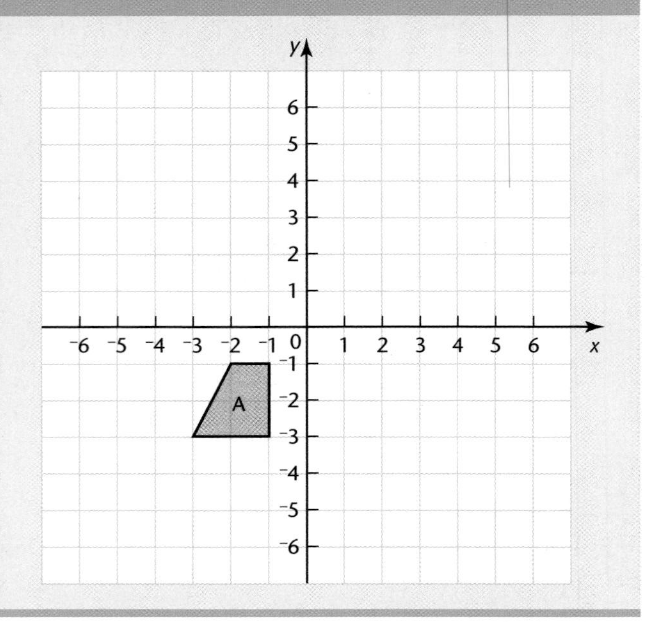

Challenge 2

- Copy the diagram.
- Translate shape A by $\begin{pmatrix} -2 \\ -6 \end{pmatrix}$. Label the image B.
- Translate shape B by $\begin{pmatrix} 9 \\ 5 \end{pmatrix}$. Label the image C.
- Translate shape C by $\begin{pmatrix} 2 \\ -4 \end{pmatrix}$. Label the image D.
- What must you translate D by to get back to A?

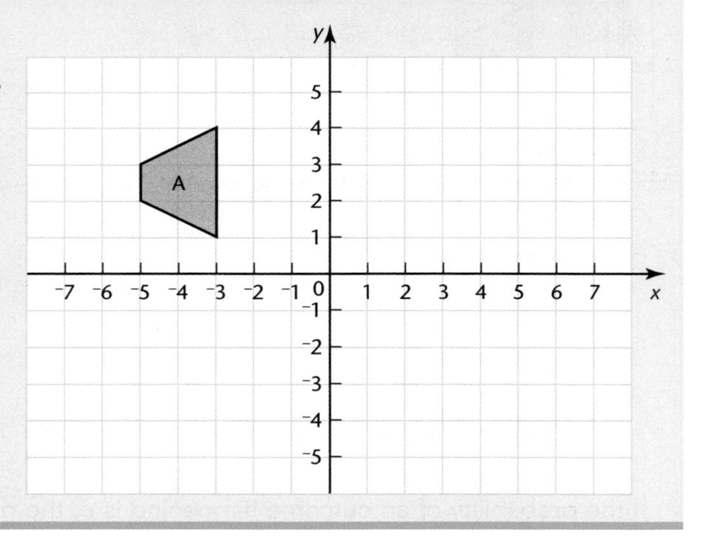

Key Ideas

- A translation moves every point on the shape the same distance, in the same direction. The image is congruent to the object.
- A translation is defined by movements to the left or right and up or down.
- A column vector can be used to describe a translation.

  Translate by $\begin{pmatrix} -3 \\ 4 \end{pmatrix}$ means 'move 3 units left and 4 units up'.

Probability 1

YOU WILL LEARN ABOUT	YOU SHOULD ALREADY KNOW
○ The probability of an event not happening ○ Mutually exclusive outcomes	○ That probabilities are expressed as fractions or decimals ○ That all probabilities lie on a scale of 0–1 ○ How to find a probability from a set of equally likely outcomes ○ How to subtract fractions and decimals from 1

The probability of an outcome not happening

Activity 1

There are ten counters in a bag. Seven are black and the rest are red.

A counter is taken without looking.

Calculate the probability that the counter is

a) black. **b)** not black.

What do you notice about your answers to **a)** and **b)**?

Activity 2

When a fair die is thrown, there are six equally likely scores or outcomes: 1, 2, 3, 4, 5 and 6.

What is the probability of

a) throwing a 6? **b)** throwing a 1, 2, 3, 4 or 5 (not a 6)?

What do you notice about your answers to **a)** and **b)**?

In both activities you should have found that a) + b) = 1.

This shows you how to find the probability that an outcome does not occur.

> If the probability of an outcome happening is p, the probability of the outcome not happening is $1 - p$.

Instead of writing 'the probability of an outcome happening', you can use the shorter form P(outcome).

So in Activity 1

P(black) = $\frac{7}{10}$ P(not black) = $1 - \frac{7}{10} = \frac{3}{10}$

Example 1

The probability that Nayim's school bus is late is $\frac{1}{8}$.

What is the probability that Nayim's bus is not late?

Solution

$$\text{P(not late)} = 1 - \text{P(late)}$$
$$= 1 - \frac{1}{8}$$
$$= \frac{7}{8}$$

Example 2

The probability that it will rain tomorrow is 0.3.
What is the probability that it will not rain tomorrow?

Solution

$$\text{P(no rain)} = 1 - \text{P(rain)}$$
$$= 1 - 0.3$$
$$= 0.7$$

Exercise 4.1

1 The probability of burning the toast is 0.8.
What is the probability of not burning the toast?

2 The probability that I will have cereal for breakfast is $\frac{2}{7}$.
What is the probability that I will not have cereal for breakfast?

3 The picture shows a fair spinner.

a) What is the probability of getting an even number with one spin?
b) What is the probability of getting an odd number with one spin?

4 The probability that Emma will pass her next maths examination is 0.85.
What is the probability that she will fail the examination?

5 Based on past experience, the probability that Kevin's school bus will be late tomorrow is 0.23.
What is the probability that it will be on time or early?

6 The probability that Jane will do the washing up tonight is $\frac{3}{8}$.
What is the probability that she will not do the washing up?

7 The probability that Chris will get a motorcycle for his sixteenth birthday is 0.001.
What is the probability that he will not get a motorcycle?

8 The probability that the gig will sell out is $\frac{1}{25}$.
What is the probability that the gig will not sell out?

9 The probability that Liam will watch City on Saturday is 0.98.
What is the probability that he will not watch City?

10 The probability that Umair will choose to study geography at GCSE is $\frac{2}{7}$.
What is the probability that he will not choose geography?

11 There are 13 diamonds in a pack of 52 playing cards.
I pick a card at random.
a) What is the probability that I pick a diamond?
b) What is the probability that I do not pick a diamond?

12 In a multiple-choice test paper five possible answers to each question are given. Only one of the answers is right. If Obaid does not know an answer to a question, he guesses.
Obaid guessed the answer to question 15.

a) What is the probability that Obaid got question 15 right?

b) What is the probability that Obaid got question 15 wrong?

13 The probability that the first ball drawn in the lottery is white is $\frac{9}{49}$.
What is the probability that it is not white?

14 The probability that Peter goes out on a Friday night is 0.995.
What is the probability that Peter stays in on a Friday night?

15 When two dice are thrown, what is the probability of not scoring a double?

Challenge 1

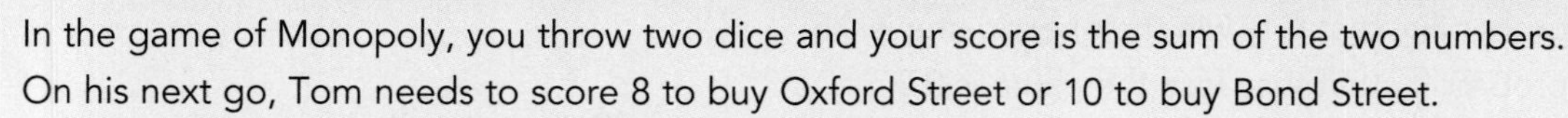

In the game of Monopoly, you throw two dice and your score is the sum of the two numbers.
On his next go, Tom needs to score 8 to buy Oxford Street or 10 to buy Bond Street.

a) What is the probability that, on his next go, Tom can buy
   (i) Oxford Street?
   (ii) Bond Street?
   (iii) either one of them?
   (iv) neither of them?

Tom gets 7 on that go but Oxford Street and Bond Street are not sold before his next go.

b) What is the probability that on his next go Tom can buy
   (i) Oxford Street?
   (ii) Bond Street?
   (iii) either one of them?
   (iv) neither of them?

Challenge 2

The weather forecast says the probability that it will be sunny tomorrow is 0.4.
Terry says this means that the probability that it will rain is 0.6.
Is Terry correct? Why?

Mutually exclusive outcomes

Mutually exclusive outcomes are outcomes which cannot happen together. For example, if you toss a coin once then the outcomes 'the coin comes down heads' and 'the coin comes down tails' are mutually exclusive, since it is impossible for the coin to come down both heads and tails.

If outcomes are mutually exclusive and cover all the possibilities then the probabilities of those outcomes must add up to 1.

So, in the above example,

P(heads) + P(tails) = 1.

This would be true even if the coin was not fair but was biased towards heads.

If P(heads) = 0.6 then P(tails) = 0.4 and P(heads) + P(tails) = 1 since you cannot throw both a head and a tail on the same coin on the same toss, and no other outcome is possible.

If A, B, C and D are the mutually exclusive outcomes of an event, then

$$P(A) + P(B) + P(C) + P(D) = 1.$$

So, for example

$$P(B) = 1 - [P(A) + P(C) + P(D)]$$
$$\text{or } P(B) = 1 - P(A) - P(C) - P(D).$$

Example 3

The probability that City win their next game is 0.5.

The probability that they lose the game is 0.2.

What is the probability that the game will be drawn?

Solution

The outcomes 'win', 'lose' and 'draw' are mutually exclusive, as it is impossible for any two or all of them to happen together.

Also, no outcomes other than 'win', 'lose' and 'draw' are possible.

Therefore P(win) + P(lose) + P(draw) = 1.

P(draw) = 1 – P(win) – P(lose) = 1 – 0.5 – 0.2 = 0.3.

Exercise 4.2

1 In a game of tennis you can only win or lose. A draw is not possible.
Qasim plays Robin at tennis. The probability that Qasim wins is 0.7.
What is the probability that Robin wins?

2 A set of traffic lights may be on red, red and amber, amber or green.
The probability that they are on red is 0.5. The probability that they are on red and amber is 0.05 and the probability that they are on amber is 0.05.
What is the probability that the lights are on green?

3 A bag contains red, white and blue balls.
I pick one ball out of the bag at random.
The probability that I pick a red one is $\frac{1}{12}$.
The probability that I pick a white one is $\frac{7}{12}$.
What is the probability that I pick a blue one?

4 Alex is choosing his GCSE options.
In one pool, he can choose History or Geography or Business Studies.
The probability that he chooses History is 0.3.
The probability that he chooses Geography is 0.45.
What is the probability that he chooses Business Studies?

5 For my next holiday I will go to Spain or France or the USA.
The probability that I will go to Spain is $\frac{7}{20}$.
The probability that I will go to France is $\frac{11}{20}$.
What is the probability that I will go to the USA?

6 A coach is selecting a baseball team. She has three pitchers to choose from: Raisa, Kimberley and Melanie.
The probability that she chooses Raisa is $\frac{3}{8}$.
The probability that she chooses Kimberley is $\frac{5}{8}$.
What is the probability that she chooses Melanie?

7 There are four breakfast cereals in the cupboard: muesli, cornflakes, Weetycrisps and frosted flakes.
Kim has decided to have cereal for breakfast.
The probability that she has muesli is 0.05.
The probability that she has Weetycrisps is 0.4.
The probability that she has frosted flakes is 0.2.
What is the probability that she will have cornflakes?

8 Greg is choosing a main course from the menu.
There are three choices: burger and chips, tuna salad and tagliatelli.
The probability that he chooses tuna salad is $\frac{1}{12}$.
The probability that he chooses tagliatelli is $\frac{5}{12}$.
What is the probability that he chooses burger and chips?

9 Max travels to school by bus, car or cycle or he walks.
The probability that he travels by bus is $\frac{1}{11}$.
The probability that he travels by car is $\frac{3}{11}$.
The probability that he cycles is $\frac{2}{11}$.
What is the probability that he walks?

10 There are red, yellow and blue beads in a bag.
The probability of choosing a red one is $\frac{4}{12}$.
The probability of choosing a yellow one is $\frac{3}{12}$.
What is the probability of choosing a blue one?

11 Outcomes A, B and C are mutually exclusive and one of them must happen.
If $P(A) = 0.47$ and $P(B) = 0.31$, what is $P(C)$?

12 The probability that Rovers will win their next game is 0.4.
Geri says this means that the probability they will lose is 0.6.
Why is she almost certainly wrong?

13 A shop has black, grey and blue dresses on a rail. Jen picks one at random.
The probability of picking a grey dress is 0.2.
The probability of picking a black dress is 0.1.
What is the probability of picking a blue dress?

14 Heather always travels to the gym by car, bus or bike.
On any day, the probability that Heather will travel by car is $\frac{3}{20}$ and the probability that she will travel by bus is $\frac{11}{20}$.
What is the probability that Heather will travel to the gym by bike?

15 The probability that the school hockey team will win their next match is 0.4.
The probability that they will lose is 0.25.
What is the probability that they will draw the match?

16 Pat has either boiled eggs, cereal or toast for breakfast.
The probability that she will have toast is $\frac{2}{11}$ and the probability that she will have cereal is $\frac{5}{11}$.
What is the probability that she will have boiled eggs?

17 The table shows the probability of getting some of the scores when a biased six-sided dice is thrown.

Score	1	2	3	4	5	6
Probability	0.27	0.16	0.14		0.22	0.1

What is the probability of getting 4?

Challenge 3

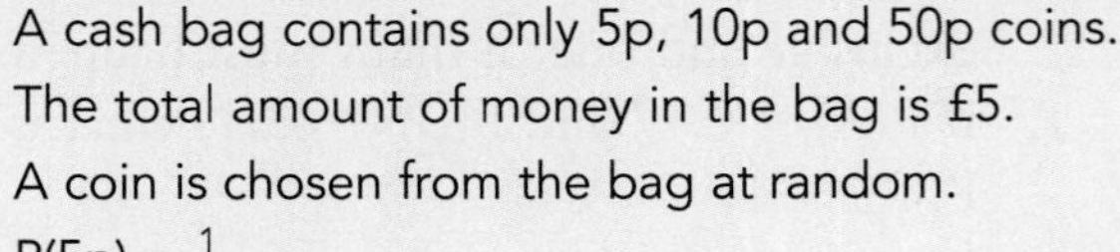

A cash bag contains only 5p, 10p and 50p coins.
The total amount of money in the bag is £5.
A coin is chosen from the bag at random.
P(5p) = $\frac{1}{2}$
P(10p) = $\frac{3}{8}$

a) Work out P(50p).
b) How many of each kind of coin are there in the bag?

Key Ideas

- If the probability of an outcome happening is p, the probability of the outcome not happening is $1 - p$.
- If two outcomes A and B cannot occur together they are mutually exclusive.
- If two outcomes A and B are mutually exclusive and cover all possible outcomes, then P(A) + P(B) = 1.

  Similarly, if events A, B and C are mutually exclusive and cover all possible outcomes then P(A) + P(B) + P(C) = 1.

Drawing triangles and other shapes

YOU WILL LEARN ABOUT	YOU SHOULD ALREADY KNOW
○ Constructing triangles using a ruler, compasses and protractor ○ Constructing regular polygons inscribed in circles ○ Constructing nets of 3-D shapes	○ How to use simple scales ○ How to use a pair of compasses and either a protractor or an angle measurer ○ The meaning of the words *parallel*, *perpendicular*, *isosceles* and *equilateral* ○ That angles in a triangle add to 180°

Constructing triangles given two sides and the included angle

The *included angle* is the angle between the two given sides.

Example 1

Make an accurate drawing of this triangle.

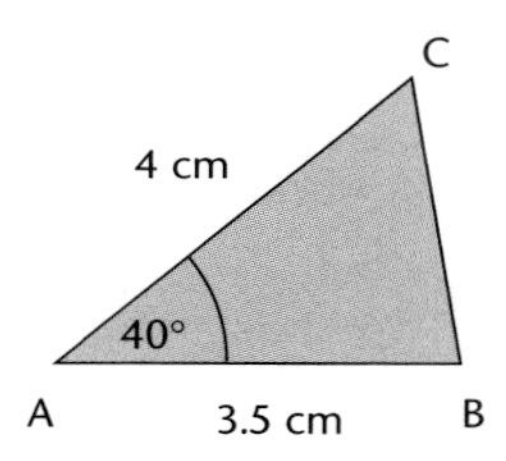

Solution

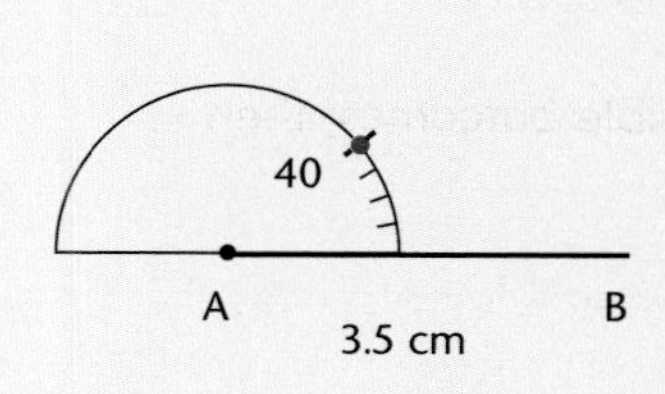

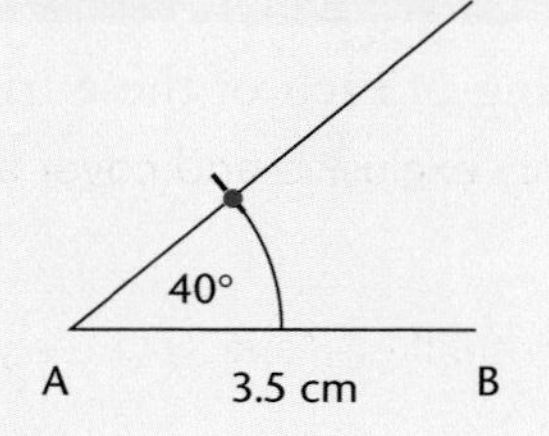

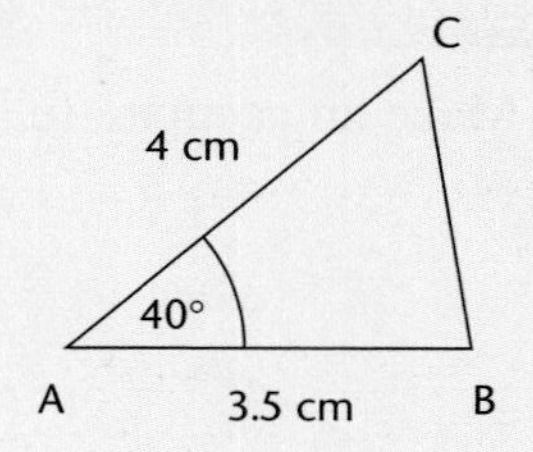

Draw the line AB 3.5 cm long. Measure and mark the 40° angle at A.
Draw a line from A through the point.
Mark the point C on this line, 4 cm from A. Join C to B.

> **Exam tip**
> Make sure your pencil is sharp before you start accurate drawings and graphs.

Constructing triangles given one side and two angles

Example 2

In triangle PQR, PQ = 6 cm, angle RPQ = 35° and angle PQR = 29°.
Make an accurate drawing of triangle PQR.

Solution

First draw a sketch.

Draw the line PQ, 6 cm long. Mark the 35° angle at P and draw a line through the point, longer than PQ.

Mark the 29° angle at Q. Draw a line to meet the previous line at R. This completes the triangle.

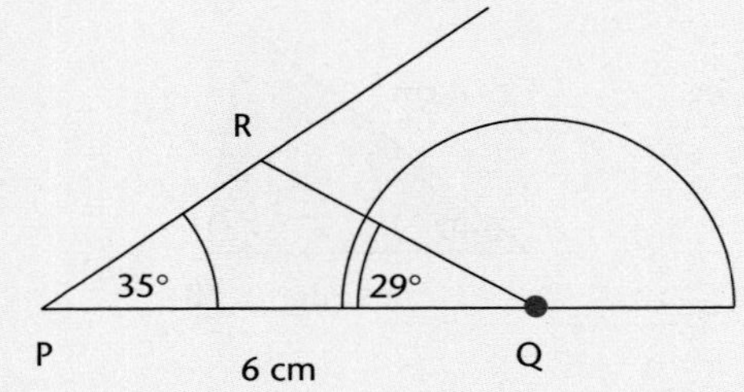

Exam tip

Draw a sketch of the triangle first. When an angle is written as three letters, the middle letter indicates the vertex (corner) of the angle.

Exam tip

If the two given angles were at R and Q, then angle P could be found by adding the two known angles and subtracting from 180°.

Exercise 5.1

1 Make an accurate full-size drawing of each of these triangles.

a)

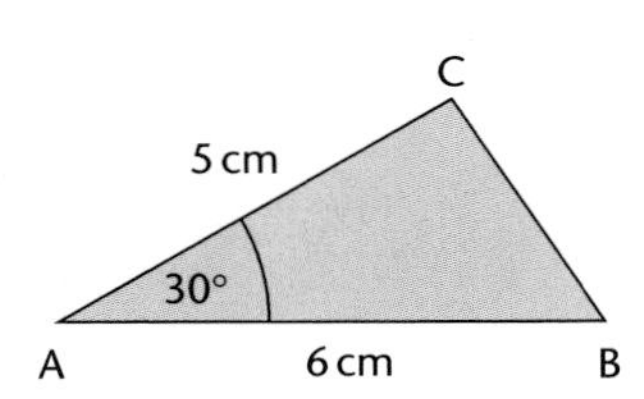

b)

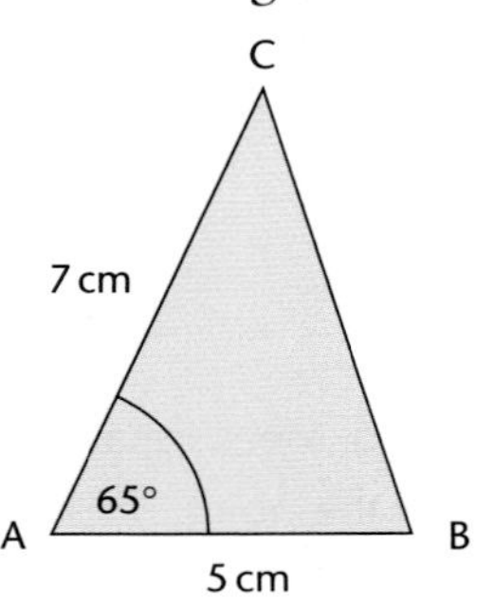

c)

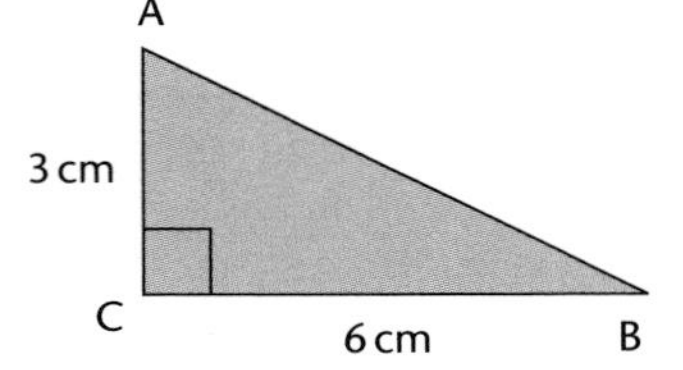

d)

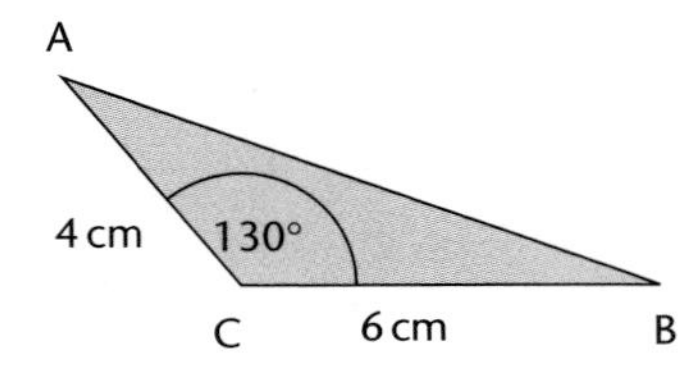

2 For each triangle in question **1**, measure the unmarked side and the other two angles on your drawing.

3 Make an accurate full-size drawing of each of these triangles.

a)

C
3.5 cm
62°
A
6 cm
B

b)

C
6.2 cm
27°
A
6.2 cm
B

c)

A
3.8 cm
118°
7.3 cm
C
B

d)

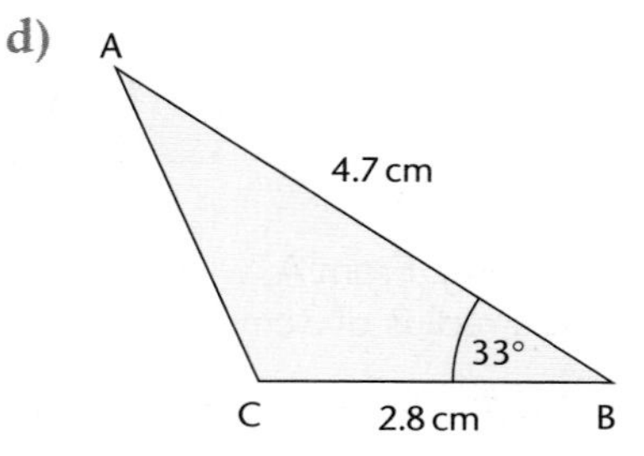

4 For each triangle in question **3**, measure the unmarked side and the other two angles on your drawing.

5 This is a sketch of Joe's garden.

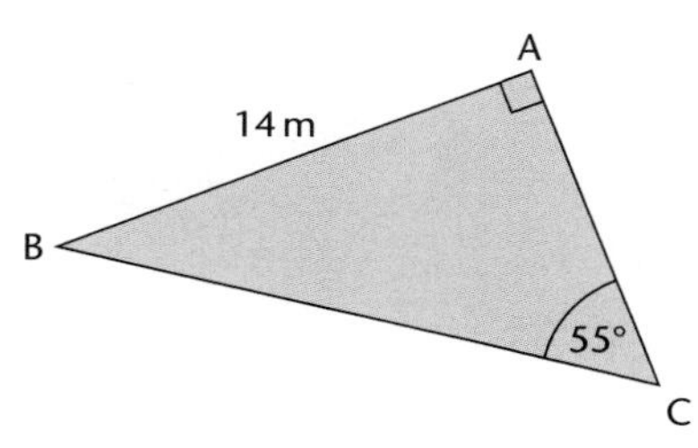

a) Make a scale drawing of the garden using a scale of 1 cm to 2 m.

b) Measure the length of each unmarked side on your drawing and hence find the length of each of the other two sides of the garden.

Constructing triangles given three sides

Example 3

Make an accurate drawing of triangle ABC, where AB = 5 cm, BC = 4 cm and AC = 3 cm.

Solution

First draw a sketch.

C
3 cm
4 cm
A
5 cm
B

Draw the line AB, 5 cm long. From A, with compasses set to a radius of 3 cm, draw an arc above the line.

A
5cm
B

From B, with compasses set to a radius of 4 cm, draw another arc to intersect the first. The point where the arcs meet is C.

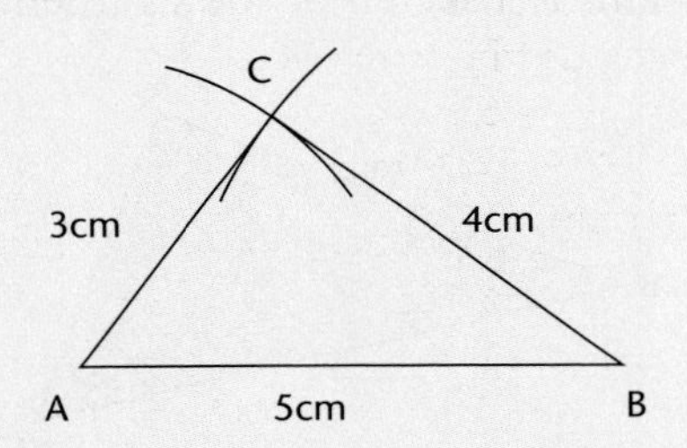

Constructing triangles given two sides and a non-included angle

When you are given two sides and one of the angles which is not between them, you need to use a pair of compasses and a protractor to draw the triangle. The following two examples show the method.

Example 4

Make an accurate drawing of this triangle.

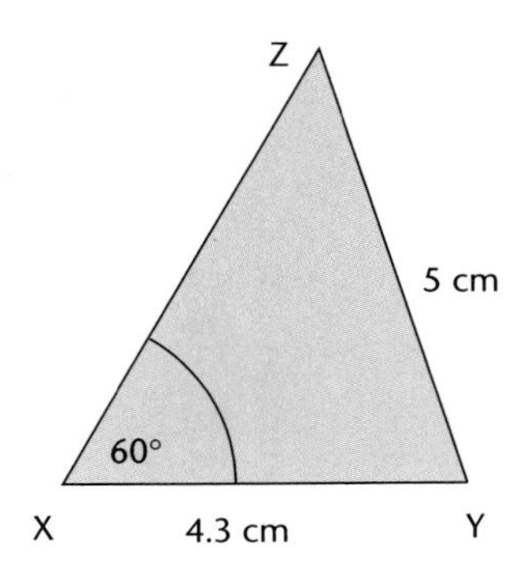

Solution

Draw the line XY 4.3 cm long.

Measure an angle of 60° at X and draw a line from X of any length, but longer than XY.

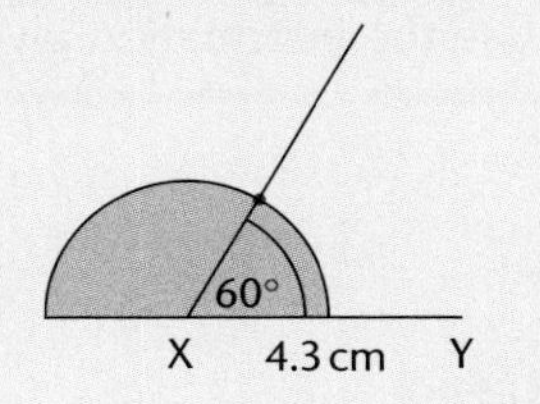

From Y, with your compasses set to a radius of 5 cm, draw an arc to intersect the line.

The point where the arc cuts the line is Z.

Join Z to Y using your ruler.

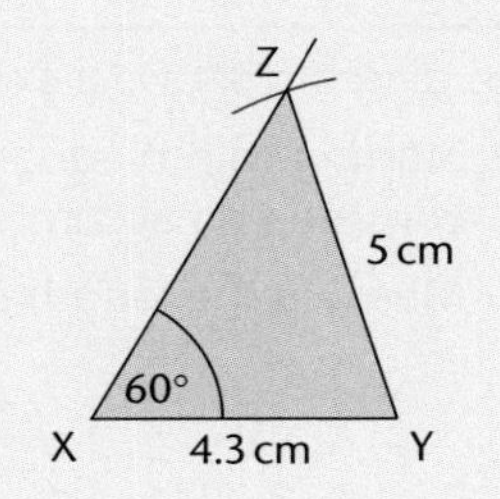

When you are not given a sketch of the triangle, it is a good idea to make one first, before you start an accurate drawing.

Example 5

Make an accurate drawing of triangle PQR where PQ = 6.3 cm, angle RPQ = 60° and QR = 6 cm.

Solution

Draw a sketch of the triangle.

Draw the line PQ 6.3 cm long.

Measure an angle of 60° at P and draw a line from P.

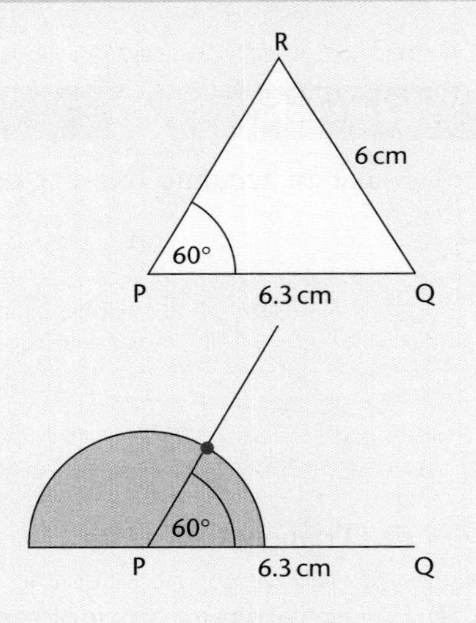

From Q, with your compasses set to a radius of 6 cm, draw an arc to intersect the line.

The point where the arc cuts the line is R.

Join R to Q using your ruler.

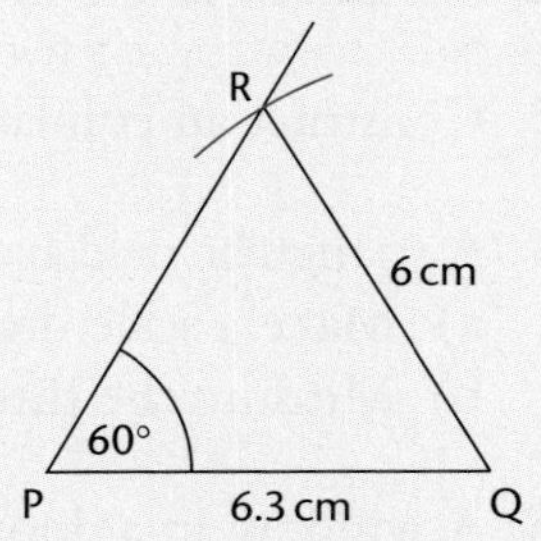

Challenge 1

Use the information given in Example 5 to draw a different triangle.

This triangle will have an obtuse angle at R.

Activity I

The diagram shows a sketch of kite ABCD.

a) Starting with diagonal AC, use a ruler and compasses to construct an accurate drawing of the kite.

b) Measure the length of the diagonal BD.

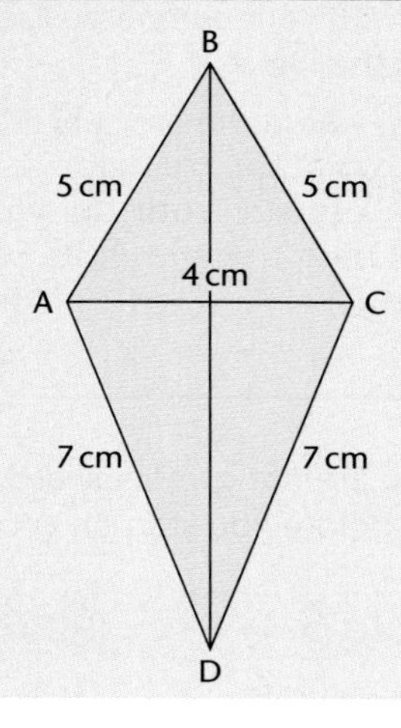

Exercise 5.2

1 Make an accurate full-size drawing of each of these triangles.

a)

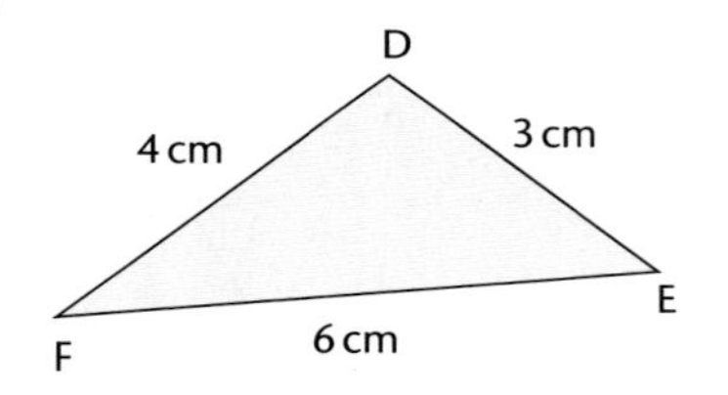

b)

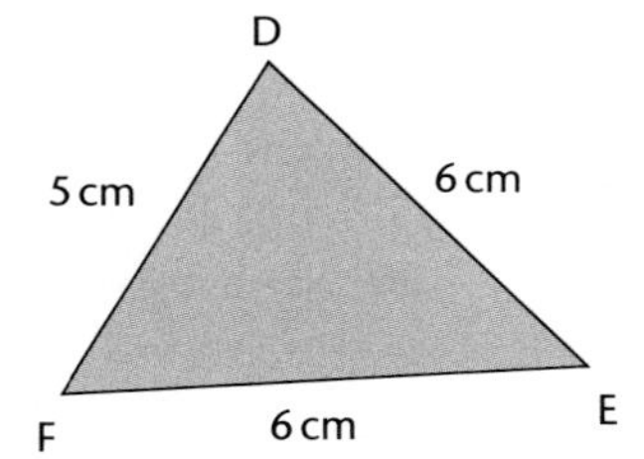

c) Triangle DEF with DE = 4.3 cm, EF = 7.2 cm and FD = 6.5 cm.

2 For each triangle in question **1**, measure the angles on your drawing.

3 Construct an equilateral triangle with sides of 4.2 cm.

4 A triangular field has sides of length 30 m, 14 m and 20 m.

a) Make a scale drawing of the field. Use a scale of 1 cm to 5 m.

b) Measure the three angles of the field.

5 A window in a door is a triangle with the side AB horizontal and 40 cm long. The other two sides meet at a point C above AB and are each 25 cm long.

a) Make a scale drawing of the window. Use a scale of 1 cm to 5 cm.

b) How far is C above AB?

6 Draw each of these triangles accurately.

a)

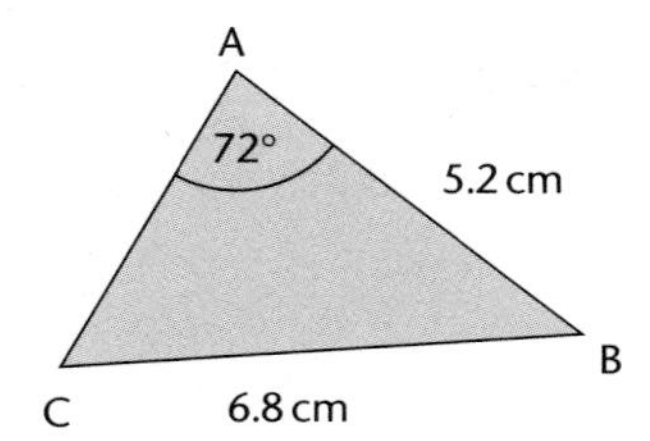

b)

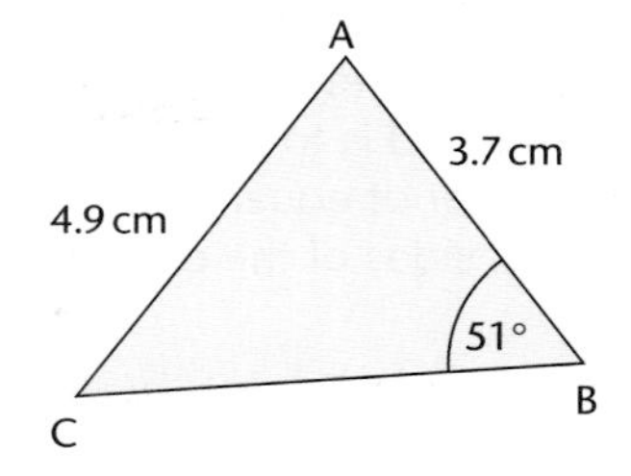

7 Measure angle C in each of the triangles you drew in question 6.

8 a) Draw the triangle ABC where AB = 5 cm, BC = 4 cm and angle BAC = 40°.
Compare your drawing with your neighbour's drawing.

b) Draw triangle ABC where
AB = 7 cm, AC = 6 cm and angle ABC = 50°.
Compare your drawing with your neighbour's drawing.

9 a) Use a ruler, protractor and pair of compasses to construct an accurate drawing of this quadrilateral.

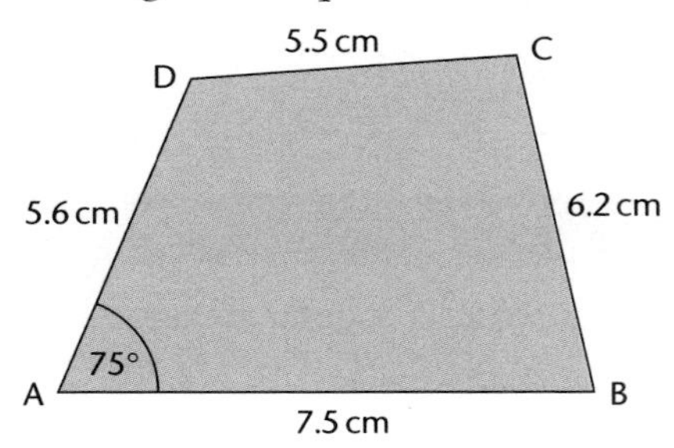

b) Measure angle D.

10 a) Draw a rhombus ABCD with diagonal AC = 3.5 cm and sides of length 5.3 cm.

b) Measure the length of diagonal BD.

11 a) Make an accurate full-size drawing of this parallelogram.

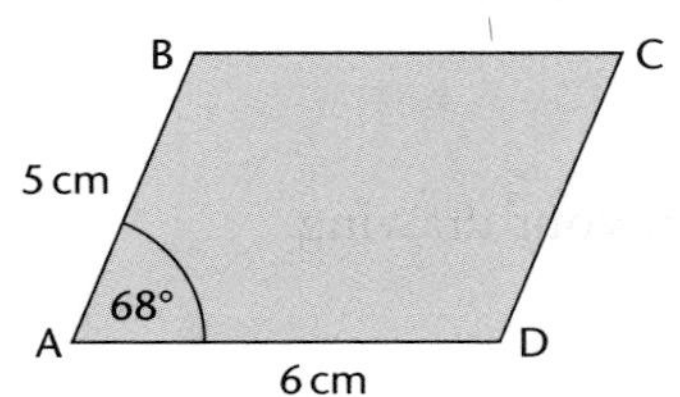

b) Measure the size of the other three angles.

12 This is a sketch of a park.

a) Using a suitable scale, make an accurate drawing of the park.

b) Use your drawing to find the distance between the gates to the park at corners B and D.

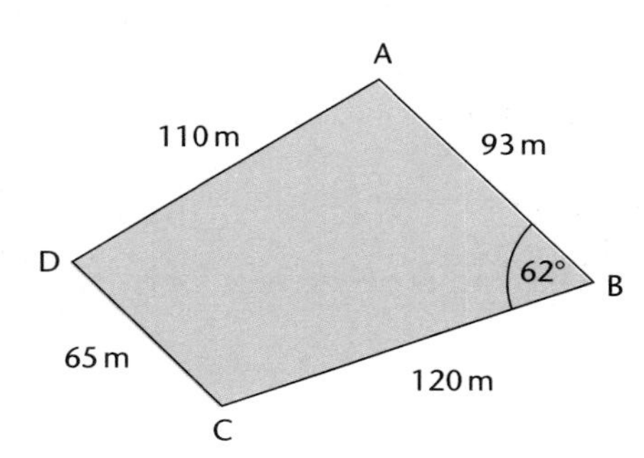

Constructing regular polygons in circles

There are several different methods to construct a regular polygon. Possibly the easiest method is to divide the angle at the centre of the circle, 360°, into a number of equal parts – the number of parts being equal to the number of sides of the polygon.

Example 6	Solution
Construct a regular pentagon.	Draw a circle with a radius of, for example, 5 cm. A pentagon has five sides, so divide the angle at the centre by five giving 360° ÷ 5 = 72°. Measure, using a protractor, 72° angles around the centre, joining the centre to the circumference. Join the points on the circumference. 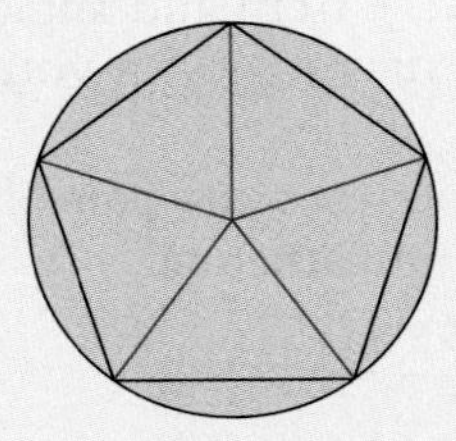

Exercise 5.3

1 Construct a regular hexagon.

2 Construct a regular octagon.

Nets

This flat shape can be folded up to make a cube – it is a **net** of the cube.

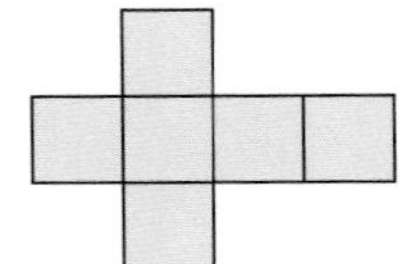

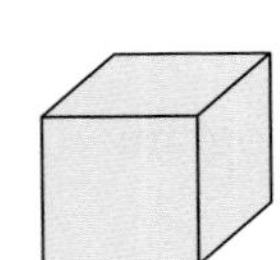

A regular **tetrahedron** is the name for a triangular-based pyramid with edges that are all the same length.

Here is a net for a regular tetrahedron.

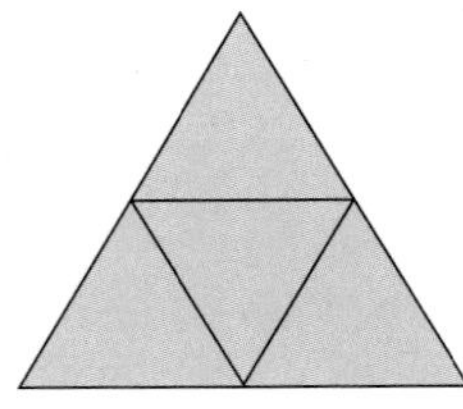

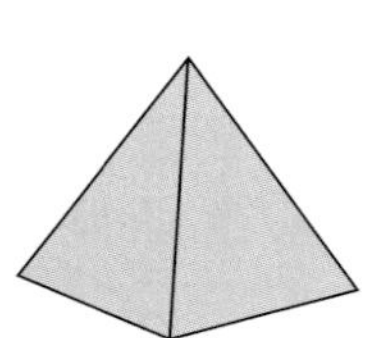

You may have to draw the nets for pyramids using compasses.

Exercise 5.4

1 Which of these shapes are a net of a cube?

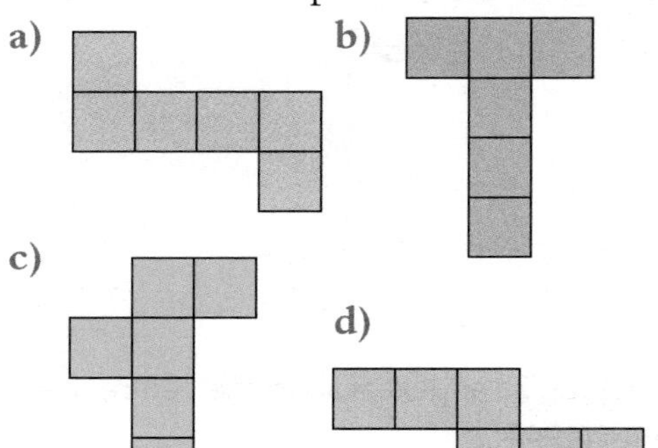

2 Which of the shapes **a)**, **b)**, or **c)** could be the net of this cuboid?

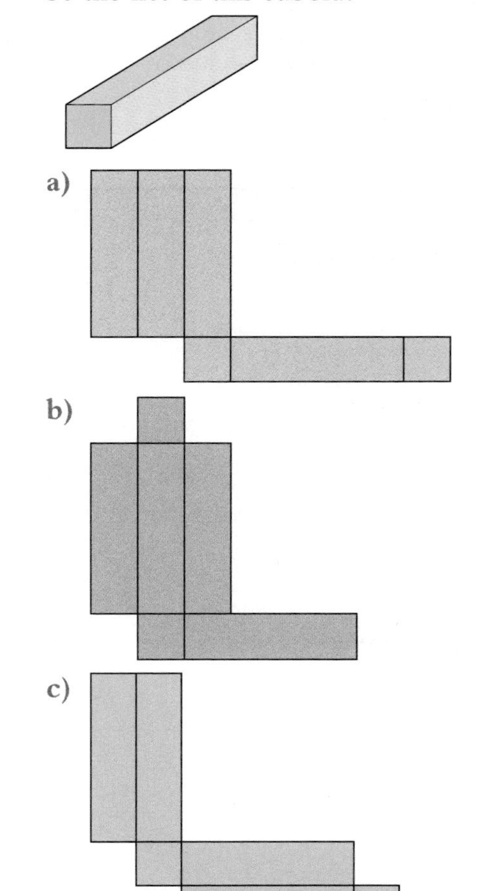

3 Construct an accurate net of the following 3-D shapes.

a) A cube of side 3 cm

b) A cuboid 2 cm by 3 cm by 5 cm

4 This is the net for a square-based pyramid.

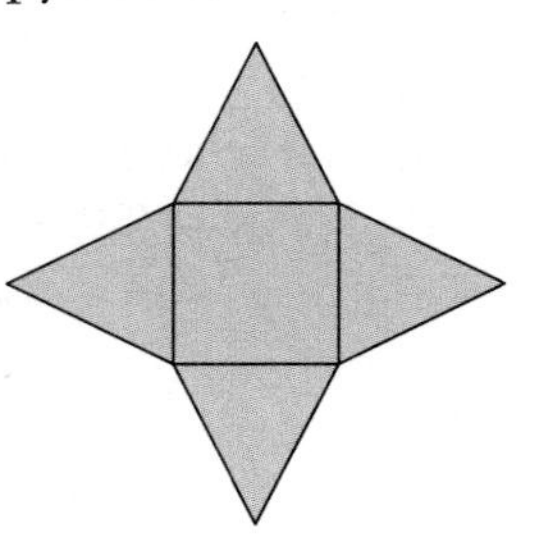

Is this a net for a different square-based pyramid?

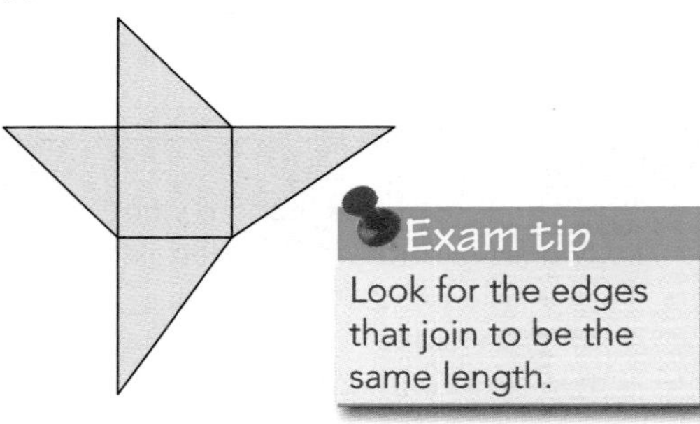

5 Construct a net of this triangular prism.

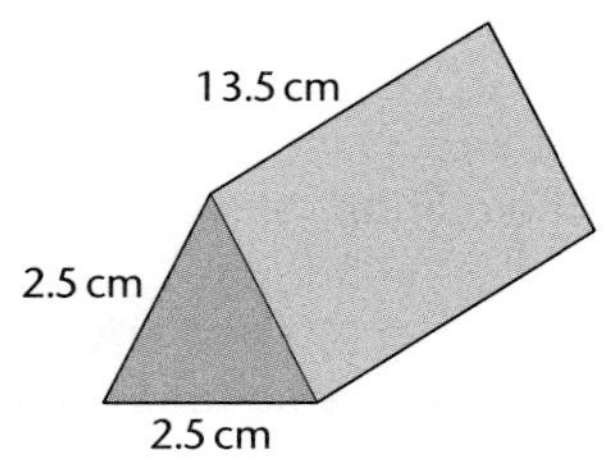

6 Construct the net of a regular tetrahedron [triangular-based pyramid] with side 5 cm.

Activity 2

Follow these instructions to make a gift box.

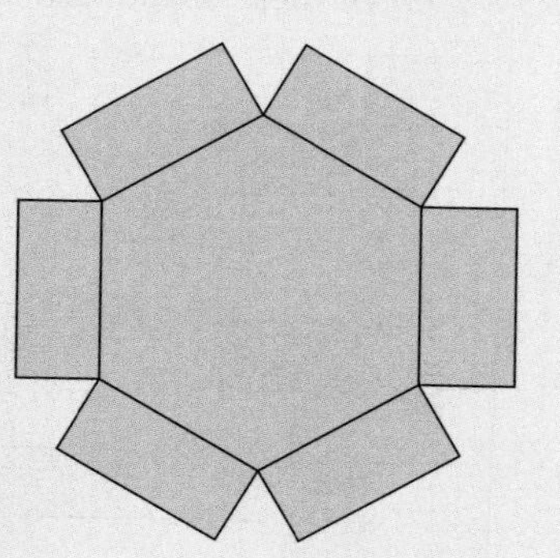

- Choose what shape you want the box to be.
- For example, this net will fold up to make a hexagonal prism.
- Choose what dimensions you want the finished box to be.
- Think where you need to add flaps to the net so that you can glue the edges together.
- Make an experimental box in paper and fold it up to check your measurements and that you have flaps in the right places.
- Make any adjustments that are necessary and then draw an accurate net on card.
- Think how you want to decorate your box. You may find it easier to decorate your box while it is flat, before you fold and glue it.
- Fold your net. Card is easier to fold if you score along the fold lines first. Then glue it.
- You can make a lid to fit your box by making the edges of the hexagon or other cross-section 1 to 2 mm larger than the base. Making the lid less deep than the box means that the lid is easier to take off.

- You can construct a triangle when you know
  - two sides and the included angle
  - one side and two angles
  - all three sides
  - two sides and an angle that is not between them.
- When you are given two sides and one of the angles which is not between them, you need to use a pair of compasses and a protractor to draw the triangle.
- You can construct a regular polygon in a circle using the fact that the angle at the centre is equal to 360 ÷ the number of sides.
- A net is a flat shape that can be folded to form a 3-D shape.

6 Working with fractions

YOU WILL LEARN ABOUT	YOU SHOULD ALREADY KNOW
o Multiplying and dividing fractions o Adding and subtracting mixed numbers	o How to find equivalent fractions o How to convert improper fractions and mixed numbers o How to add and subtract simple fractions

Multiplying fractions

Multiplying a fraction by a whole number

In this diagram $\frac{1}{8}$ is red.

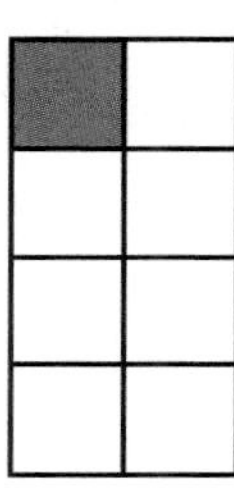

In this diagram five times as much is red, so $\frac{1}{8} \times 5 = \frac{5}{8}$.

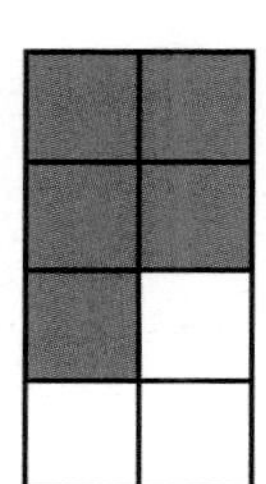

This shows that to multiply a fraction by an integer, you multiply the numerator by the integer. Then simplify by cancelling and changing to a mixed number if possible.

Example 1

Work out these.

a) $\frac{3}{10} \times 3$ **b)** $\frac{2}{3} \times 9$ **c)** $\frac{3}{8} \times 2$ **d)** $5 \times \frac{1}{2}$

Solution

a) $\frac{3}{10} \times 3 = \frac{9}{10}$ Multiply the numerator by 3.

b) $\frac{2}{3} \times 9 = \frac{18}{3} = \frac{6}{1} = 6$ $\frac{6}{1}$ means $6 \div 1 = 6$.

c) $\frac{3}{8} \times 2 = \frac{6}{8} = \frac{3}{4}$ Give $\frac{6}{8}$ in its lowest terms.

d) $5 \times \frac{1}{2} = \frac{5}{2} = 2\frac{1}{2}$ Change $\frac{5}{2}$ to a mixed number.

Multiplying a fraction by a fraction

In this diagram, $\frac{1}{4}$ is red.

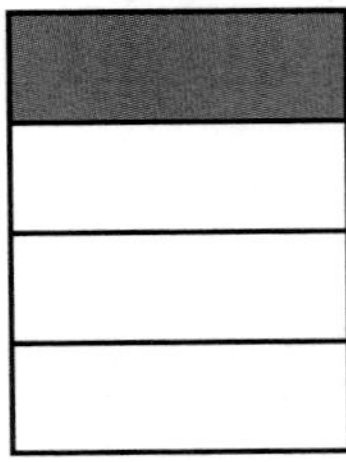

In this diagram, $\frac{1}{3}$ of the red is dotted, so $\frac{1}{3} \times \frac{1}{4} = \frac{1}{12}$.

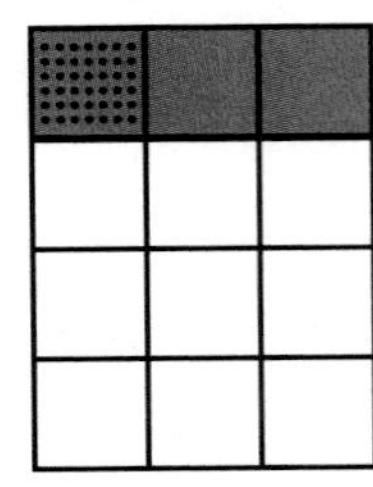

To multiply fractions, multiply the numerators and multiply the denominators, then simplify if possible.

Example 2

Work out these.

a) $\frac{1}{3} \times \frac{1}{2}$ **b)** $\frac{3}{5} \times \frac{1}{2}$ **c)** Find $\frac{3}{4}$ of $\frac{6}{7}$.

Solution

a) $\frac{1}{3} \times \frac{1}{2} = \frac{1}{6}$ Multiply the numerators and multiply the denominators.

b) $\frac{3}{5} \times \frac{1}{2} = \frac{3}{10}$

c) $\frac{3}{4} \times \frac{6}{7} = \frac{18}{28}$ 'of' means the same as '×'.

$\frac{18}{28} = \frac{9}{14}$ Simplify by dividing the numerator and the denominator by 2.

$\frac{3}{\cancel{4}_2} \times \frac{\cancel{6}^3}{7} = \frac{9}{14}$ Another way is to 'cancel' a term in the numerator with one in the denominator. This can make the arithmetic simpler. Divide both 4 and 6 by 2, then multiply the numerators and denominators.

Exam tip

Note: $\frac{1}{2} \times \frac{1}{3} = \frac{1}{6}$
A common error is to multiply 1 × 1 and get 2.

Exam tip

When cancelling, cancel a term in the numerator with one in the denominator.

Exercise 6.1

Work out these. Where necessary, write your answers as proper fractions or mixed numbers in their lowest terms.

1 $\frac{1}{2} \times 4$

2 $7 \times \frac{1}{2}$

3 $9 \times \frac{1}{3}$

4 $\frac{3}{4} \times 12$

5 $\frac{2}{5} \times 5$

6 $24 \times \frac{5}{12}$

7 $\frac{2}{3} \times 4$

8 $\frac{4}{9} \times 2$

9 $\frac{4}{5} \times 3$

10 $\frac{1}{5} \times 3$

11 $\frac{1}{4} \times \frac{2}{3}$

12 $\frac{2}{3} \times \frac{3}{5}$

13 $\frac{4}{9} \times \frac{1}{2}$ 14 $\frac{1}{3} \times \frac{2}{3}$ 15 $\frac{5}{6} \times \frac{3}{5}$ 16 $\frac{3}{7} \times \frac{7}{9}$

17 $\frac{1}{2} \times \frac{5}{6}$ 18 $\frac{3}{10} \times \frac{5}{11}$ 19 $\frac{2}{3} \times \frac{5}{8}$ 20 $\frac{3}{5} \times \frac{5}{12}$

21 Jane makes 120 cakes for a car boot sale. She sells $\frac{5}{6}$ of them. How many does she sell?

22 After everyone has helped themselves to part of a pizza, $\frac{2}{5}$ of the pizza is left over. For a second helping, Dom has $\frac{1}{4}$ of what is left. What fraction of the whole pizza does Dom have for a second helping? Give your answer in its lowest terms.

Dividing fractions

When you work out 6 ÷ 3, you are finding how many 3s there are in 6.

Finding $6 \div \frac{1}{3}$ is the same as finding how many $\frac{1}{3}$s there are in 6.

In this diagram, $\frac{1}{3}$ of the rectangle is shaded red.

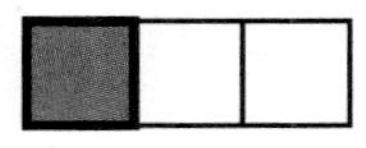

You can see that 18 of the red squares would fit in 6 of these rectangles.

So $6 \div \frac{1}{3} = 6 \times 3 = 18$

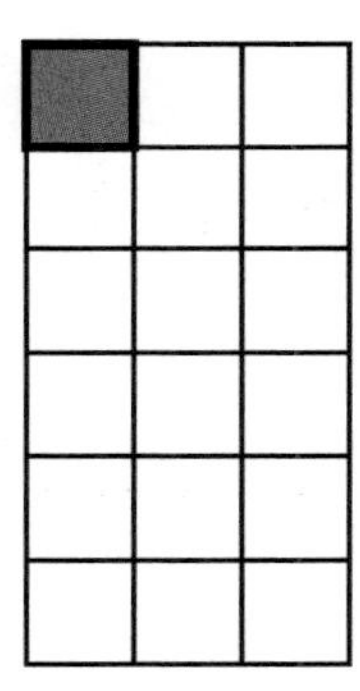

This can be extended. For example, finding $4 \div \frac{2}{3}$.

This is the same as finding how many $\frac{2}{3}$s there are in 4.

In this diagram $\frac{2}{3}$ of the rectangle is shaded red.

You can see that 6 of these 2-square shapes fit in the 4 rectangle rows.

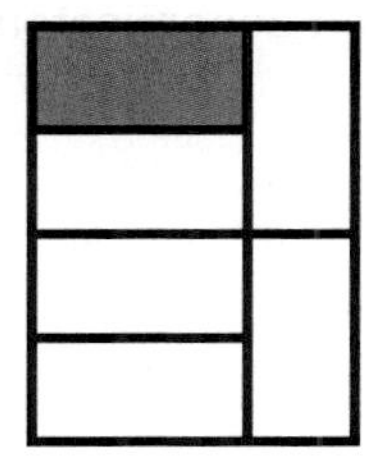

You can also think of it in two steps:

$4 \div \frac{1}{3} = 4 \times 3 = 12$ (there are 12 small squares in the 4 rectangles)

So $4 \div \frac{2}{3} = 12 \div 2 = 6$

Or $4 \div \frac{2}{3} = 4 \times \frac{3}{2} = \frac{12}{2} = 6$.

Here is a rule that reminds you what to do:

When dividing fractions, turn the second fraction upside-down and then multiply.

Example 3

Work out these.

a) $\frac{8}{9} \div \frac{1}{3}$ **b)** $\frac{8}{9} \div 2$ **c)** $\frac{1}{2} \div \frac{1}{3}$

Solution

a) $\frac{8}{9} \div \frac{1}{3} = \frac{8}{9} \times \frac{3}{1}$ Turn the second fraction upside-down and multiply.

$= \frac{8}{{}_3\cancel{9}} \times \frac{\cancel{3}^1}{1}$ Divide 9 and 3 by 3.

$= \frac{8}{3} \times 1$

$= \frac{8}{3}$

$= 2\frac{2}{3}$ Change to a mixed number.

b) $\frac{8}{9} \div 2 = \frac{8}{9} \div \frac{2}{1}$ Write 2 as $\frac{2}{1}$.

$= \frac{8}{9} \times \frac{1}{2}$ Turn the second fraction upside-down and multiply.

$= \frac{{}^4\cancel{8}}{9} \times \frac{1}{\cancel{2}_1}$ Divide 8 and 2 by 2.

$= \frac{4}{9} \times 1$

$= \frac{4}{9}$

c) $\frac{1}{2} \div \frac{1}{3} = \frac{1}{2} \times \frac{3}{1}$ Turn the second fraction upside-down and multiply.

$= \frac{3}{2}$

$= 1\frac{1}{2}$ Change to a mixed number.

Example 4

Work out these.

a) $\frac{4}{5} \div \frac{3}{10}$ **b)** $\frac{9}{10} \div \frac{3}{4}$

Solution

a) $\frac{4}{5} \div \frac{3}{10} = \frac{4}{5} \times \frac{10}{3}$ Turn the second fraction upside-down and multiply.

$= \frac{4}{{}_1\cancel{5}} \times \frac{\cancel{10}^2}{3}$ Divide 10 and 5 by 5.

$= \frac{4}{1} \times \frac{2}{3}$

$= \frac{8}{3}$

$= 2\frac{2}{3}$ Change to a mixed number.

b) $\frac{9}{10} \div \frac{3}{4} = \frac{9}{10} \times \frac{4}{3}$

$= \frac{{}^{3}\cancel{9}}{{}_{5}\cancel{10}} \times \frac{\cancel{4}^{2}}{\cancel{3}_{1}}$ Divide 9 and 3 by 3 and 4 and 10 by 2.

$= \frac{3}{5} \times \frac{2}{1}$

$= \frac{6}{5}$

$= 1\frac{1}{5}$ Change to a mixed number.

Exercise 6.2

Work out these. Where necessary, write your answers as proper fractions or mixed numbers in their lowest terms.

1 $9 \div \frac{1}{3}$	**2** $12 \div \frac{3}{4}$	**3** $\frac{3}{4} \div \frac{1}{2}$	**4** $\frac{2}{3} \div 3$
5 $\frac{2}{3} \div 5$	**6** $\frac{1}{3} \div \frac{3}{4}$	**7** $\frac{4}{9} \div 2$	**8** $\frac{3}{8} \div \frac{1}{4}$
9 $\frac{4}{5} \div 4$	**10** $\frac{2}{3} \div \frac{1}{3}$	**11** $\frac{2}{3} \div \frac{1}{6}$	**12** $\frac{5}{6} \div 10$
13 $\frac{7}{9} \div \frac{1}{9}$	**14** $\frac{3}{4} \div \frac{1}{8}$	**15** $\frac{4}{5} \div \frac{3}{10}$	**16** $\frac{1}{4} \div \frac{3}{8}$
17 $\frac{3}{4} \div \frac{5}{6}$	**18** $\frac{2}{5} \div \frac{1}{15}$	**19** $\frac{2}{5} \div \frac{7}{10}$	**20** $\frac{1}{6} \div \frac{3}{4}$
21 $\frac{3}{8} \div \frac{1}{4}$	**22** $\frac{2}{3} \div \frac{5}{8}$	**23** $\frac{3}{4} \div \frac{3}{8}$	**24** $\frac{2}{3} \div \frac{4}{15}$
25 $\frac{3}{5} \div \frac{9}{10}$	**26** $\frac{1}{8} \div \frac{5}{12}$	**27** $\frac{5}{7} \div 3$	**28** $6 \div \frac{2}{3}$
29 $\frac{2}{3} \div 5$	**30** $\frac{4}{9} \times \frac{1}{3}$	**31** $\frac{5}{6} \times \frac{1}{4}$	**32** $\frac{11}{12} \div \frac{2}{3}$
33 $\frac{4}{5} \div \frac{1}{2}$	**34** $\frac{5}{7} \times \frac{3}{4}$	**35** $\frac{8}{9} \times \frac{1}{6}$	**36** $\frac{7}{10} \div \frac{4}{5}$
37 $\frac{8}{9} \times \frac{5}{6}$	**38** $\frac{7}{15} \div \frac{3}{10}$	**39** $\frac{4}{9} \div \frac{1}{12}$	**40** $\frac{7}{20} \times \frac{5}{8}$

41 Out of Peter's 'take home' pay, he gives his Mum £50 for food etc. He spends $\frac{3}{8}$ of the rest and saves what is left over. How much is his 'take home' pay in a week when he saves £60?

42 In an election, the Green candidate got 240 votes, which was $\frac{3}{8}$ of the votes. The Independent candidate got $\frac{1}{5}$ of the votes. How many votes did the Independent candidate get?

Challenge 1

a) Stephen and Vicky bought a house together. It cost £245 000. Vicky paid three-fifths of the cost. How much did she pay?

b) Iain gave one-tenth of his income to charity. Two-thirds of his gift went to overseas aid. What fraction of his income did he give to overseas aid?

Adding and subtracting mixed numbers

Reminder: when adding or subtracting fractions, change them to equivalent fractions with the same denominator, then add or subtract the numerators.

To add or subtract mixed numbers, deal with the whole numbers first.

Example 5

Work out these.

a) $1\frac{1}{6} + 2\frac{1}{3}$ **b)** $2\frac{3}{4} + \frac{3}{5}$ **c)** $3\frac{2}{3} - \frac{1}{6}$ **d)** $4\frac{1}{5} - 1\frac{1}{2}$

Solution

a) $1\frac{1}{6} + 2\frac{1}{3} = 3 + \frac{1}{6} + \frac{1}{3}$

$= 3 + \frac{1}{6} + \frac{2}{6}$

$= 3\frac{3}{6}$

$= 3\frac{1}{2}$

Add the whole numbers, then deal with the fractions in the normal way.

b) $2\frac{3}{4} + \frac{3}{5} = 2 + \frac{15}{20} + \frac{12}{20}$

$= 2 + \frac{27}{20}$

$= 2 + 1 + \frac{7}{20}$

$= 3\frac{7}{20}$

You end up with an improper fraction which you have to change to a mixed number, and then add the whole numbers.

c) $3\frac{2}{3} - 1\frac{1}{6} = 3 - 1 + \frac{2}{3} - \frac{1}{6}$

$= 2 + \frac{4}{6} - \frac{1}{6}$

$= 2\frac{3}{6}$

$= 2\frac{1}{2}$

Subtract the whole numbers and then the fractions.

d) $4\frac{1}{5} - 1\frac{1}{2} = 3 + \frac{2}{10} - \frac{5}{10}$

$= 2 + \frac{10}{10} + \frac{2}{10} - \frac{5}{10}$

$= 2\frac{7}{10}$

Working out $\frac{2}{10} - \frac{5}{10}$ gives an answer of negative $\frac{3}{10}$. Change one of the whole numbers into $\frac{10}{10}$, then subtract.

Exercise 6.3

1 Add these fractions. Write your answers as simply as possible.

a) $\frac{3}{10} + 1\frac{2}{5}$ b) $1\frac{1}{10} + \frac{3}{5}$

c) $2\frac{1}{5} + 1\frac{1}{10}$ d) $4\frac{1}{2} + 2\frac{3}{10}$

e) $1\frac{3}{10} + \frac{2}{5}$

2 Add these fractions. Write your answers as simply as possible.

a) $1\frac{5}{8} + \frac{1}{8}$ b) $\frac{1}{8} + 2\frac{3}{8}$

c) $1\frac{3}{4} + 2\frac{1}{8}$ d) $3\frac{1}{2} + \frac{7}{8}$

e) $4\frac{5}{8} + \frac{3}{4}$

3 Subtract these fractions. Write your answers as simply as possible.

a) $2\frac{3}{10} - \frac{1}{10}$ b) $3\frac{5}{10} - 1\frac{3}{10}$

c) $4\frac{1}{5} - 2\frac{1}{10}$ d) $3\frac{1}{2} - \frac{3}{10}$

e) $6\frac{7}{10} - 6\frac{2}{5}$

4 Subtract these fractions. Write your answers as simply as possible.

a) $1\frac{3}{8} - \frac{1}{8}$ b) $5\frac{5}{8} - 3\frac{3}{8}$

c) $2\frac{1}{4} - 1\frac{1}{8}$ d) $4\frac{1}{2} - 3\frac{3}{8}$

e) $5\frac{5}{8} - 1\frac{1}{4}$

5 Add these fractions. Write your answers as simply as possible.

a) $\frac{5}{12} + \frac{1}{12} + \frac{3}{4}$ b) $\frac{7}{12} + \frac{1}{4} + \frac{2}{3}$

c) $\frac{3}{4} + \frac{5}{12} + \frac{1}{6}$ d) $\frac{1}{12} + \frac{5}{6} + \frac{3}{4}$

e) $\frac{1}{5} + \frac{17}{20} + \frac{3}{10}$

6 Add these fractions. Write your answers as simply as possible.

a) $\frac{1}{8} + \frac{3}{8} + \frac{7}{8}$ b) $\frac{7}{8} + \frac{3}{8} + \frac{1}{4}$

c) $\frac{3}{4} + 1\frac{1}{8} + \frac{1}{8}$ d) $1\frac{1}{2} + \frac{7}{8} + \frac{3}{4}$

e) $2\frac{1}{2} + \frac{1}{4} + \frac{3}{8}$

7 Subtract these fractions. Write your answers as simply as possible.

a) $1\frac{3}{8} - \frac{5}{8}$ b) $2\frac{5}{8} - 1\frac{3}{8}$

c) $3\frac{1}{4} - 1\frac{5}{8}$ d) $3\frac{1}{2} - 1\frac{7}{8}$

e) $3\frac{1}{8} - \frac{1}{4}$

8 Subtract these fractions. Write your answers as simply as possible.

a) $5\frac{1}{10} - \frac{7}{10}$ b) $1\frac{5}{6} - \frac{2}{3}$

c) $1\frac{1}{5} - \frac{7}{10}$ d) $1\frac{1}{2} - \frac{3}{4}$

e) $2\frac{7}{10} - 1\frac{4}{5}$

9 Work out these.

a) $3\frac{1}{2} + 2\frac{1}{5}$ b) $4\frac{7}{8} - 1\frac{3}{4}$

c) $4\frac{2}{7} + \frac{1}{2}$ d) $6\frac{5}{12} - 3\frac{1}{3}$

e) $4\frac{3}{4} + 2\frac{5}{8}$ f) $5\frac{5}{6} - 1\frac{1}{4}$

10 Work out these.

a) $4\frac{7}{9} + 2\frac{5}{6}$ b) $4\frac{7}{13} - 4\frac{1}{2}$

c) $3\frac{5}{7} + 2\frac{1}{3}$ d) $7\frac{2}{5} - 1\frac{3}{4}$

e) $5\frac{2}{7} - 3\frac{1}{2}$ f) $4\frac{1}{12} - 3\frac{1}{4}$

11 Write down two fractions that add up to $\frac{3}{4}$.

12 Gurdeep had a packet of sweets. He ate $\frac{1}{2}$ of them himself and gave Torin $\frac{1}{8}$. What fraction did he have left?

13 Faisal cut two pieces of wood $3\frac{3}{8}$ inches and $5\frac{1}{4}$ inches long from a piece 10 inches long. How long was the piece that was left?

14 A table $31\frac{5}{8}$ inches high was raised by putting each of its legs on a block $1\frac{3}{4}$ inches high. What was the new height of the table?

15 Luci baked two apple pies of the same size. Two-fifths of one of them was eaten at lunch time. One and a quarter pies were eaten at supper time. What fraction of a pie was left over?

Challenge 2

Make up a fraction 'story problem' similar to questions 12 to 15 in Exercise 6.3.
Solve it yourself then swap stories with a partner.
Solve your partner's problem.
Check that you get the same answers.

Key Ideas

- To multiply a fraction by an integer, multiply just the numerator by the integer. Simplify the answer if possible.
- When multiplying fractions, multiply the numerators and multiply the denominators and simplify if possible.
- When dividing fractions, turn the second fraction upside-down and multiply.
- To add or subtract fractions, change them to equivalent fractions all with the same denominator and add or subtract the numerators.
- To add or subtract mixed numbers, deal with the whole numbers separately.

CHAPTER 7

Plans and elevations

YOU WILL LEARN ABOUT	YOU SHOULD ALREADY KNOW
o Plans and elevations	o How to draw accurately using a pencil and ruler o How to draw triangles using a ruler and a protractor o How to draw triangles using a ruler and a pair of compasses o How to make isometric drawings of shapes

Any one view of a three-dimensional object will not show all its features. To get the full picture you need three different views from three perpendicular directions.

The view from above is called the **plan**. The views from the sides are called the **front elevation** and the **side elevation.**

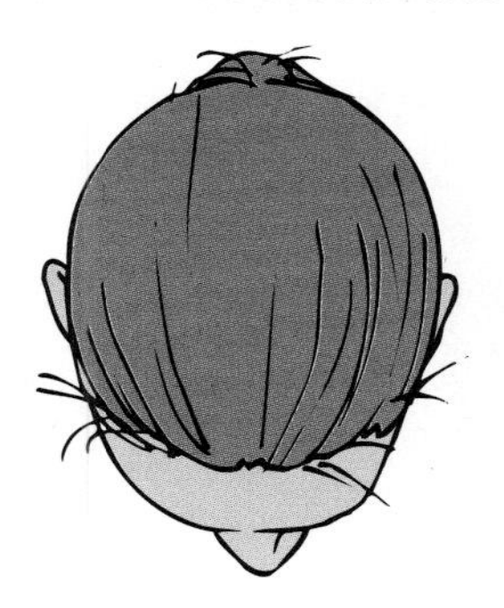

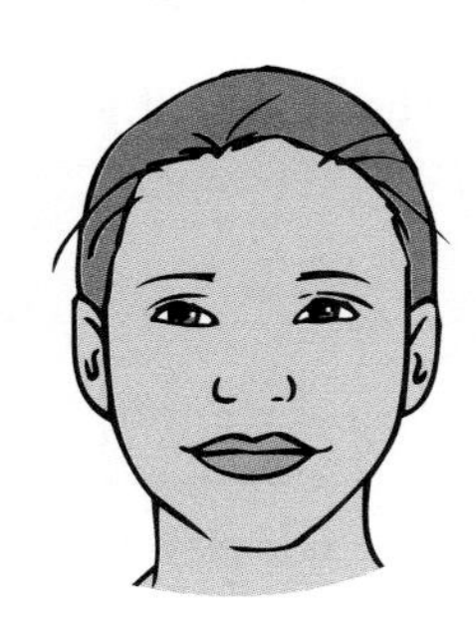

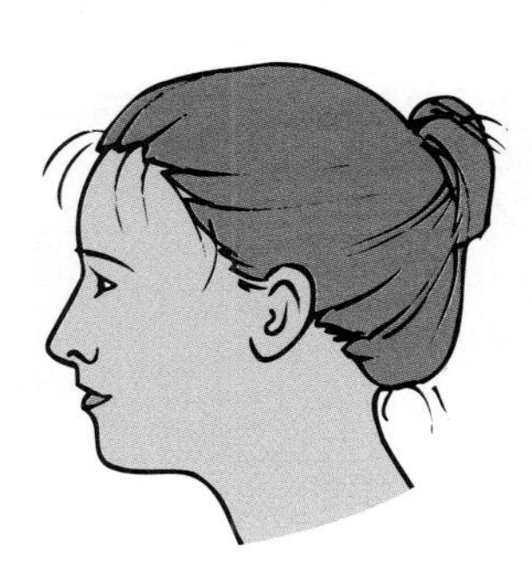

Example 1

Sketch the plan and elevations of this solid.

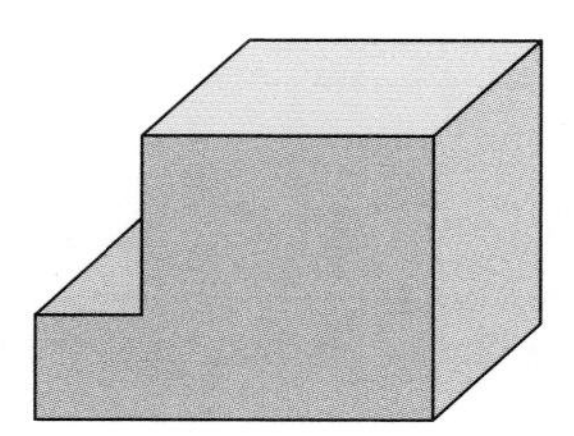

Solution

The plan view of the solid shape is the view of the surfaces that you get if you look at the solid directly from above. This is sometimes referred to as the 'bird's-eye' view.

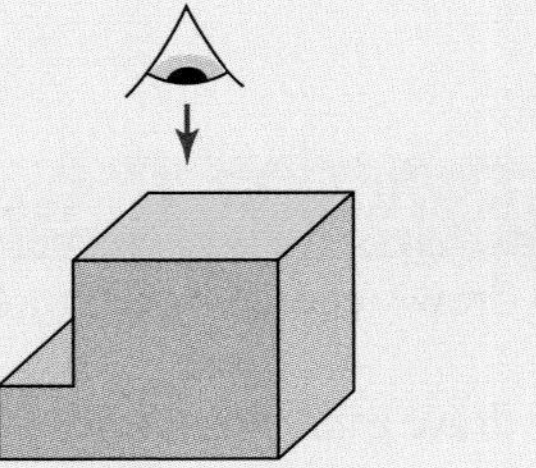

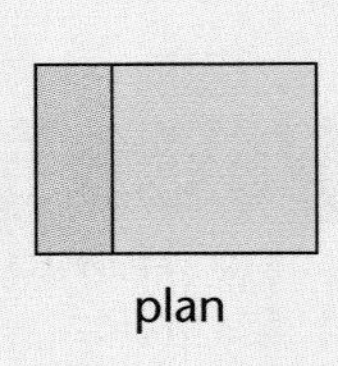

plan

Notice that the only indication of the step in the solid is a line on the plan. This view will not tell you how high the step in the solid is.

The front elevation, or front view, of the solid shape is the view of the surfaces that you get if you look at the solid directly from the front.

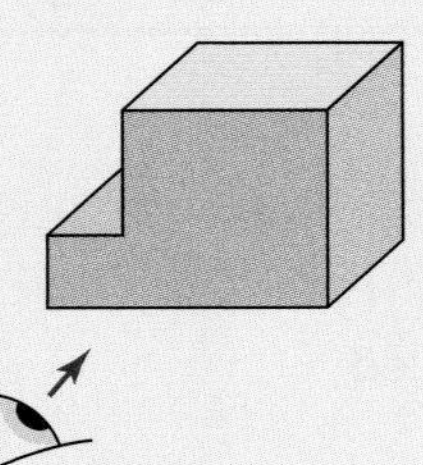

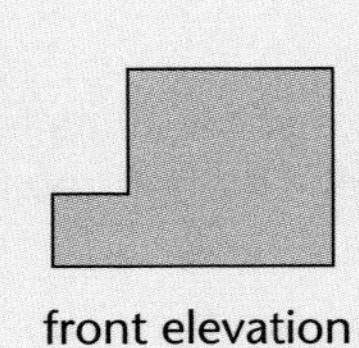

front elevation

This view will not tell you how deep the solid is.

The side elevation, or side view, of the solid shape is the view of the surfaces that you get if you look at the solid directly from the side.

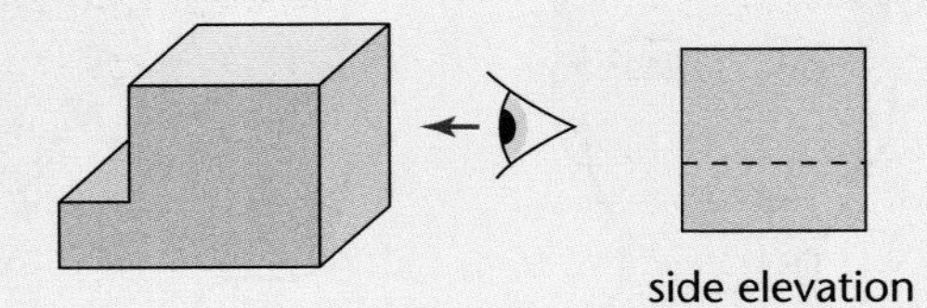

side elevation

The step in the solid is hidden from view from this direction. To show it is there, but cannot be seen, we use a dotted line. Dotted lines are always used to indicate hidden features of the solid.

The side elevation can be viewed from either side of the solid. From the opposite side, the view would be the same but with a solid line instead of a dotted line: the step in the solid *can* be seen from this direction.

Activity 1

Make a shape from multilink cubes (use five to eight cubes) and put it on your table or desk.

On centimetre-squared paper, draw its plan view and its elevations from all four sides.

Challenge 1

Swap the drawings you made in Activity 1 with a partner.
Use your partner's drawings to construct the shape from multilink cubes.
Check with your partner that you both have made the same shape.

Example 2

Sketch the front elevation, F, the plan, P, and the side elevation, S, of this ring.

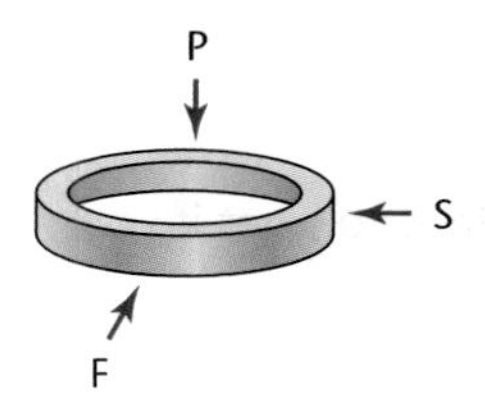

Solution

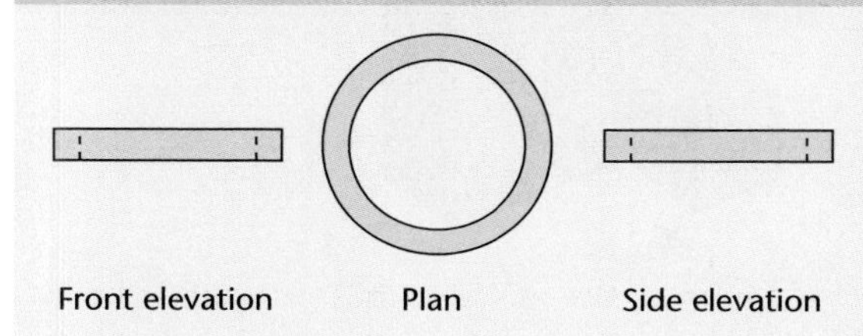

Notice that the front and side elevations are the same. The two dotted lines indicate the hidden edge of the inside of the ring. The curved part of the ring appears only as a rectangle in the front and the side elevations.

Example 3

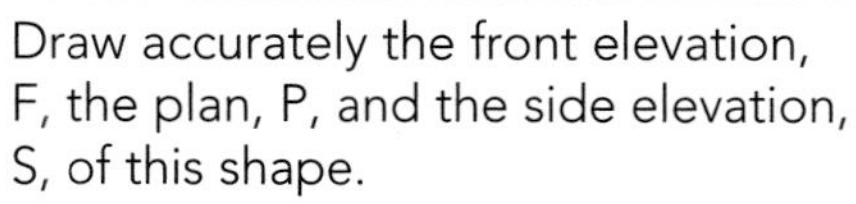

Draw accurately the front elevation, F, the plan, P, and the side elevation, S, of this shape.

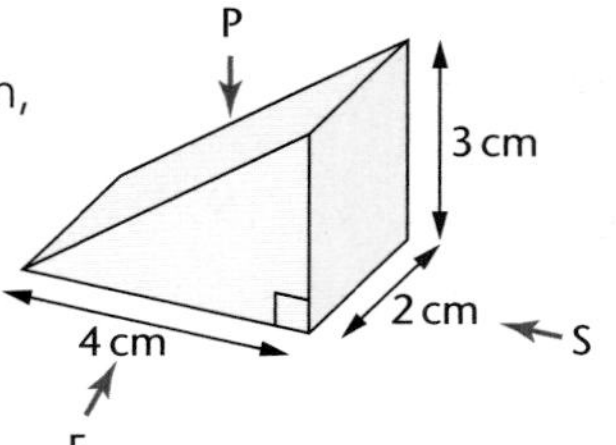

Exam tip

If you do not have squared paper, you need to draw the right angles using a protractor or set square. Sometimes you may need to use compasses.

Solution

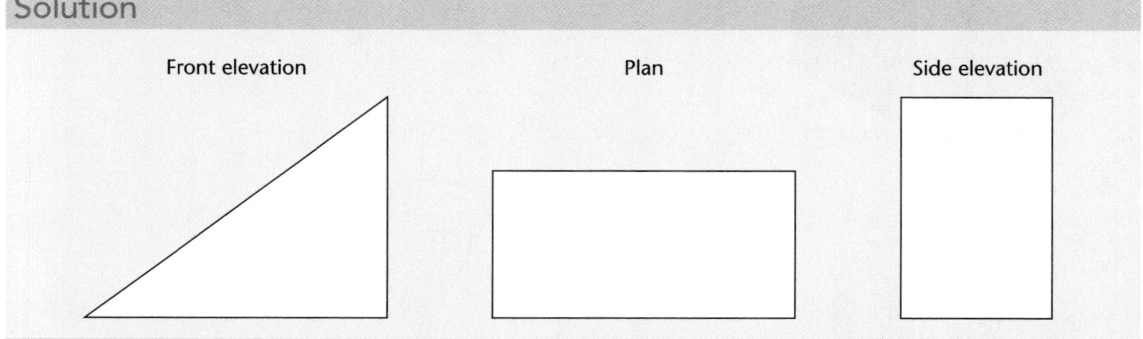

Exercise 7.1

Sketch the plan and elevations of each of the solid shapes in questions **1** to **15**.

The arrows indicate the direction of the plan, P, the front elevation, F, and the side elevation, S.

1

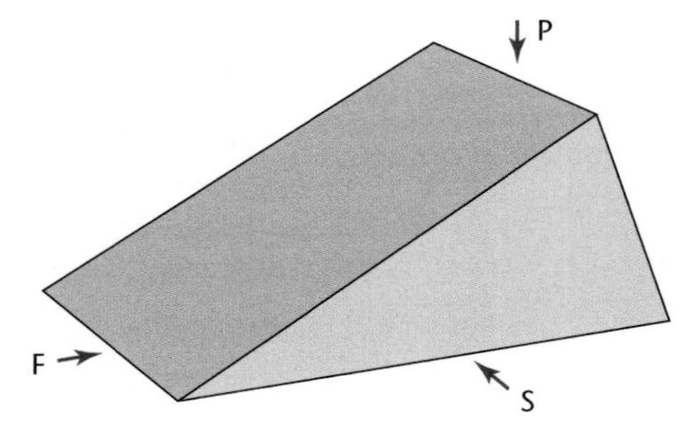

2

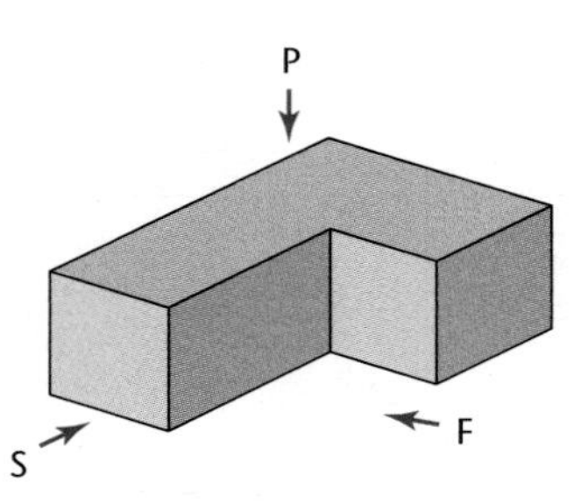

3

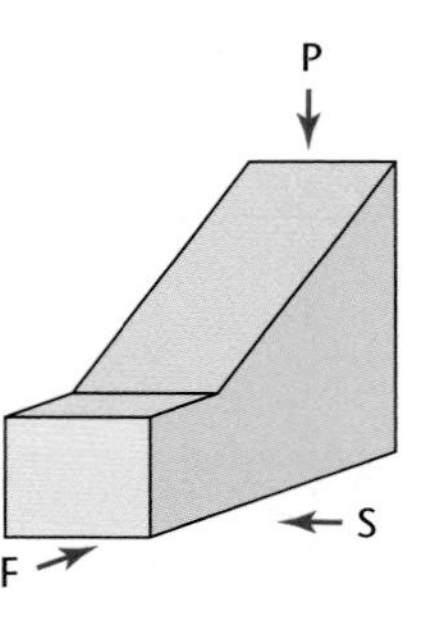

4

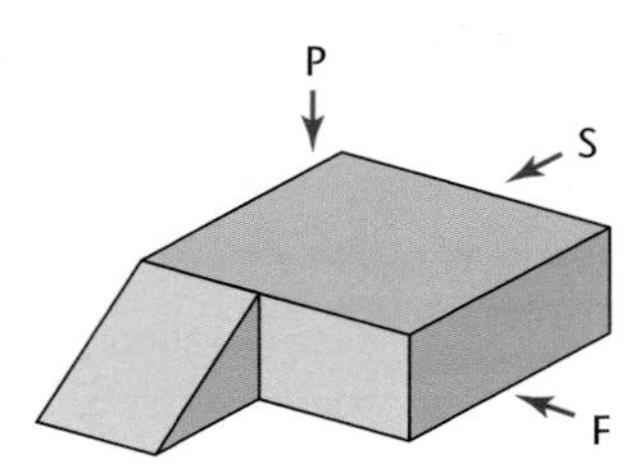

5

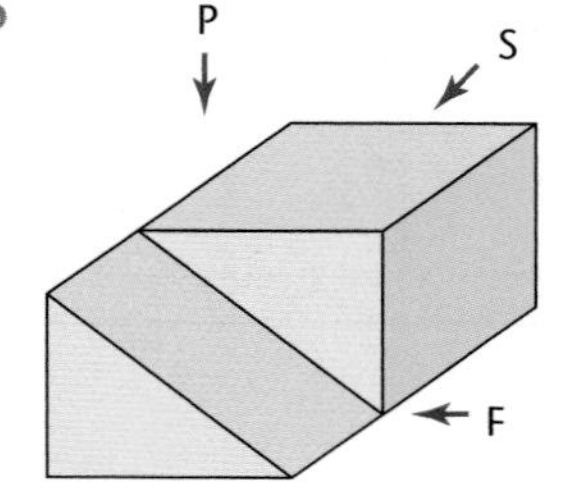

6

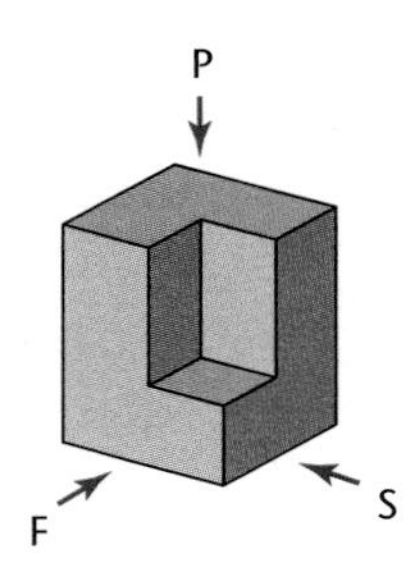

7

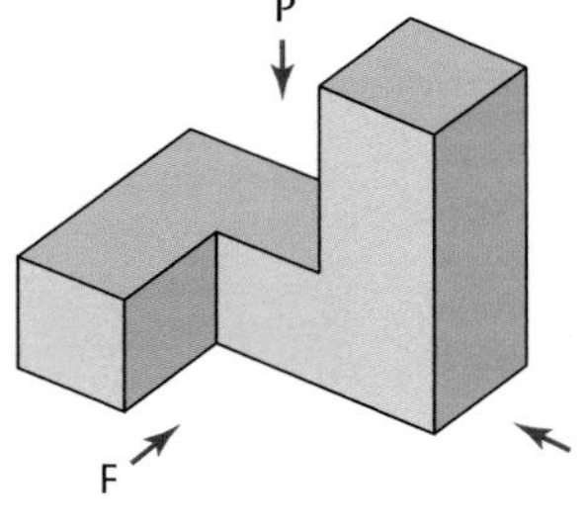

8

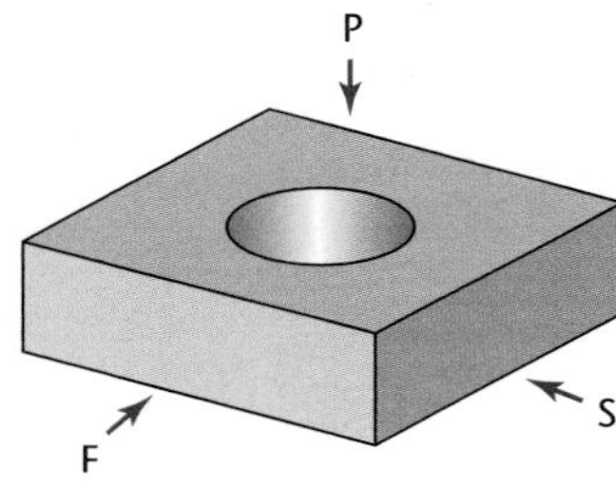

9

P

F

S

10

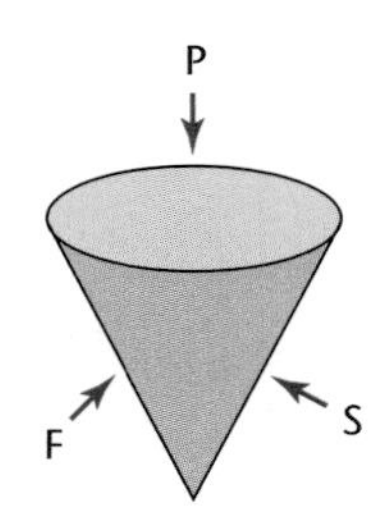

11

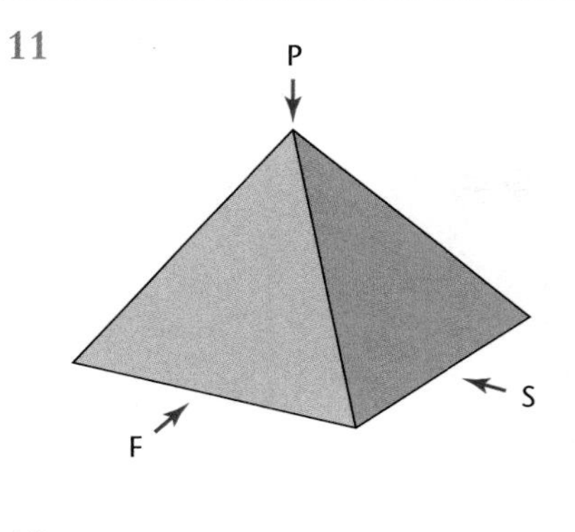

12

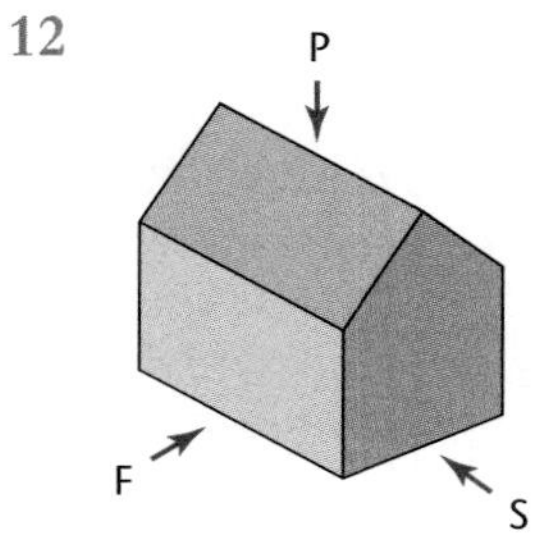

13

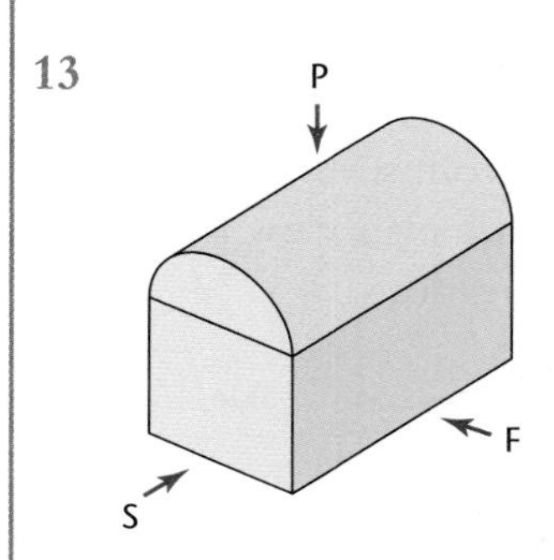

14

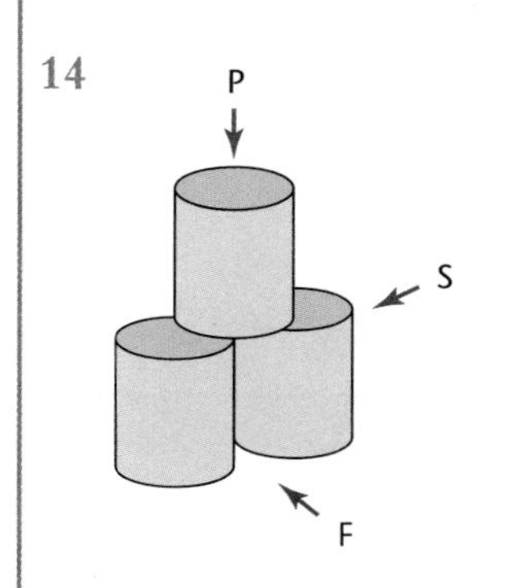

15

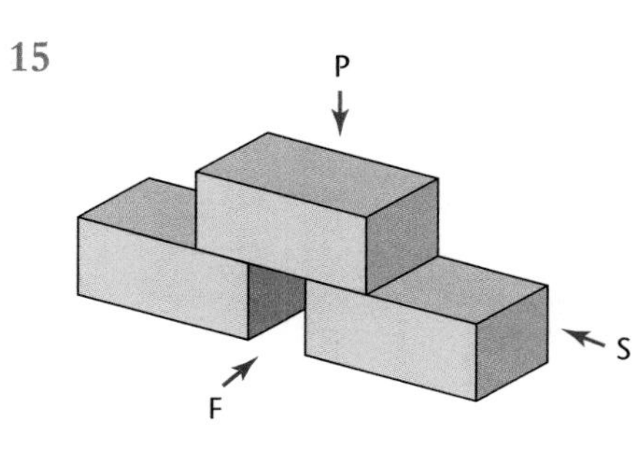

Draw accurately the plan and elevations of each of the solid shapes in questions **16** to **24**.

Use a protractor and compasses as required.

16

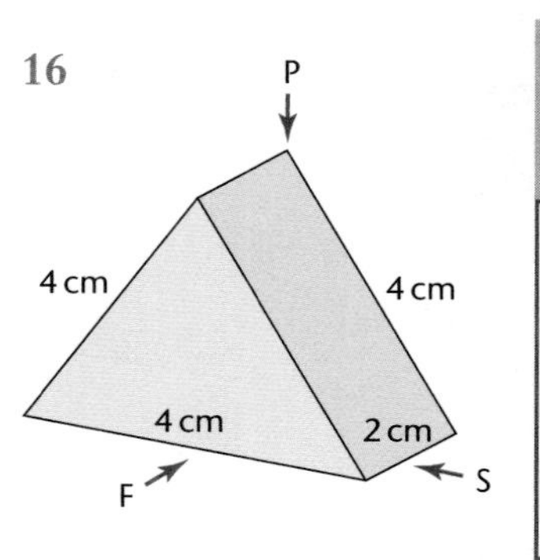

You will need to measure the height of the triangle in the front elevation to draw the side elevation.

17

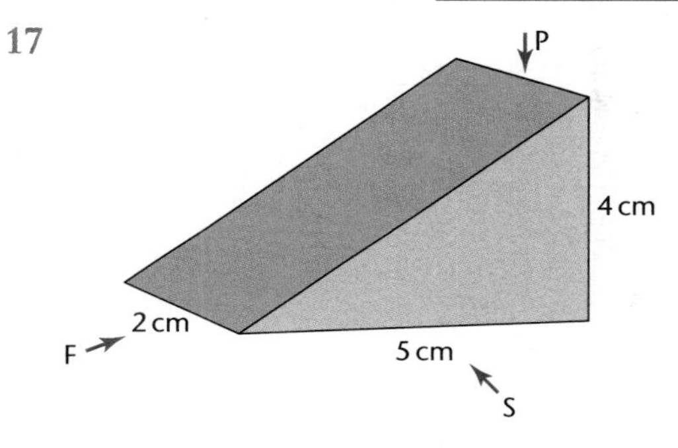

18

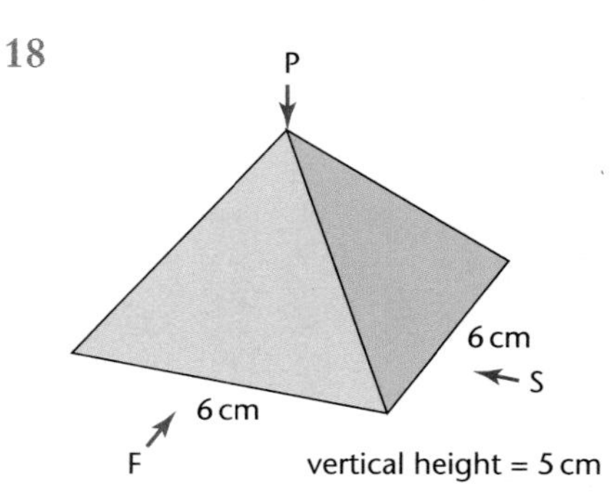

19

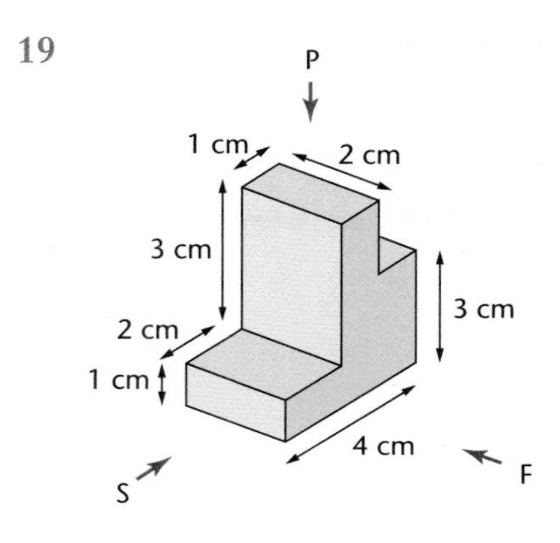

20

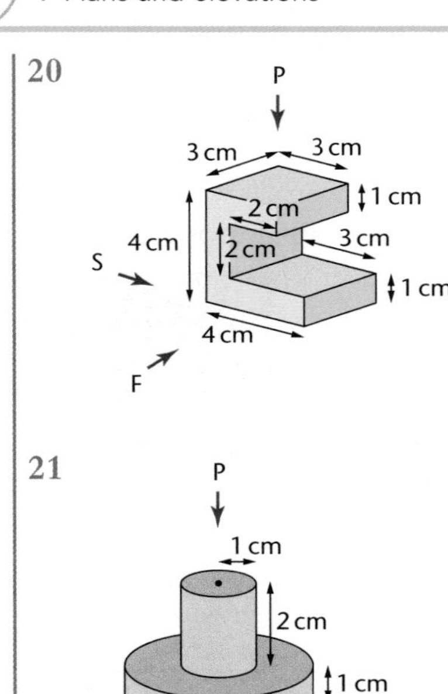

21

P
1 cm
2 cm
1 cm
F
2 cm
S

22

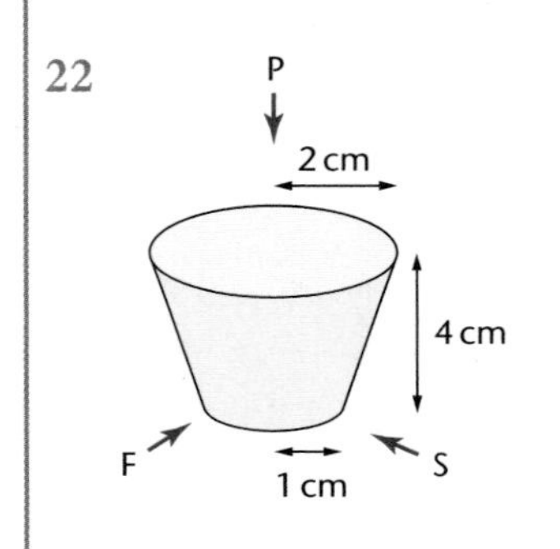

23

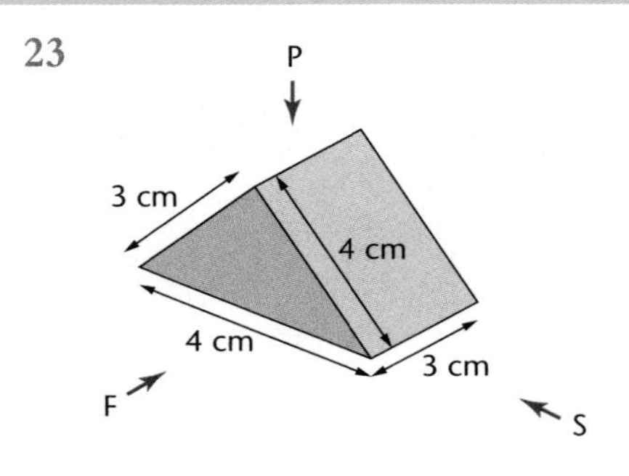

24

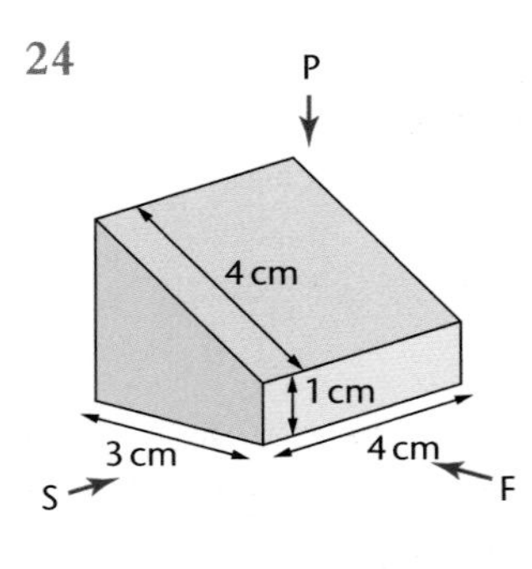

You will need to draw the side elevation before you draw the front elevation. You will need to use compasses to draw the side elevation.

25 A shape is made from seven centimetre cubes. Here are the plan view and the front and side elevations.

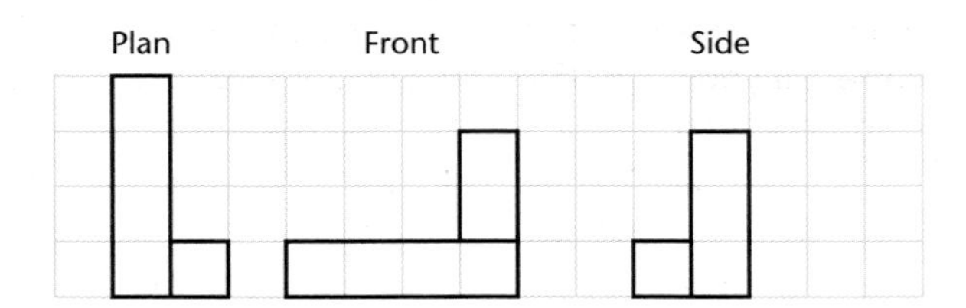

On isometric paper, make an isometric drawing of the shape.

Key Ideas

- A plan view of an object is the shape of the object viewed from above.
- An elevation of an object is the shape of the object viewed from the front, back or side.

Expressions and formulae

YOU WILL LEARN ABOUT	YOU SHOULD ALREADY KNOW
○ Word formulae ○ Writing a formula in letters ○ Substituting numbers for letters in a formula or expression	○ Basic algebra (letters and simplifying) ○ How to add, subtract, multiply and divide with decimals, fractions and negative numbers

Letters for unknowns

Imagine you had a job where you were paid by the hour. You would receive the same amount for each hour you worked.

How could you work out how much you will earn in a week?

You would need to work it out as the number of hours you worked multiplied by the amount you are paid for each hour.

This is a formula in words.

If you work 35 hours at £4.50 an hour, it is easy to work out $35 \times £4.50$, but what if the numbers change? The calculation '$35 \times £4.50$' is only correct if you work 35 hours. Suppose you move to another job where you are paid more for each hour.

You need a simple formula that *always* works. You can use symbols to stand for the numbers that can change.

You could use ? or □, but it is less confusing to use letters.

Let the number of hours worked be N
the amount you are paid each hour be £P
the amount you earn in a week be £W.

Then $W = N \times P$.

Example 1

Use the formula $W = N \times P$ to find the amount earned in a week when $N = 40$ and $P = 5$.

Solution

$W = N \times P$
$= 40 \times 5$
$= 200$

The amount earned in a week is £200.

Some rules of algebra

- You do not need to write the $\times$ sign.
  $4 \times t$ is written $4t$.
- In multiplications, the number is always written in front of the letter.
  $p \times 6 - 30$ is written $6p - 30$.
- You often start a formula with the single letter you are finding.
  $2 \times l + 2 \times w = P$ is written $P = 2l + 2w$.
- When there is a division in a formula, it can be written as a fraction.
  $y = k \div 6$ can be written $y = \frac{k}{6}$.

Example 2

Write a formula for the mean height, M cm, of a group of n people whose heights total h cm.

Solution

To find the mean height, you divide the total of the heights by the number of people, so the formula for M is

$M = h \div n$ or $M = \frac{h}{n}$.

Exam tip

If you are not sure whether to multiply or divide, try an example with numbers first.

Example 3

The cost (£C) of hiring a car is a fixed charge (£f), plus the number of days (n) multiplied by the daily rate (£d). Write a formula for C.

Solution

$C = f + n \times d$ or $C = f + nd$

Exercise 8.1

Write a formula for each of these, using the letters given.

1 The cost (C pence) of x pencils at y pence each.

2 The area (A cm^2) of a rectangle m cm long and n cm wide.

3 The height (h cm) of a stack of n tins each t cm high.

4 The temperature in °F (F) is 32, plus 1.8 times the temperature in °C (C).

5 My gas bill (£B) is a fixed charge (£s), plus the number (n) of units used multiplied by the cost (£u) of each unit.

6 The mileage performance (R), the number of miles per litre of a car is the number (m) of miles travelled divided by the number (p) of litres of petrol used.

7 The time (T minutes) to cook a turkey is 40 minutes for each kilogram (k) plus 30 minutes.

8 The area (A cm^2) of a triangle is half the base (b cm) times the height (h cm).

9 The number of dollars (d) is 1.65 times the number of pounds (p).

10 The current in a circuit (I) is the voltage (V) divided by the resistance (R).

11 The cost (£C) of hiring a car for n days at a rate of £40 per day plus a basic charge of £12.

12 The cost (£C) of n units of electricity at 12p per unit, plus a standing charge of £6.

13 The hire charge (£C) for a car is £28 multiplied by the number of days (d), plus a fixed charge of £15.

14 The cost of petrol (£C) is the number of litres (n) multiplied by the price of petrol per litre (£p).

15 The total of the wages (£w) in a supermarket is the number of workers (n) multiplied by the weekly wage (£q).

16 The perimeter (p cm) of a quadrilateral is the sum of the lengths in centimetres of its sides (a, b, c, d).

17 The cost (£P) of n books at q pounds each.

18 The number of books (N) that can fit on a shelf is the length (L cm) of the shelf divided by the thickness (t cm) of each book.

19 The number of euros (E) is 1.5 times the number of pounds (P).

20 The number of posts (Q) for a fence is the length of the fence (R metres) divided by 2, plus 1.

21 The number of eggs (n) in a box a eggs across, b eggs along and c eggs up.

22 The cost (£C) of n CDs at £7.50 each and v DVDs at £12 each.

23 The cost (£C) of n minutes of mobile phone use at 30p per minute, plus a rental charge of £14.

24 The area (A cm^2) of a kite is half the product of the width (w cm) and the length (m cm).

Before you do Exercise 8.2, check your answers to Exercise 8.1

Exercise 8.2

Use the formulae from Exercise 8.1 to find each of these.

1 C when $x = 15$ and $y = 12$

2 A when $m = 7$ and $n = 6$

3 h when $n = 20$ and $t = 17$

4 F when $C = 40$

5 B when $s = 9.80$, $n = 234$ and $u = 0.065$

6 R when $m = 320$ and $p = 53.2$

7 T when $k = 9$

8 A when $b = 5$ and $h = 6$

9 d when $p = 200$

10 I when $V = 13.6$ and $R = 2.5$

11 C when $n = 5$

12 C when $n = 532$

13 C when $d = 5$

14 C when $n = 50$ and $p = 70$

15 w when $n = 200$ and $q = 150$

16 p when $a = 7$, $b = 5$, $c = 8$ and $d = 2$

17 P when $n = 25$ and $q = 7$

18 N when $L = 90$ and $t = 3$

19 E when $P = 50$

20 Q when $R = 36$

21 n when $a = 12$, $b = 20$ and $c = 6$

22 C when $n = 2$ and $v = 3$

23 C when $n = 80$

24 A when $w = 40$ and $m = 60$

Substituting numbers into algebraic expressions

Numbers that can be substituted into algebraic expressions can be positive, negative, decimals or fractions.

Example 4	Solution
a) Find the value of $4x + 3$ when $x = 2$.	**a)** $4x + 3 = 4 \times 2 + 3$ $= 8 + 3$ $= 11$
b) Find the value of $3x^2 + 4$ when $x = 3$.	**b)** $3x^2 + 4 = 3 \times 3^2 + 4$ $= 3 \times 9 + 4$ $= 27 + 4$ $= 31$
c) Find the value of $2x^2 + 6$ when $x = {}^-2$.	**c)** $2x^2 + 6 = 2 \times ({}^-2)^2 + 6$ $= 2 \times 4 + 6$ $= 8 + 6$ $= 14$

Exam tip

Work out each term separately and then collect together.

Exam tip

Remember $3x^2$ means $3 \times x \times x$ not $3 \times x \times 3 \times x$.

Exam tip

Take special care when negative numbers are involved.

Exercise 8.3

1 Find the value of these expressions when $a = 5$, $b = 4$ and $c = 2$.

a) $a + b$
b) $b + c$
c) $a - c$
d) $a + b + c$
e) $2a$
f) $3b$
g) $5c$
h) $3a + b$
i) $3c - b$
j) $a + 6c$
k) $4a + 2b$
l) $2b + 3c$
m) $a - 2c$
n) $8c - 2b$
o) bc
p) $4ac$
q) abc
r) $ac + bc$
s) $ab - bc - ca$
t) a^2
u) $\frac{a + b}{3}$
v) $\frac{ab}{c}$
w) $b^2 + c^2$
x) $3c^2$
y) a^2b
z) c^3

2 Find the value of these expressions when $t = 3$.

a) $t + 2$
b) $t - 4$
c) $5t$
d) $4t - 7$
e) $2 + 3t$
f) $10 - 2t$
g) t^2
h) $10t^2$
i) $t^2 + 2t$
j) t^3

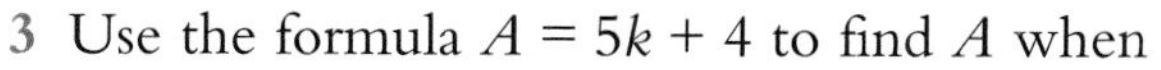

3 Use the formula $A = 5k + 4$ to find A when

a) $k = 3$.
b) $k = 7$.
c) $k = 0$.
d) $k = \frac{1}{2}$.
e) $k = 2.1$.

4 Use the formula $y = mx + c$ to find y when

a) $m = 3, x = 2, c = 4$.
b) $m = \frac{1}{2}, x = 8, c = 6$.
c) $m = 10, x = 15, c = 82$.

5 Use the formula $D = \frac{m}{v}$ to find D when

a) $m = 24, v = 6$.
b) $m = 150, v = 25$.
c) $m = 17, v = 2$.
d) $m = 4, v = \frac{1}{2}$.

Challenge 1

Find the value of these expressions when $x = {}^{-}2$, $y = 3$ and $z = {}^{-}4$.

a) $x + y$
b) $y - z$
c) $3y + z$
d) $5z + 2x$
e) yz
f) xz

Challenge 2

Formula codes

Give each letter of the alphabet a number, in order, from 1 to 26.

$a = 1 \quad b = 2 \quad c = 3 \quad d = 4 \quad e = 5 \quad \ldots \quad x = 24 \quad y = 25 \quad z = 26$

a) Substitute the value of the letters into each of the formulae below.
Find the letters for each of the answers and they will spell a word.
The first one has been done for you.

$a + c = 1 + 3 = 4 \rightarrow d$

$t - s$

$2g$

$\frac{u}{c}$

$2p - 3i$

$dt + b - h^2$

b) Make some messages of your own.
Exchange your messages with a friend and decipher them.

c) Try a different numbering of the letters of the alphabet such as $a = 26$, $b = 25$, $c = 24$, ...
and write some messages using these.
Exchange your messages with a friend and see if you can 'crack' the new code.

Using harder numbers when substituting

You need to be able to work with positive and negative numbers, decimals or fractions, and to do so with or without a calculator.

Example 5

Find the value of

a) $a + 5b^3$ when $a = 2.6$ and $b = 3.2$

b) $5cd$ when $c = 2.4$ and $d = 3.2$

Solution

a) $a + 5b^3 = 2.6 + 5 \times 3.2^2$
$= 2.6 + 5 \times 10.24$
$= 2.6 + 51.2$
$= 53.8$

The working has been shown here to remind you of the order of operations, but you should be able to do this calculation in one step on your calculator.

b) $5cd = 5 \times 2.4 \times 3.2$
$= 12 \times 3.2$
$= 10 \times 3.2 + 2 \times 3.2$
$= 32 + 6.4$
$= 38.4$

You could work out 2.4×3.2 first, but this would be harder – use your number facts to help you see the best way to do non-calculator calculations.

Example 6

$P = ab + 4b^2$. Find P when $a = \frac{4}{5}$ and $b = \frac{3}{8}$, giving your answer as a fraction.

Solution

$P = \left(\frac{4}{5} \times \frac{3}{8}\right) + \left(4 \times \frac{3}{8} \times \frac{3}{8}\right) = \frac{3}{10} + \frac{9}{16} = \frac{24}{80} + \frac{45}{80} = \frac{69}{80}$

Exercise 8.4

Work out each of the formulae in questions **1** to **10** for the values given.
Do not use your calculator.

1 $V = ab - ac$ when $a = 3$, $b = {}^-2$ and $c = 5$

2 $P = 2rv + 3r^2$ when $r = 5$ and $v = {}^-2$

3 $T = 5s^2 - 2t^2$ when $s = {}^-2$ and $t = 3$

4 $M = 2a(3b + 4c)$ when $a = 5$, $b = 3$ and $c = {}^-2$

5 $R = \dfrac{2qv}{q + v}$ when $q = 3$ and $v = {}^-4$

6 $L = 2n + m$ when $n = \frac{2}{3}$ and $m = \frac{5}{6}$

7 $D = a^2 - 2b^2$ when $a = \frac{4}{5}$ and $b = \frac{2}{5}$

8 $A = a^2 + b^2$ when $a = 5$ and $b = {}^-3$

9 $P = 2c^2 - 3cd$ when $c = 2$ and $d = {}^-5$

10 $B = p^2 - 3q^2$ when $p = {}^-4$ and $q = {}^-2$

You may use your calculator for questions **11** to **15**.

11 Find the value of $M = \dfrac{ab}{(2a + b^2)}$ when $a = 2.75$ and $b = 3.12$.
Give your answer correct to 2 decimal places.

12 The distance S metres fallen by a pebble is given by the formula
$S = \frac{1}{2}gt^2$, where t is in seconds.

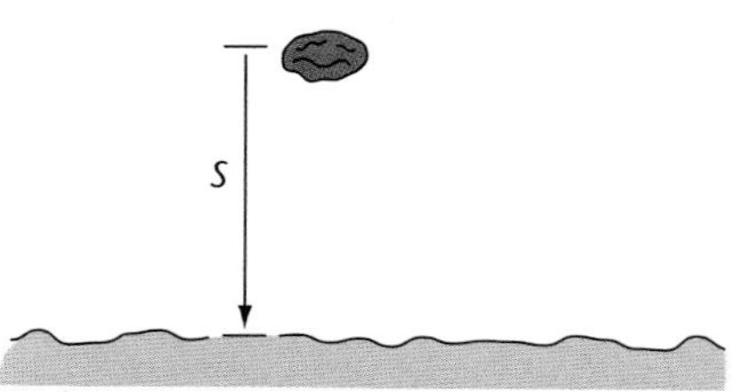

Find S in each of these cases.

a) $g = 10$ and $t = 12$

b) $g = 9.8$ and $t = 2.5$

13 The surface area (A cm^2) of a cuboid with sides x, y and z cm is given by the formula
$A = 2xy + 2yz + 2xz$.

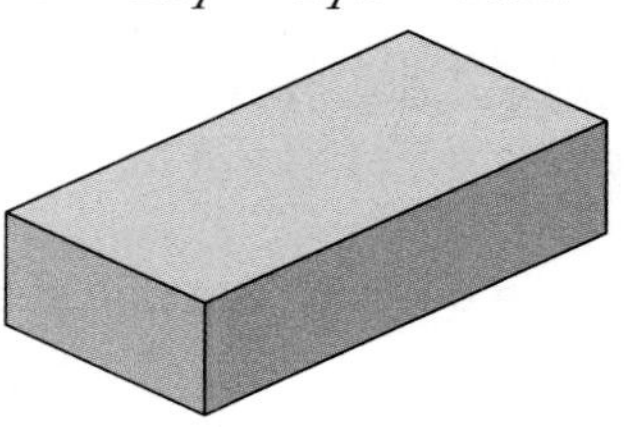

Find A when $x = 5$, $y = 4.5$ and $z = 3.5$.

14 The elastic energy of an elastic string is given by the formula $E = \dfrac{\lambda x^2}{2a}$.
Find E when $\lambda = 3.4$, $x = 5.7$ and $a = 2.5$. Give your answer correct to 1 decimal place.

15 The focal length of a lens is given by the formula $f = \dfrac{uv}{u + v}$.
Find the focal length when $u = 6$ and $v = {}^{-}7$.

Key Ideas

- When you have substituted numbers in a formula and are working out the result, remember the order of operations – brackets and indices (powers), then divide and multiply, then add and subtract.

Exam tip

Remembering the word BIDMAS may help.

Percentages 1

YOU WILL LEARN ABOUT	YOU SHOULD ALREADY KNOW
○ Expressing one quantity as a fraction of another ○ Expressing one quantity as a percentage of another ○ Increasing and decreasing an amount by a percentage	○ How to change a fraction to a percentage ○ How to find a percentage of a quantity

Expressing one quantity as a fraction or a percentage of another

To write one quantity as a fraction of another, write the first quantity as the numerator of the fraction and the second quantity as the denominator. Cancel the fraction, if possible, to give the fraction in its lowest terms.

Example 1	Solution
Express 14 as a fraction of 60.	$\frac{14}{60} = \frac{7}{30}$ Write the fraction in its lowest terms.

To express one quantity as a percentage of another, start by writing the first quantity as a fraction of the second. Then use the methods already learned to change the fraction to a percentage.

Example 2	Solution
Express 4 as a percentage of 5.	First write 4 as a fraction of 5. $\frac{4}{5}$ Then change $\frac{4}{5}$ to a percentage. $\frac{4}{5} = 0.8 = 80\%$

Example 3	Solution
Express £5 as a percentage of £30.	$\frac{5}{30} = 0.167 = 16.7\%$ to 1 decimal place

Example 4	Solution
Express 70 cm as a percentage of 2.3 m.	To do this, both quantities must be in the same units. First change 2.3 m to centimetres. 2.3 m = 230 cm $\frac{70}{230} = 0.304 = 30.4\%$ to 1 decimal place Another way is to change 70 cm to metres. 70 cm = 0.7 m $\frac{0.7}{2.3} = 0.304 = 30.4\%$ to 1 decimal place

Exercise 9.1

Do not use a calculator for questions 1 to 13.

1 What fraction is
a) 7 of 14?
b) 5 of 15?
c) 8 of 18?
d) 12 of 30?
e) 16 of 24?
f) 11 of 55?
g) 6 of 54?
h) 12 of 64?
Write each fraction in its lowest terms.

2 Find these.
a) 12 as a percentage of 100
b) 4 as a percentage of 50
c) 80 as a percentage of 200
d) £5 as a percentage of £20
e) 4 m as a percentage of 10 m
f) 30p as percentage of £2

3 In each case, express the first quantity as a percentage of the second.

	First quantity	Second quantity
a)	16	100
b)	12	50
c)	1 m	4 m
d)	£3	£10
e)	73p	£1
f)	8p	£1
g)	£1.80	£2
h)	40 cm	2 m
i)	50p	£10
j)	£2.60	£2

4 In class 7B, 14 of the 25 students are boys. What percentage is this?

5 A sailor has a rope 50 m long. He cuts off 12 m.
What percentage of the rope has he cut off?

6 In a sales promotion a company offers an extra 150 ml of cola free. The bottles usually hold 1 litre.
What percentage of 1 litre is 150 ml?

7 Annie scores 14 out of 20 in a test.
What fraction is this?
Write your answer in its lowest terms.

8 Thelma spent £4 out of her £20 spending money on a visit to the cinema.
What percentage is this?

9 At the keep-fit club, 8 out of the 25 members are men.
What percentage is this?

10 In the Mayflower Hotel there are 15 double rooms and 5 single rooms.
What percentage of the rooms are double?

11 United won 19, lost 15 and drew 6 of the games in a season.
What percentage of the games did they draw?

12 Stephen drove 68 miles of a 200-mile journey before stopping.
What fraction of the journey was that?
Give your answer in its lowest terms.

13 Derek scored 44 runs out of his team's total of 80.
What percentage of the runs did he score?

You may use a calculator for questions 14 to 25. Give your answers either exactly or to 1 decimal place.

14 Work out these.
a) 25 as a percentage of 200
b) 40 as a percentage of 150
c) 76p as a percentage of £1.60
d) £1.70 as a percentage of £2
e) 19 as a percentage of 24
f) 213 as a percentage of 321

15 In each case, express the first quantity as a percentage of the second.

	First quantity	Second quantity
a)	11	16
b)	9	72
c)	1 m	7 m
d)	£3	£18
e)	73p	£3
f)	17p	£2.50
g)	£2.60	£7
h)	40 cm	2.6 m
i)	£3.72	£12.96
j)	£2.85	£2.47

16 Penny scores 17 out of 40 in a test.
What percentage is this?

17 Salma was given a discount of £2.50 on a meal that would have cost £17.
What percentage discount was she given?

18 David won £7500 in a lottery. He gave £2300 of the money to his son.
What percentage is this?

19 On a flight to New York, 184 out of the 240 passengers were women.
What fraction of the passengers were women?
Give your answer in its lowest terms.

20 During the 30 days of November in 2005, it rained on 28 of the days.
On what percentage of the days did it rain?

21 In a radar speed check, 45 out of the 167 motorists checked were found to be exceeding the speed limit.
What percentage were exceeding the speed limit?

22 Last Saturday 6198 fans booked online for a festival and 4232 fans booked by phone.
What percentage of the fans made online bookings?

23 Holly gets £5 a week from her grandad, £7 from her dad, £3 from her mum and earns £11 from a paper round.
What percentage of her weekly income does she earn?

24 A school has 857 students. 70 of them go on a school holiday.
What percentage is this?

25 A mathematics test has 45 questions.
Abigail gets 42 right.
What percentage does Abigail get right?

26 A photo frame 20 cm by 30 cm contains a photo 14 cm by 18 cm.
What percentage of the whole area is the photo?

Percentage increases and decreases

To find the amount after an increase, work out the increase and add it on to the original amount.

To find the amount after a decrease, work out the decrease and subtract it from the original amount.

Example 5

A shop increased its annual sales of computers by 20% in 2006.
In 2005 it sold 1200 computers.

Without using your calculator, work out how many computers the shop sold in 2006.

Solution

20% of 1200 = $0.2 \times 1200 = 240$ 20% = 0.2

Sales in 2006 = $1200 + 240 = 1440$

Example 6

An engineer receives a 3% increase in her annual salary.
She earned £24 000 before the increase.

What is her new salary?

Solution

Increase = $24\,000 \times 0.03 = 720$ 3% = 0.03

New salary = £24 000 + £720 = £24 720

Example 7

All prices are reduced by 15% in a sale.

Find the sale price of an article previously priced at £17.60.

Solution

Reduction = 17.60×0.15 15% = 0.15
= 2.64

New price = £17.60 – £2.64
= £14.96

Exercise 9.2

Do not use a calculator for questions **1** to **7**.

1 Increase £400 by each of these percentages.
a) 20%   b) 45%
c) 6%   d) 80%

2 Decrease £200 by each of these percentages.
a) 30%   b) 15%
c) 3%   d) 60%

3 The rent for a shop on the High Street was £25 000 a year and went up by 20%.
What is the new rent?

4 Simon earns £12 000 per year.
He receives a salary increase of 4%.
Find his new salary.

5 Train fares have gone up by 8%.
What is the new price of a ticket which used to cost £3?

6 Prices are reduced by 20% in a sale.
The original price of a top was £13.
What is the sale price?

7 The Shaws' luggage weighed 80 kg. They reduced the weight by 15% because it exceeded their baggage allowance. What did their baggage weigh after they had repacked it?

You may use a calculator for questions 8 to 15.

8 Mick puts £1200 into a savings account. 7.5% interest is added to the account at the end of the year. How much is in the account at the end of the year?

9 A car loses 12% of its value in one year. It is worth £15 000 when it is new. What will it be worth at the end of the year?

10 To test the strength of a piece of wire, it is stretched by 12% of its original length. A piece of wire was originally 1.5 m long. How long will it be after stretching?

11 Rushna pays 6% of her pay into a pension fund. She earns £185 per week. What will her pay be after taking off her pension payments?

12 A drum kit costs £280 before VAT is added on. What will it cost after VAT at 17.5% has been added on?

13 The price of a football club's season ticket went up by 3.5%. The old price was £840. What is the new price?

14 A savings account paid 6.2% per annum. David put £2500 in this account and left it for a year. How much was in the account at the end of the year?

15 To travel on holiday, Jane went by train then plane. When she researched her holiday, the train fare was £32 and the plane fare was £75. When she booked, her train fare had gone up by 4% and the plane fare had reduced by 3%. Calculate her total travel cost.

Challenge 1

Paul invests £10 000 in a savings account that pays 5% interest on the amount in the account each year.
Find the amount in the account after

a) 1 year. b) 2 years. c) 5 years. d) 10 years.

Challenge 2

A car loses 15% of its value each year.
When it is new it is worth £16 000.
Find its value after

a) 1 year. b) 2 years. c) 5 years. d) 10 years.

Challenge 3

Peter's annual salary is £30 000. He gets a pay rise of 5%. Then in a recession he gets a pay cut of 5%.
Peter thinks his salary is back where it started. Show that Peter is wrong!

Key Ideas

- To find one quantity as a fraction of another, write the first quantity as the numerator of the fraction and the second quantity as the denominator. Cancel the fraction, if possible, to give it in its lowest terms.
- To find one quantity as a percentage of another, write the first quantity as a fraction of the second. Then change to a percentage. (Make sure that both quantities are in the same units.)
- To find a percentage of a quantity, change the percentage to a decimal and multiply.
- To increase a quantity by a percentage, work out the increase and add it on to the original amount.
- To decrease a quantity by a percentage, work out the decrease and subtract it from the original amount.

CHAPTER 10 Areas and perimeters

YOU WILL LEARN ABOUT	YOU SHOULD ALREADY KNOW
○ Finding the area of a triangle, a parallelogram and a trapezium ○ Finding the perimeter and area of shapes made from triangles and rectangles ○ Using rectangles and triangles to find the surface area of shapes	○ The terms *perimeter* and *area* ○ The units for these quantities: cm, cm^2, or m, m^2 ○ How to find the area of a rectangle

The area of a triangle

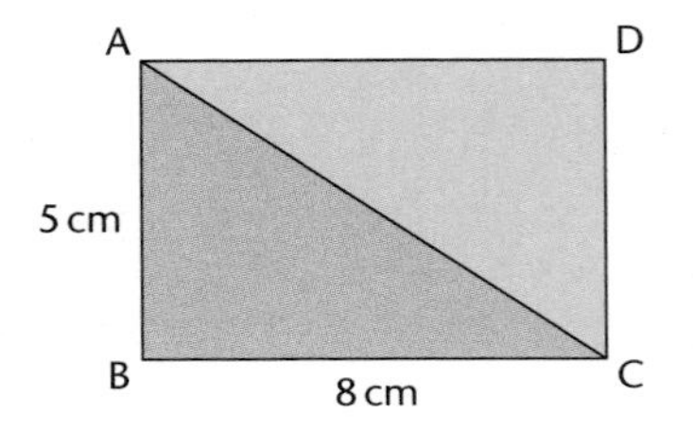

The area of the rectangle ABCD is $8 \times 5 = 40\,cm^2$.

The area of the triangle ABC is half the area of the rectangle ABCD and so is equal to $20\,cm^2$.

So the area of the right-angled triangle

$= \frac{1}{2} \times 8 \times 5$

$= 20\,cm^2$.

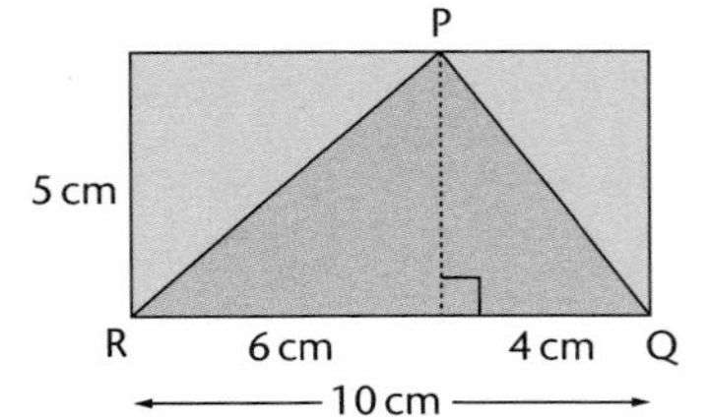

Triangle PQR is not right-angled but has been split into two right-angled triangles.

Each right-angled triangle is half a rectangle.

The area of the left-hand triangle is $\frac{1}{2} \times 6 \times 5 = 15\,cm^2$.

The area of the right-hand triangle is $\frac{1}{2} \times 4 \times 5 = 10\,cm^2$.

Therefore, the total area of triangle PQR is $15 + 10 = 25\,cm^2$.

Notice that the area of triangle PQR is half the area of the large rectangle.

So the area of the triangle is $\frac{1}{2} \times 10 \times 5 = 25\,cm^2$.

This shows that whether the triangle is right-angled or not, the area can be found by this formula.

Area of triangle = $\frac{1}{2}$ × base × perpendicular height or $A = \frac{1}{2}bh$

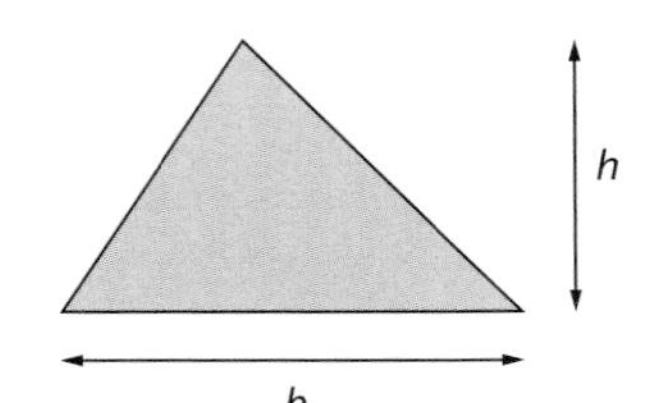

Example 1

a) Find the area of this triangle.

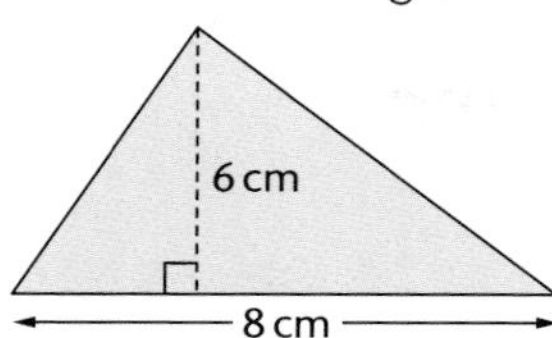

b) The area of this triangle is $20\,\text{cm}^2$.
Find the perpendicular height of the triangle.

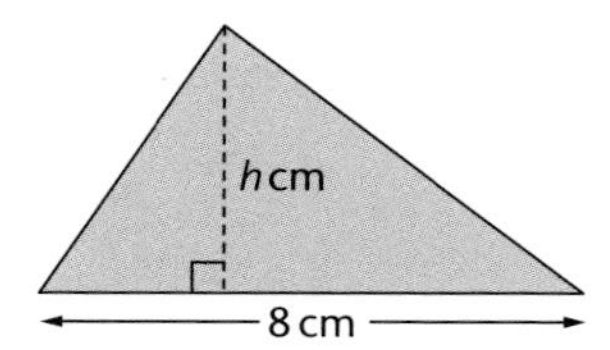

Solution

a) $\text{Area} = \frac{1}{2}bh$

$= \frac{1}{2} \times 8 \times 6$

$= 24\,\text{cm}^2$

b) $\text{Area} = \frac{1}{2}bh$

$20 = \frac{1}{2} \times 8 \times h$

$20 = 4h$

$h = 5$

So the height is 5 cm.

Remember that the units of area are always square units such as square centimetres or square metres, written cm^2 or m^2.

When using the formula, you can use any of the sides of the triangle as the base, provided you use the perpendicular height that goes with it.

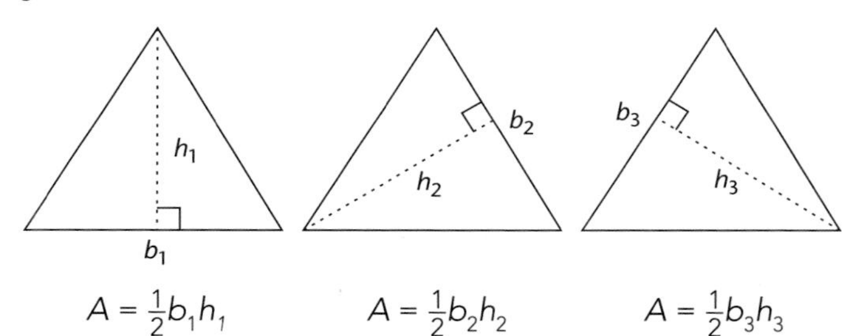

$A = \frac{1}{2}b_1h_1$ $A = \frac{1}{2}b_2h_2$ $A = \frac{1}{2}b_3h_3$

Exam tip

Always use the perpendicular height of the triangle, never the slant height. In this triangle, area $= \frac{1}{2} \times 6 \times 3 = 9\,\text{cm}^2$.

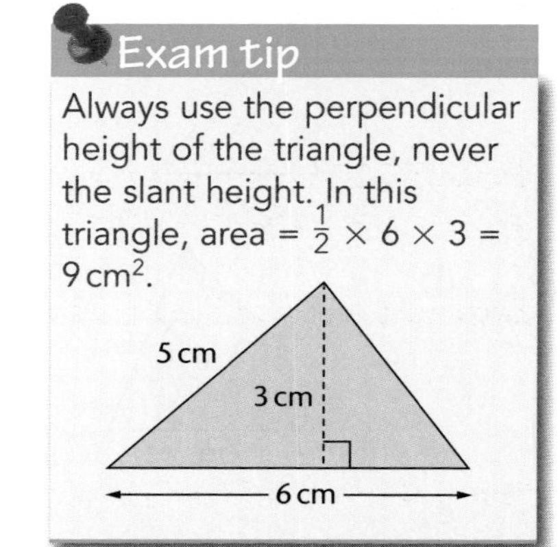

Activity 1

On squared paper, draw at least six different triangles with an area of $10\,\text{cm}^2$.

Exercise 10.1

1 Find the area of each of these triangles.

a)
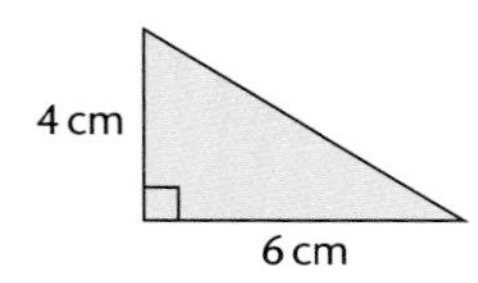

b)
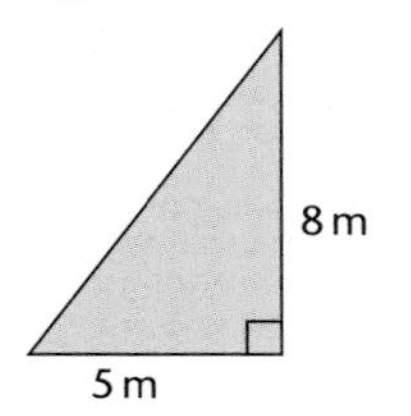

c)
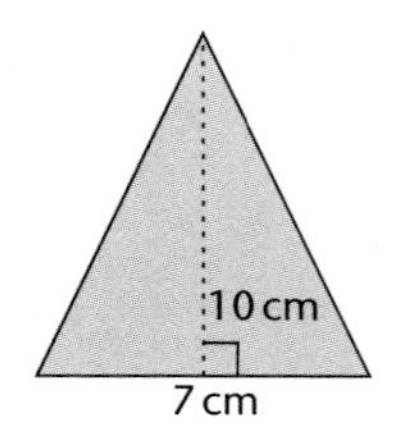

d)
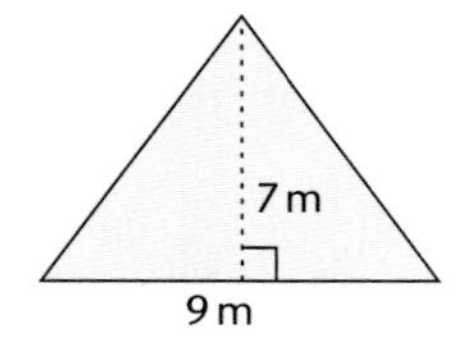

e)
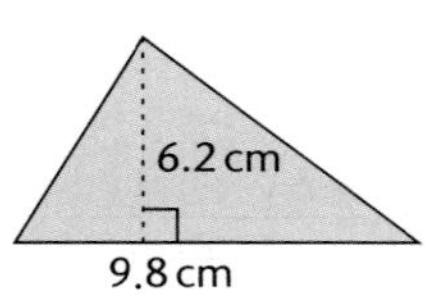

f)
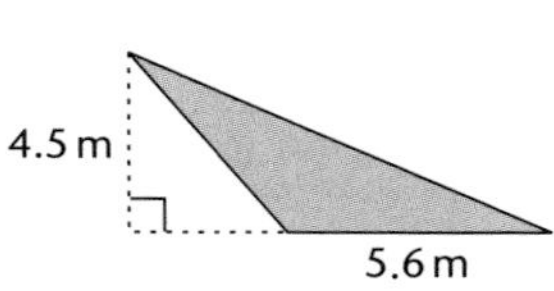

g)
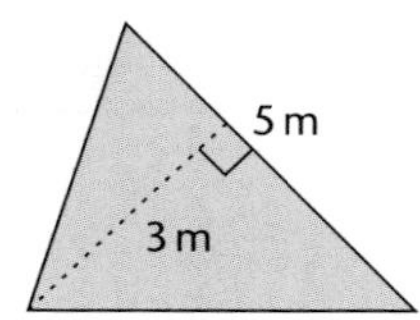

h)
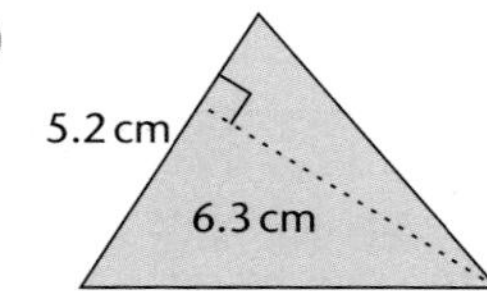

i)
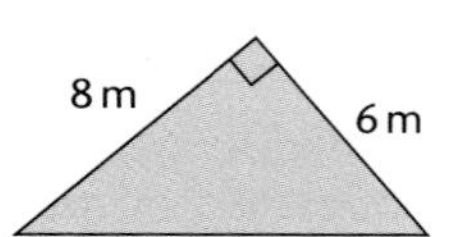

2 Find the area of each of these triangles.

a)
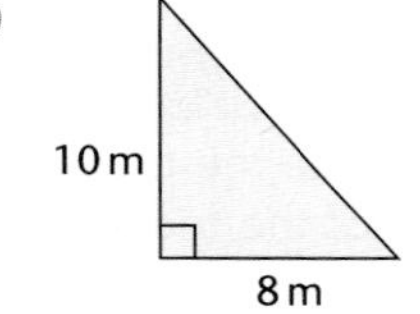

b)
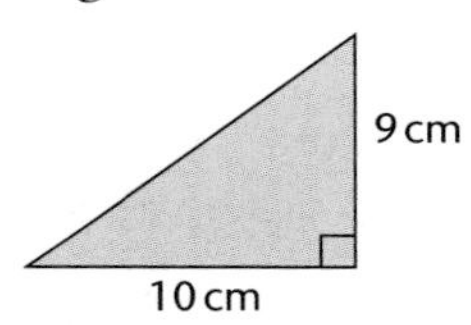

c)
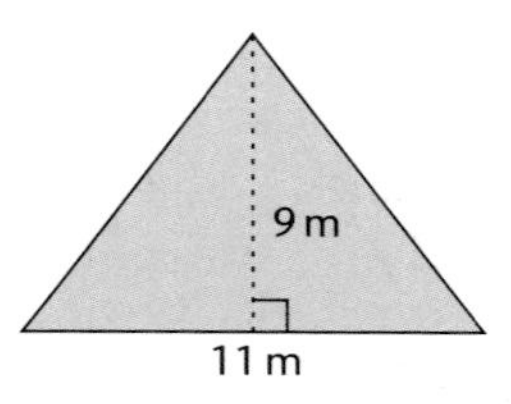

d)
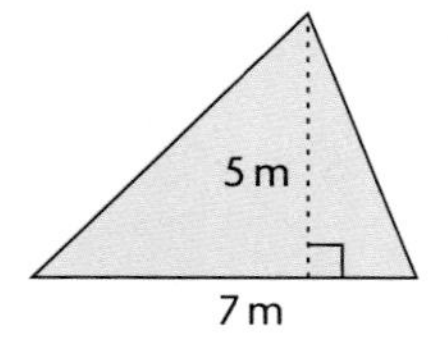

e)
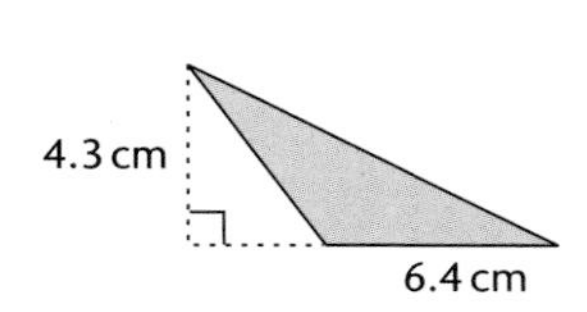

f)
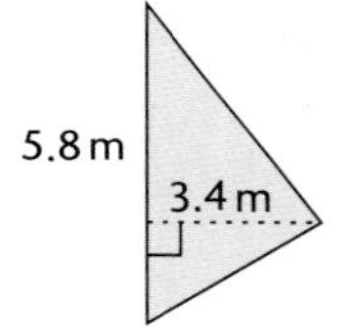

g)
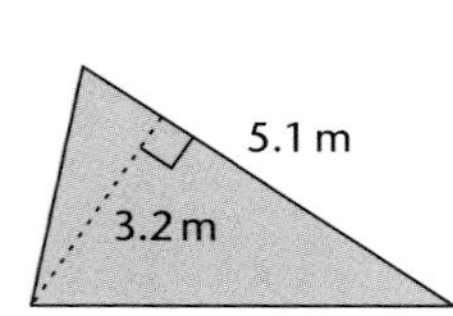

h)
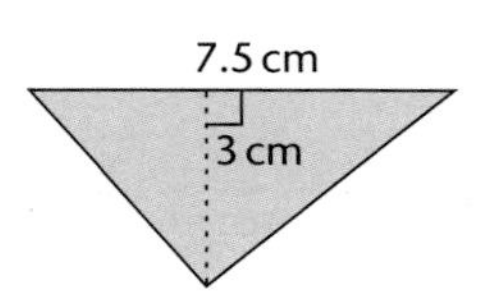

i)
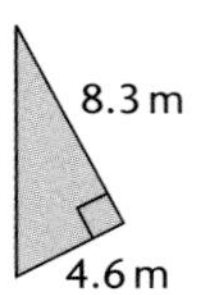

3 The vertices of a triangle are at A(2, 1), B(5, 1) and C(5, 7). Draw triangle ABC on squared paper and find its area.

4 The vertices of a triangle are at D(2, 1), E(2, 7) and F(5, 3). Draw triangle DEF on squared paper and find its area.

5 The vertices of a triangle are at K(2, 2), L(7, 2) and M(4, 6). Draw triangle KLM on squared paper and find its area.

6 The vertices of a triangle are at P($^-2$, 2), Q(3, 2) and R(5, 6). Draw triangle PQR on squared paper and find its area.

7 The vertices of a triangle are at U(⁻2, ⁻1), V(⁻2, 4) and W(3, 6).
Draw triangle UVW on squared paper and find its area.

8 Using a ruler and compasses, draw an equilateral triangle with sides of 5 cm.
Measure its perpendicular height and find its area.

9 In triangle ABC, AB = 6 cm, BC = 8 cm and angle ABC = 40°.
Draw the triangle accurately and find its area.

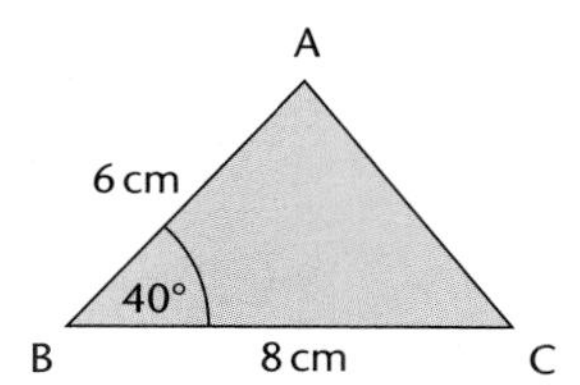

10 In triangle XYZ, XY = 7 cm, YZ = 10 cm and ZX = 6 cm.
Using a ruler and compasses, draw the triangle accurately and find its area.

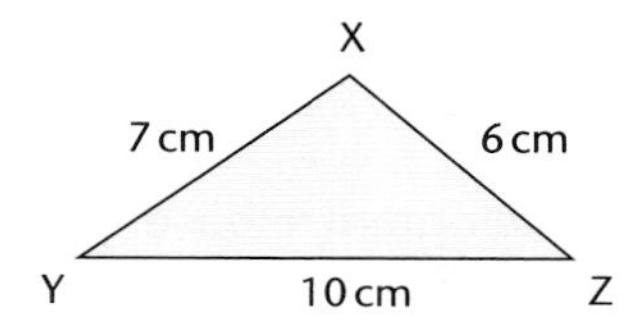

11 A triangle has an area of 12 cm^2 and a base of 4 cm.
Find the perpendicular height associated with this base.

12 In triangle ABC, AB = 6 cm, BC = 8 cm and AC = 10 cm.
Angle ABC = 90°.

a) Find the area of the triangle.

b) Find the perpendicular height BD.

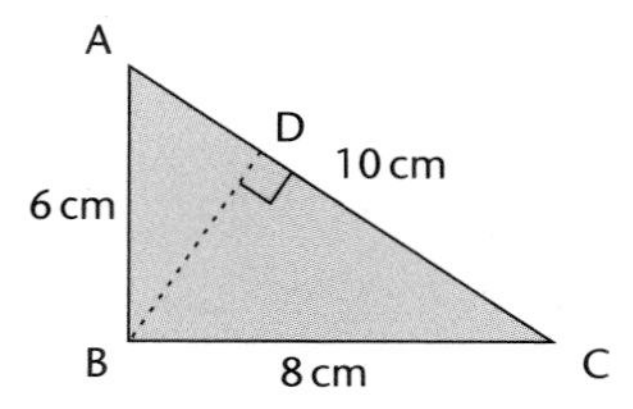

Challenge 1

Work in groups of three or four.

- Each draw a circle of radius 5 cm.
- Then draw some regular polygons with vertices on the circle – a pentagon, a hexagon, an octagon and, if there are four people, a decagon (ten sides).
- Measure carefully the perpendicular height and base of the triangles which make up your polygons.
- Then calculate the area of each polygon.
- Compare your results.

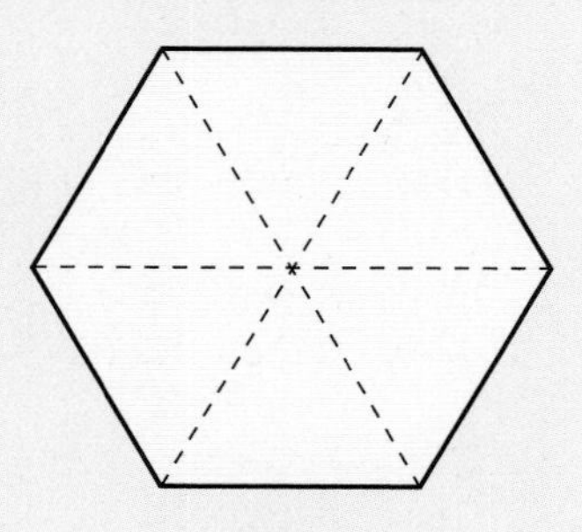

The area of a parallelogram

A parallelogram may be cut up and rearranged to form a rectangle or two congruent triangles.

Area of a rectangle = base $\times$ height

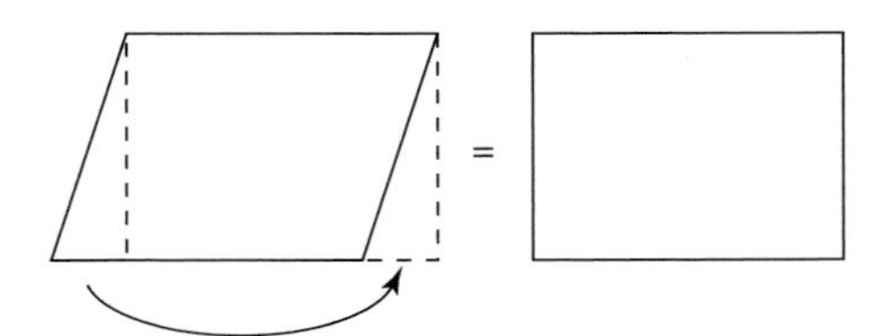

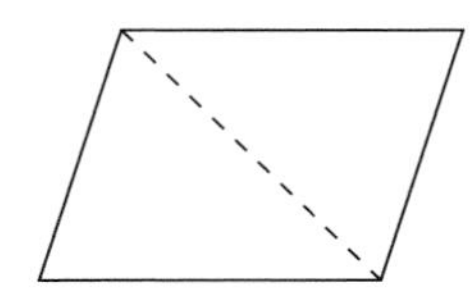

Area of each triangle = $\frac{1}{2} \times$ base $\times$ height

Both these ways of splitting a parallelogram show how to find its area.

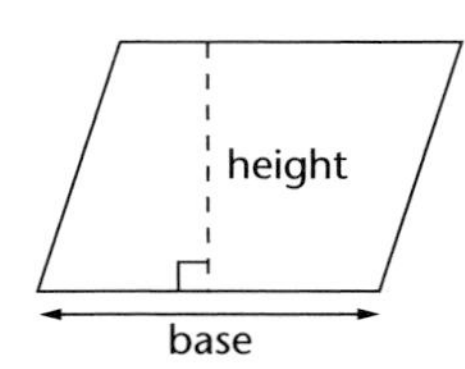

Area of a parallelogram = base $\times$ height

> **Exam tip**
>
> Make sure you use the perpendicular height and not the sloping edge when finding the area of a parallelogram.

Example 2

Find the area of this parallelogram.

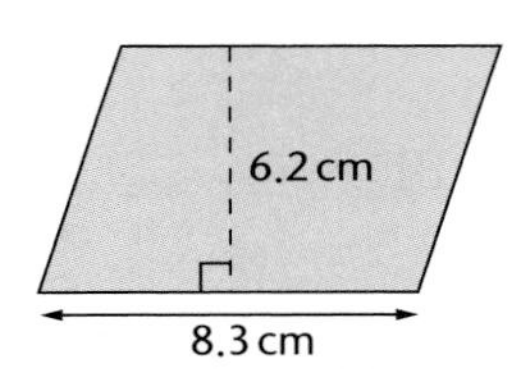

Solution

Area of a parallelogram = base $\times$ height

$= 8.3 \times 6.2$

$= 51.46\,\text{cm}^2$

$= 51.5\,\text{cm}^2$ to 1 decimal place

> **Exam tip**
>
> Always give your final answer to a suitable degree of accuracy, but don't use rounded answers in your working.

Exercise 10.2

1 Find the area of each of these parallelograms.

a)

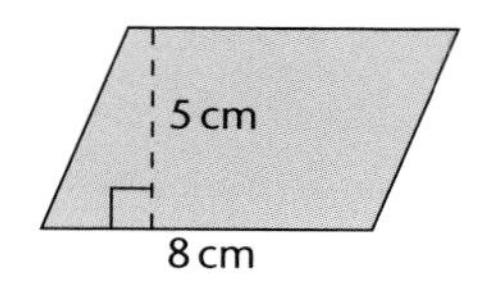

b)

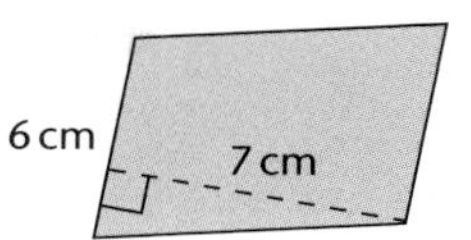

c)

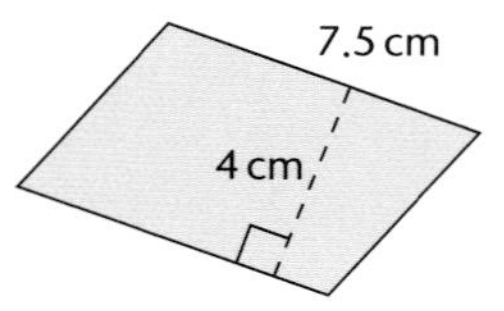

d)

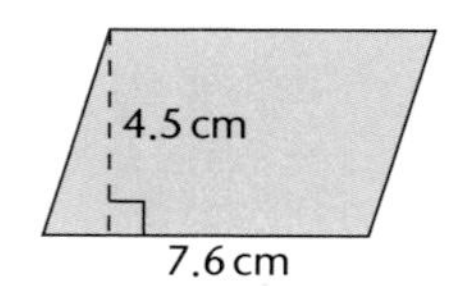

e)

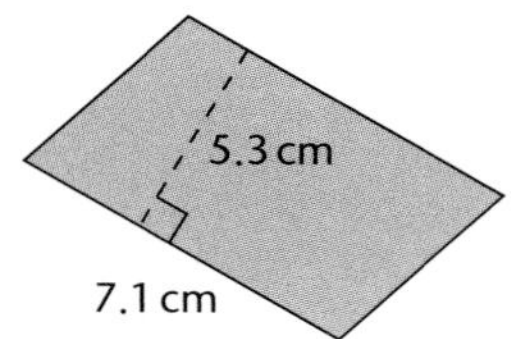

f)

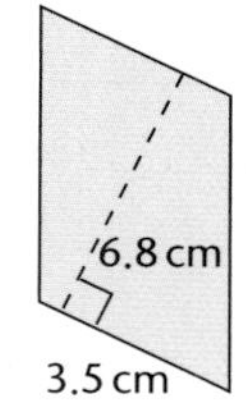

2 Find the area of each of these parallelograms. The lengths are in centimetres.

a)

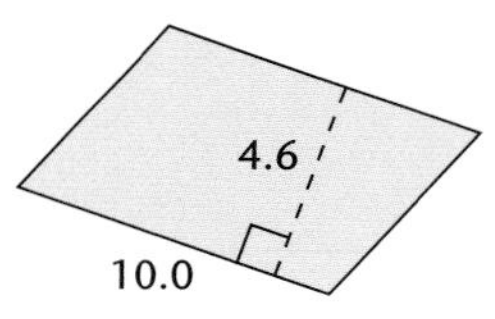

b)

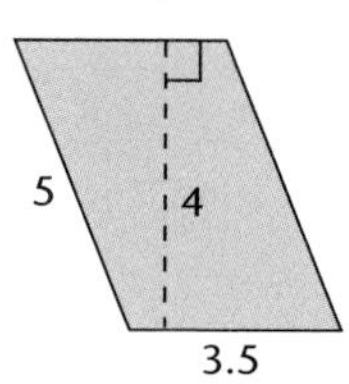

c)

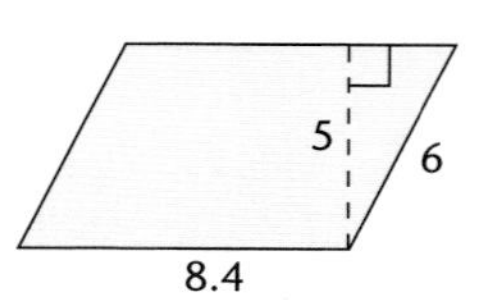

d)

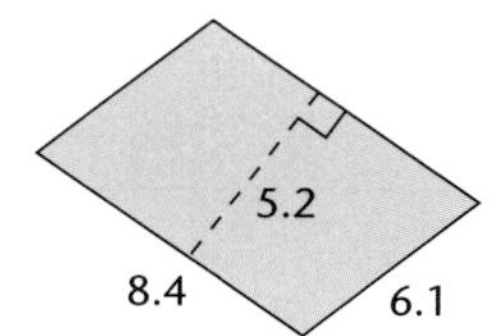

e)

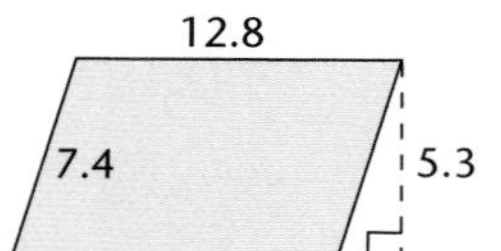

f)

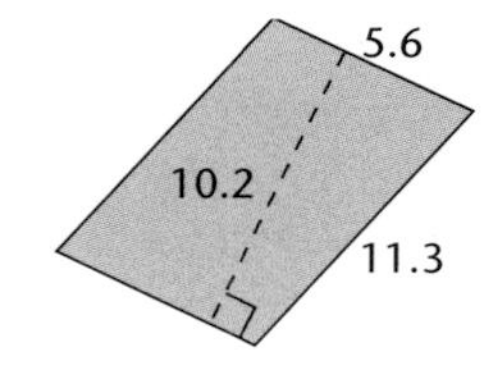

3 Measure the base and height and calculate the area of each of these parallelograms.

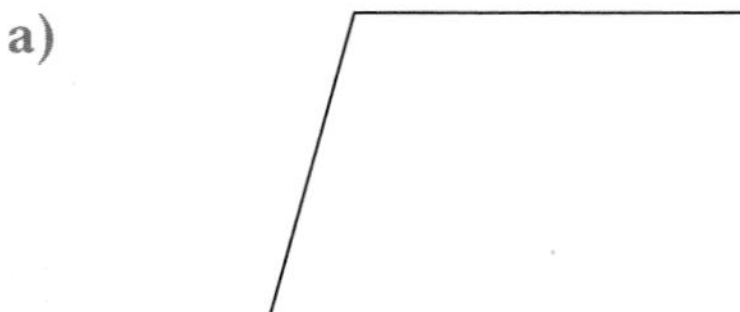

a)

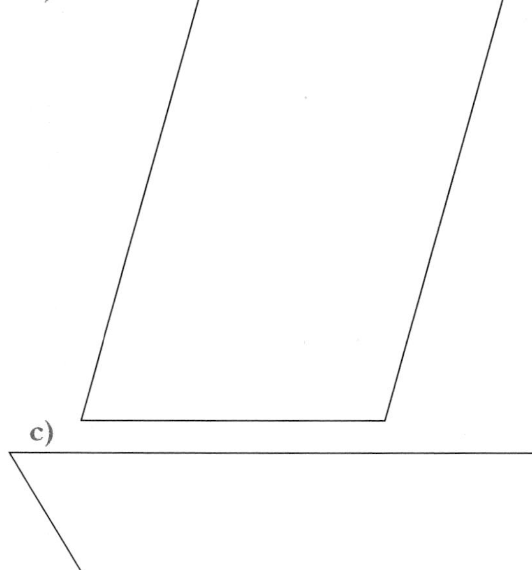

b)

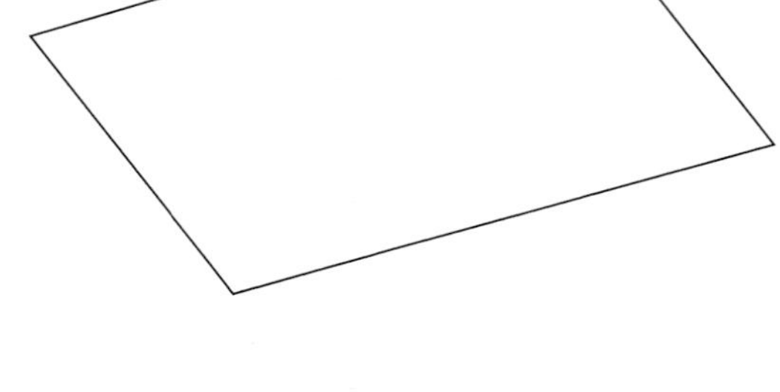

c)

d)

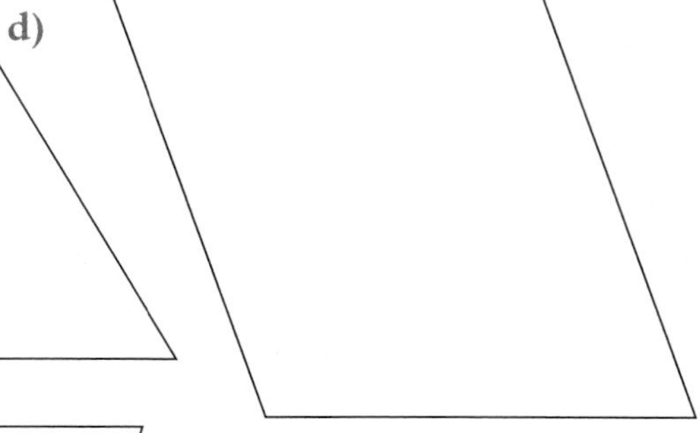

e)

f)

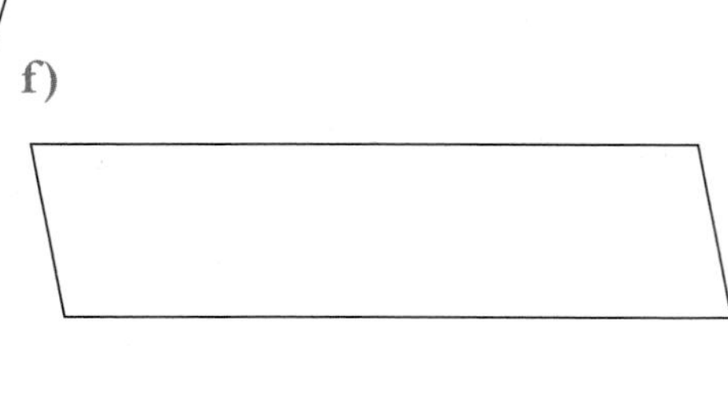

4 Find the values of a, b and c. The lengths are in centimetres.

a)

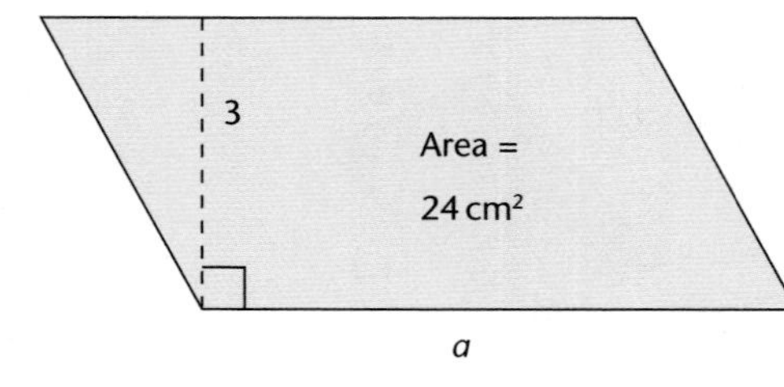

b)

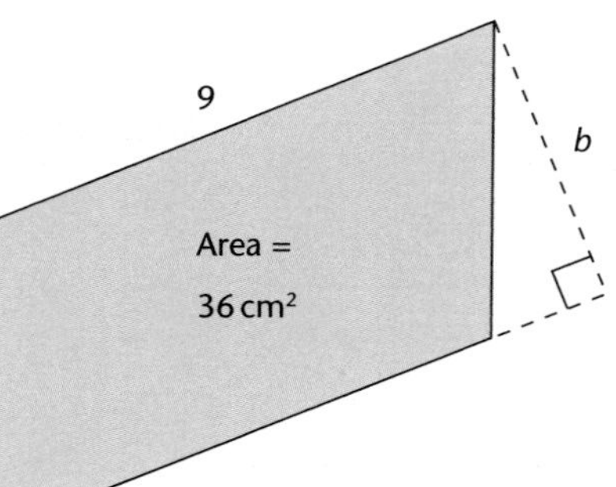

c)

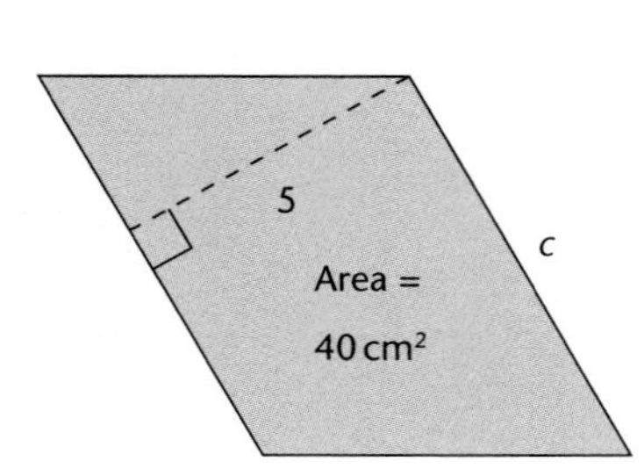

Foundation Silver/ Higher Initial

5 Find the values of x, y and z.

a)

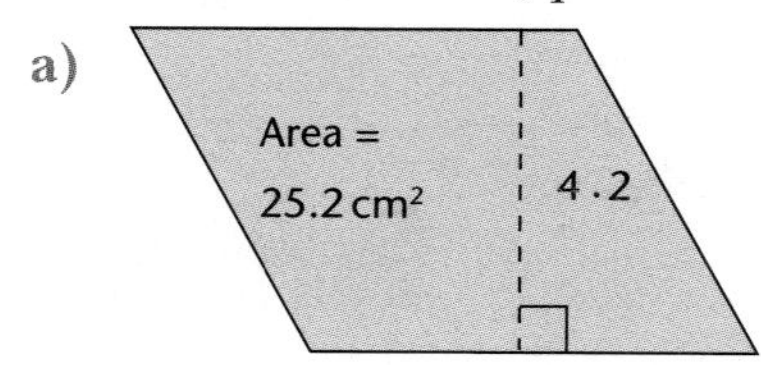

b)

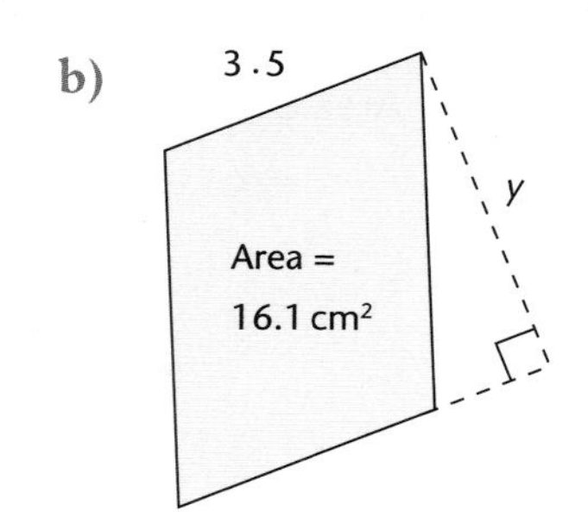

c)

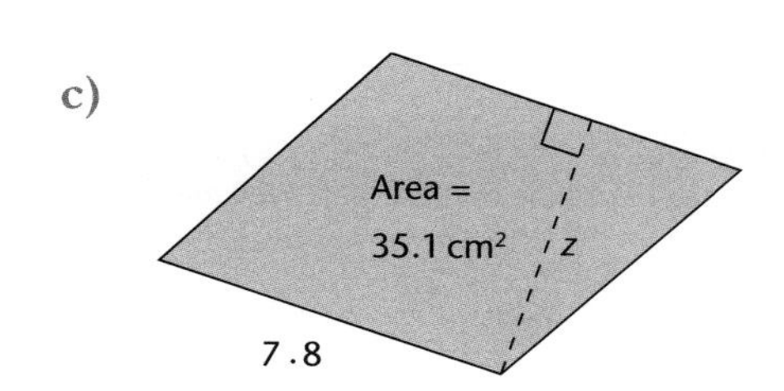

The area of a trapezium

A trapezium has one pair of opposite sides parallel.
It can be split into two triangles.

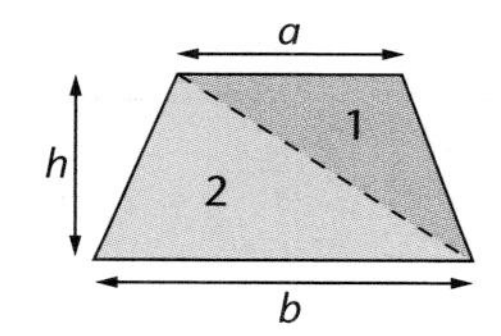

Area of triangle 1 $= \frac{1}{2} \times a \times h$

Area of triangle 2 $= \frac{1}{2} \times b \times h$

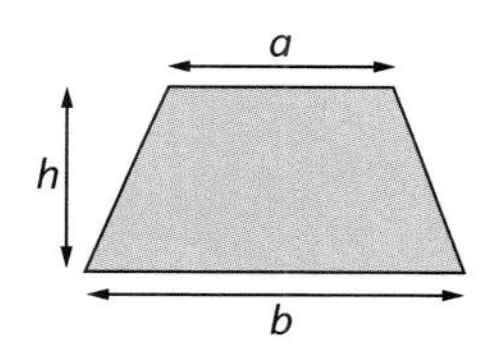

$$\begin{aligned}\text{Area of trapezium} &= \tfrac{1}{2} \times a \times h + \tfrac{1}{2} \times b \times h \\ &= \tfrac{1}{2} \times (a + b) \times h \\ &= \tfrac{1}{2}(a + b)h\end{aligned}$$

You can remember the formula in words.

Area of a trapezium = half the sum of the parallel sides × the height

Example 3

Calculate the area of this trapezium.

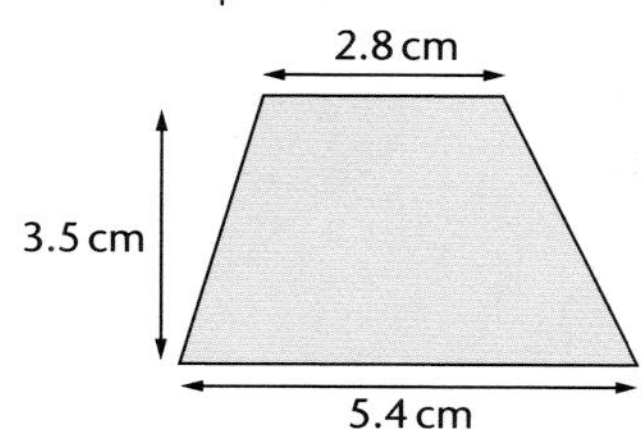

Solution

$$\begin{aligned}\text{Area of trapezium} &= \tfrac{1}{2}(a + b)h \\ &= \tfrac{1}{2} \times (2.8 + 5.4) \times 3.5 \\ &= 14.35\ \text{cm}^2 \\ &= 14.4\ \text{cm}^2 \text{ to 1 decimal place}\end{aligned}$$

Exam tip

Use the brackets function on your calculator. Without a calculator, work out the brackets first.

Example 4

Find the area of this trapezium.

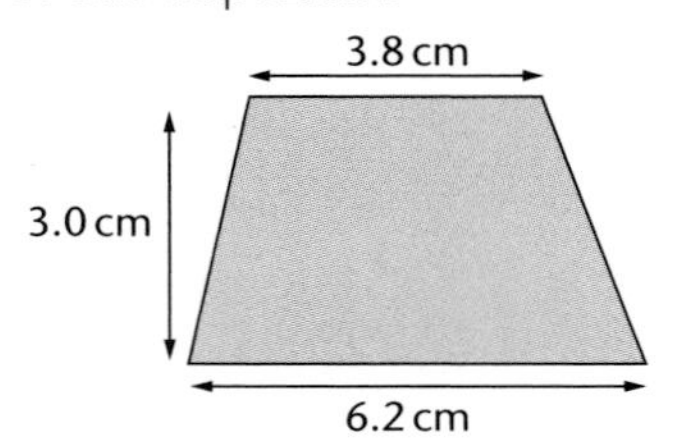

Solution

Area of a trapezium $= \frac{1}{2}(a + b)h$

$= \frac{1}{2} \times (3.8 + 6.2) \times 3$

$= \frac{1}{2} \times 10 \times 3$

$= 15\,\text{cm}^2$

Exam tip

Use the area formula to find the area of a parallelogram or trapezium as it is quicker.

Exercise 10.3

1 Find the area of each of these trapezia.

a)

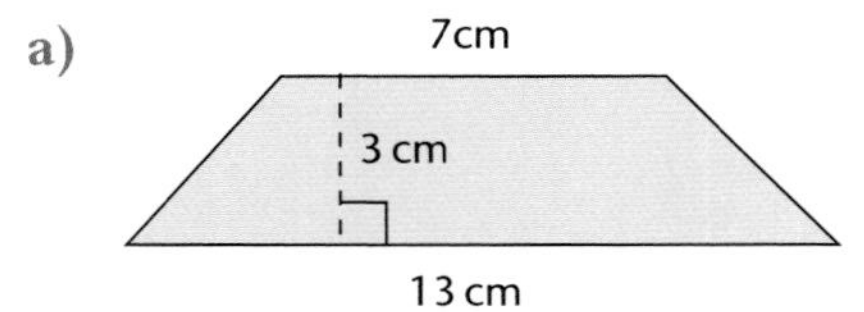

b)

3 cm
4 cm
5 cm

c)

4 cm
3 cm
6 cm

d)

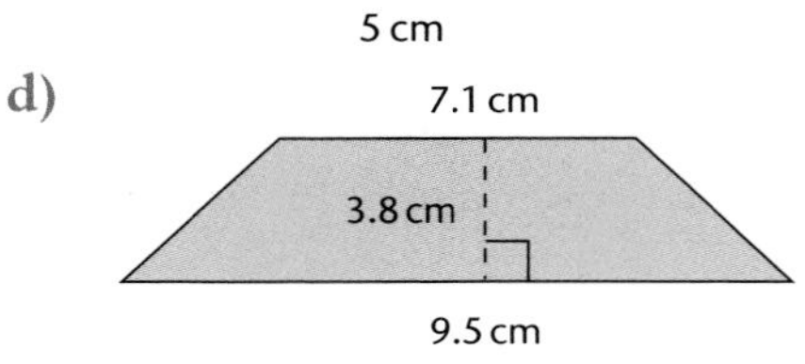

e)

2.8 cm
6.9 cm
4.7 cm

f)

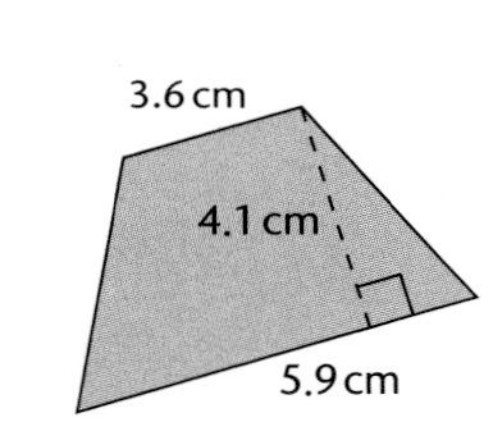

2 Find the area of each of these trapezia.

a)

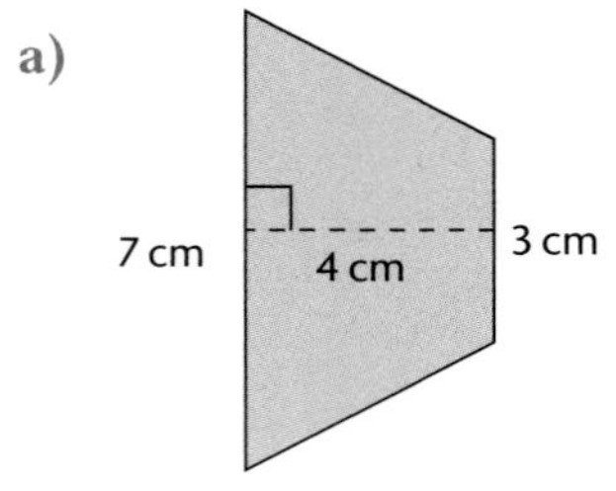

b)

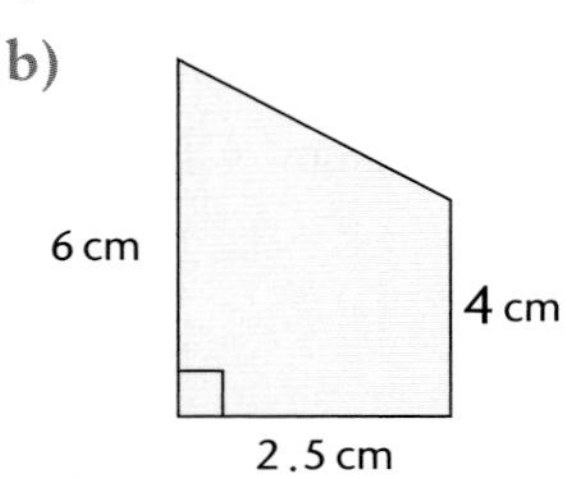

c)

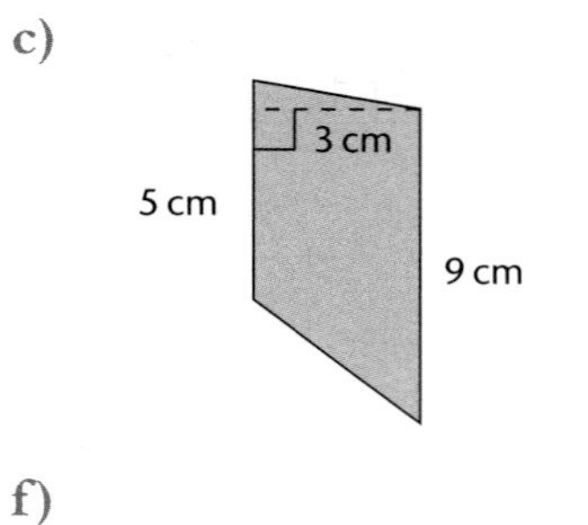

d)

3.7 cm
1.8 cm
2.1 cm

e)

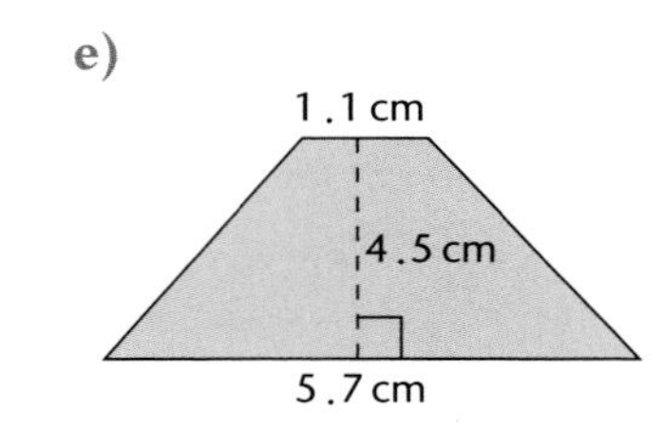

f)

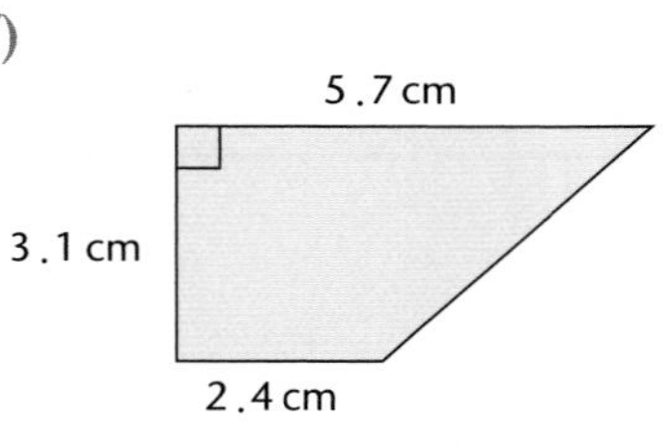

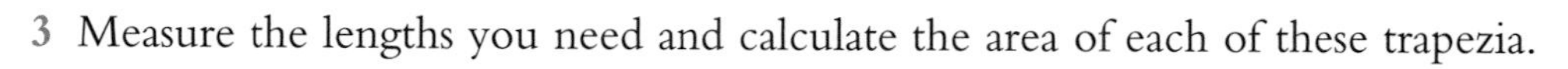

3 Measure the lengths you need and calculate the area of each of these trapezia.

a)

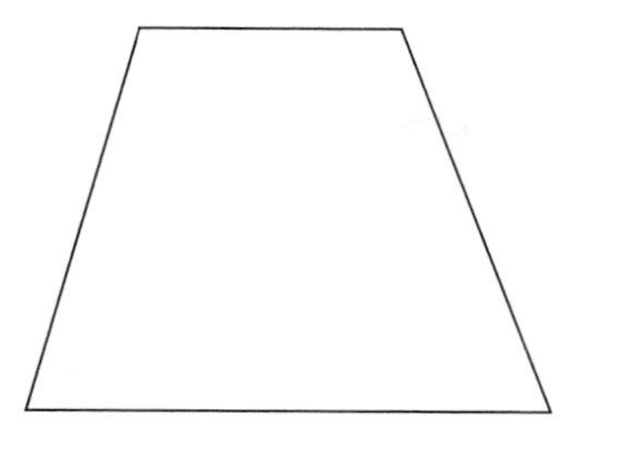

b)

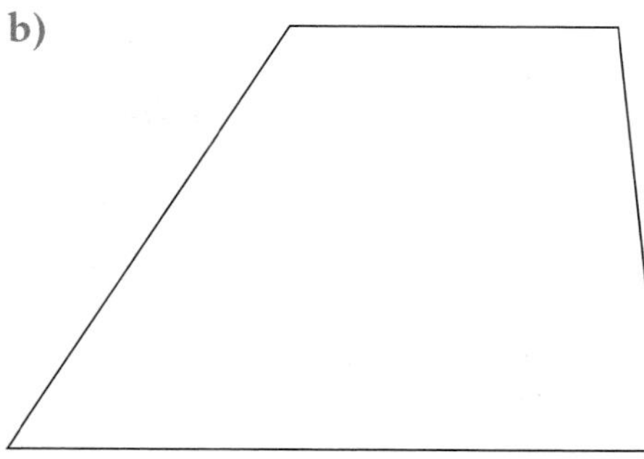

c)

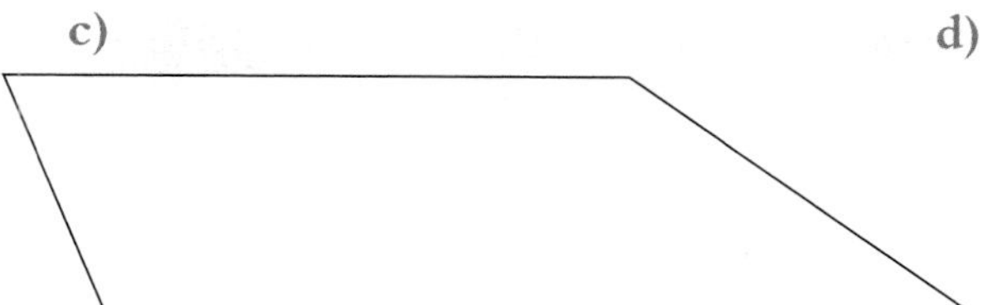

d)

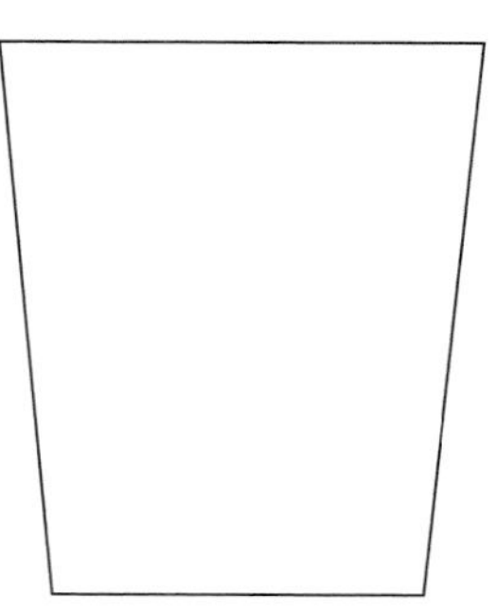

e)

f)

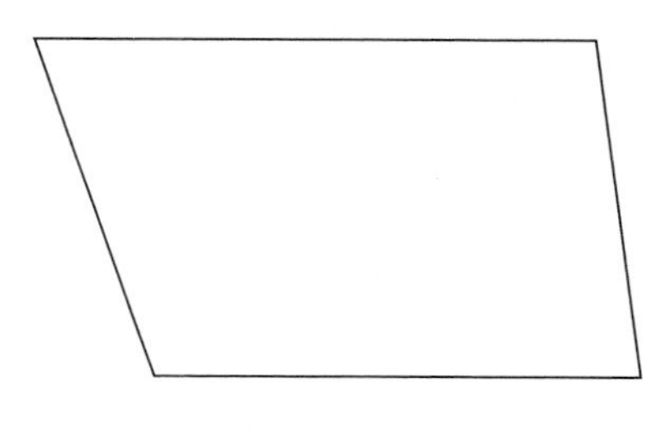

4 Find the values of a, b and c in these trapezia. The lengths are in centimetres.

a)

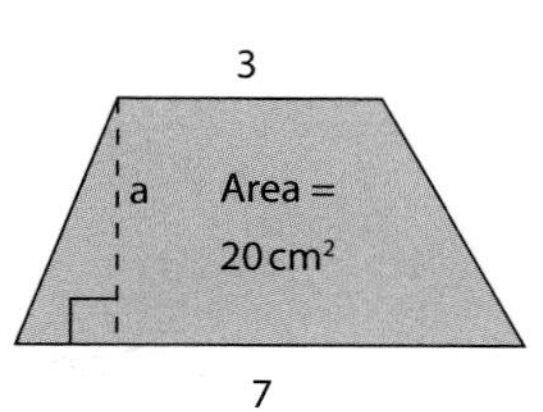

b)

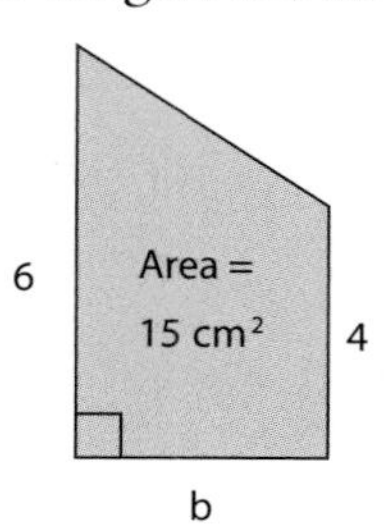

c)

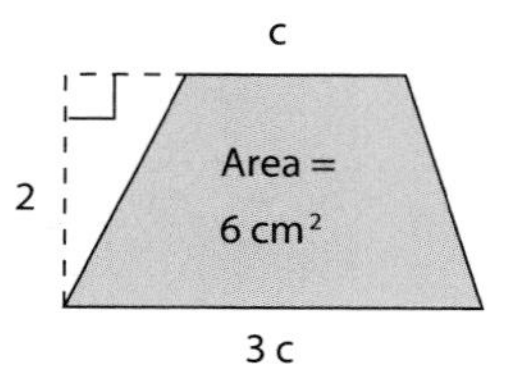

5 Find the values of x, y and z in these trapezia. The lengths are in centimetres.

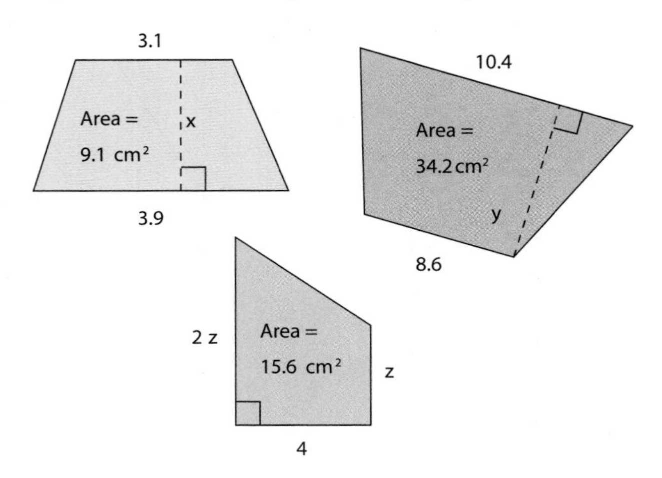

6 A trapezium has height 4 cm and area 28 cm^2.
One of its parallel sides is 5 cm long.
How long is the other parallel side?

7 A trapezium has height 6.6 cm and area 42.9 cm^2.
One of its parallel sides is 5 cm long.
How long is the other parallel side?

The perimeter and area of compound shapes

The **perimeter** of a shape is the distance all the way round the outside of the shape.

A way to find the **area** of many **compound** shapes is to split them up into rectangles and triangles.

Example 5

Work out **(i)** the perimeter and **(ii)** the area of each these shapes.
All lengths are in centimetres.

a)

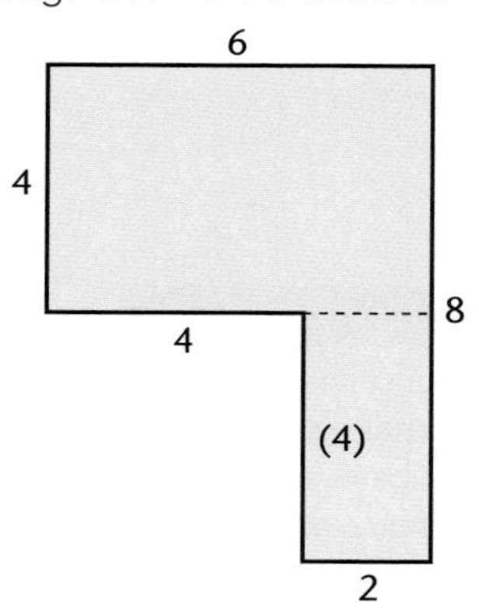

b)

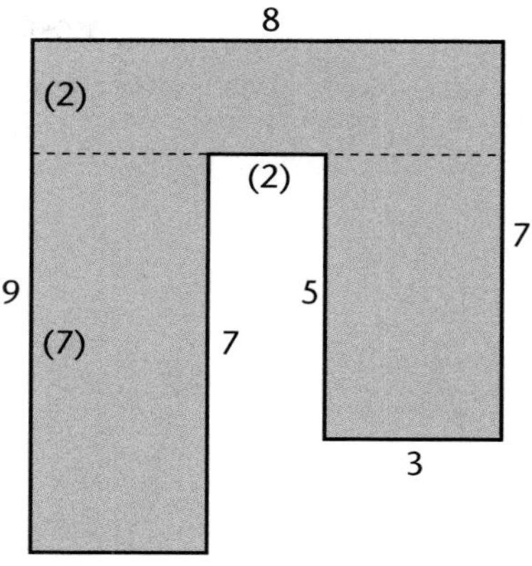

(The lengths marked in brackets were not all given in the question, but are worked out. The dashed lines have been added as part of the answer.)

Solution

a) **(i)** The perimeter of the shape is $6 + 8 + 2 + 4 + 4 + 4 = 28$ cm.

(ii) The shape has been split into two rectangles by a horizontal dotted line.
It could have been split in a different way, by a vertical line.

The area is $6 \times 4 + 2 \times 4 = 24 + 8$
$= 32\,\text{cm}^2$.

b) **(i)** The perimeter of the shape is
$9 + 8 + 7 + 3 + 5 + 2 + 7 + 3 = 44$ cm.

(ii) The shape has been split into three rectangles.

The area is $8 \times 2 + 7 \times 3 + 5 \times 3 = 16 + 21 + 15$
$= 52\,\text{cm}^2$.

Exam tip

A common error is to split the shape correctly but then multiply the wrong numbers to get the area.

Example 6

Work out the area of this shape.

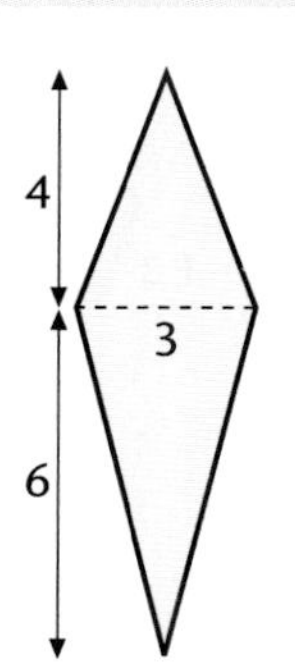

Solution

The shape has been split by the horizontal line into two triangles.

$$\text{Area} = \tfrac{1}{2} \times 4 \times 3 + \tfrac{1}{2} \times 6 \times 3$$
$$= 6 + 9$$
$$= 15\,\text{cm}^2$$

Example 7

a) On squared paper, plot the points A(1, 3), B(2, [illegible] D(6, 2) and E(3, 1). Join them up to make the pentagon ABCDE.

b) Find the area of the pentagon by splitting it into rectangles and triangles.

Solution

a)

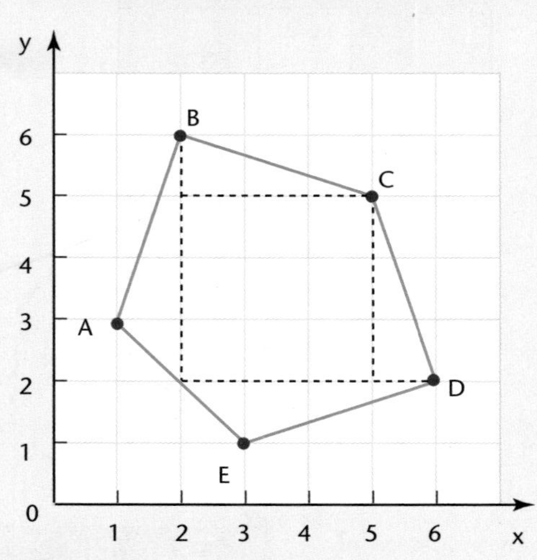

Exam tip

When splitting shapes on a grid, try to make sure that all triangles have an exact number of squares for the base and the height.

b) The pentagon has been split into one square and four triangles.
This is not the only way it could be split.

$$\text{Area} = 3 \times 3 + \tfrac{1}{2} \times 4 \times 1 + \tfrac{1}{2} \times 3 \times 1 + \tfrac{1}{2} \times 3 \times 1 + \tfrac{1}{2} \times 4 \times 1$$
$$= 9 + 2 + 1.5 + 1.5 + 2$$
$$= 16 \text{ squares}$$

Exercise 10.4

1 Find the perimeter of each of these shapes. All lengths are in centimetres.

a)

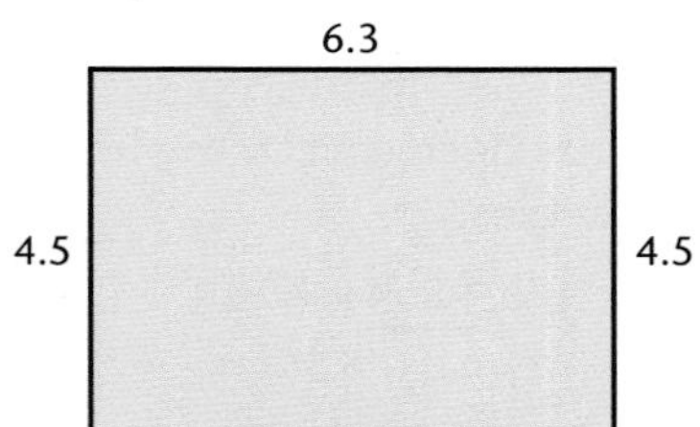

b)

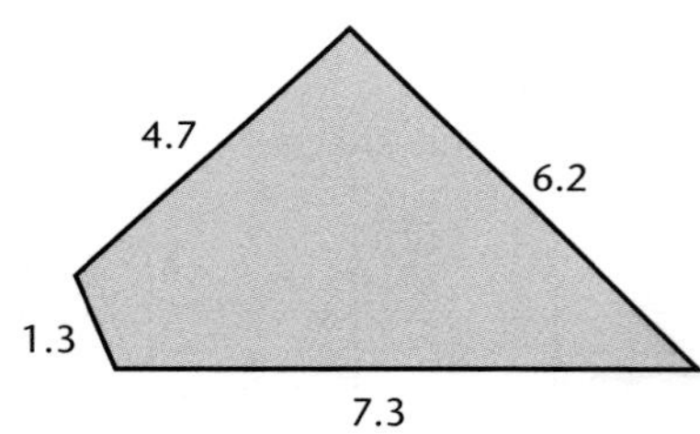

c)

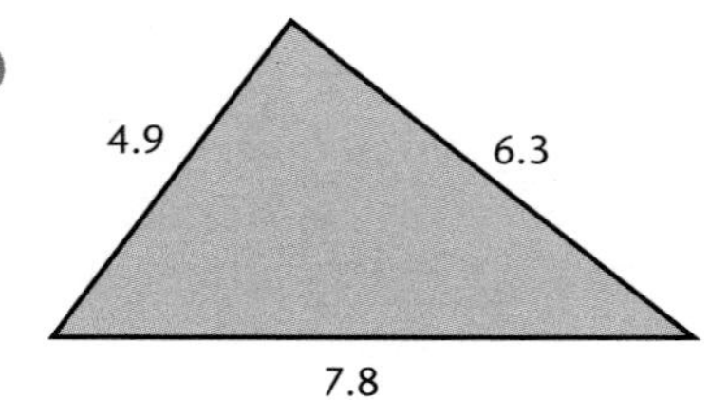

d)

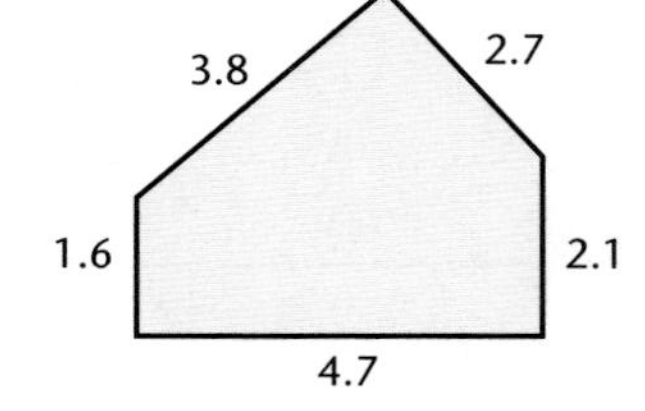

e)

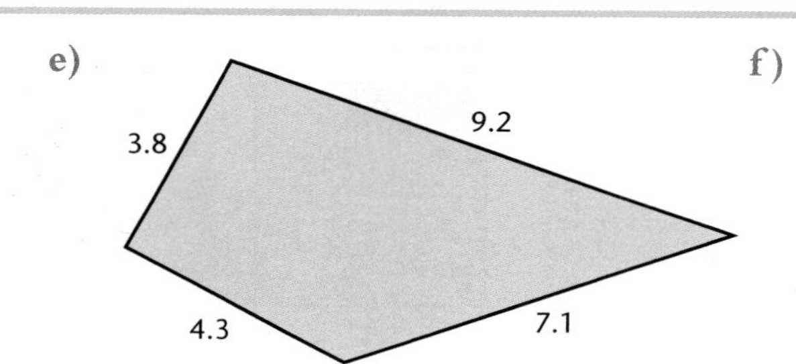

f)

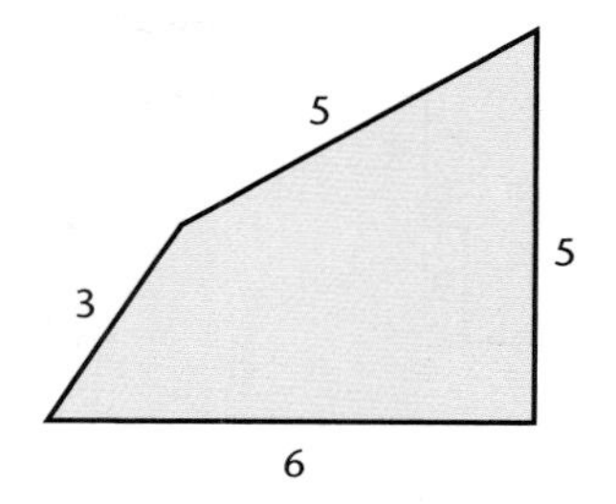

g)

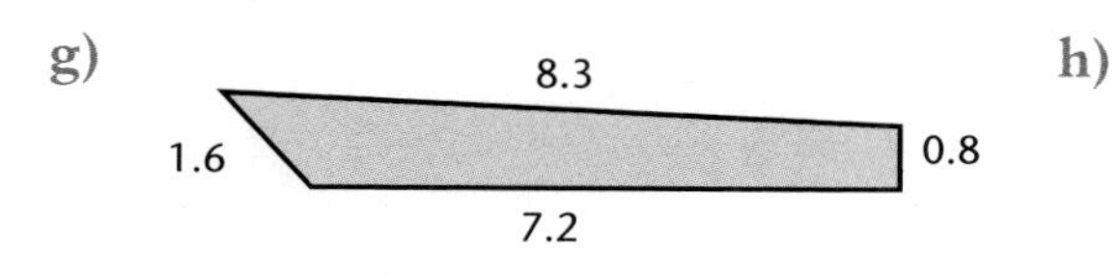

h)

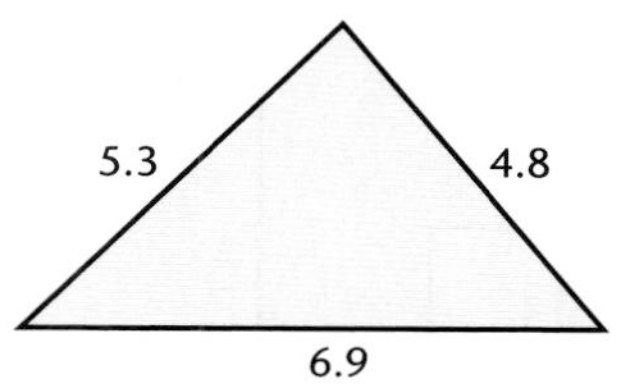

i)

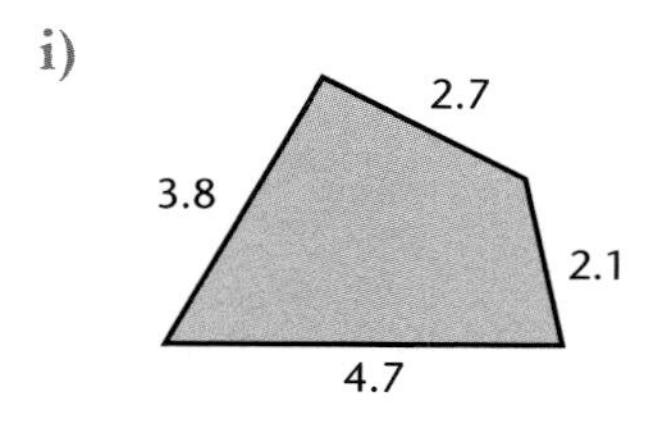

j)

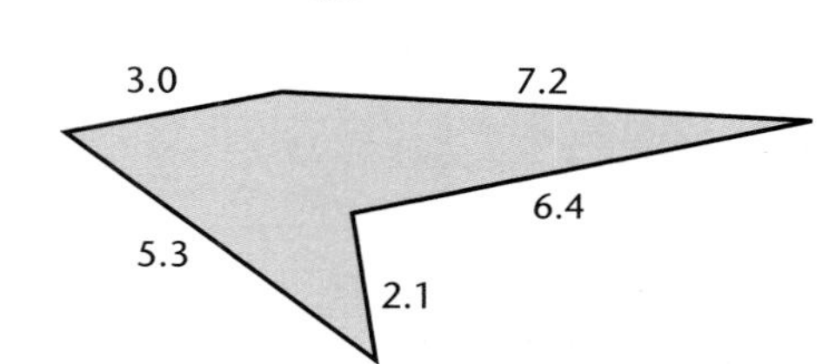

2 Work out **(i)** the perimeter and **(ii)** the area of each of these shapes. All lengths are in centimetres.

a)

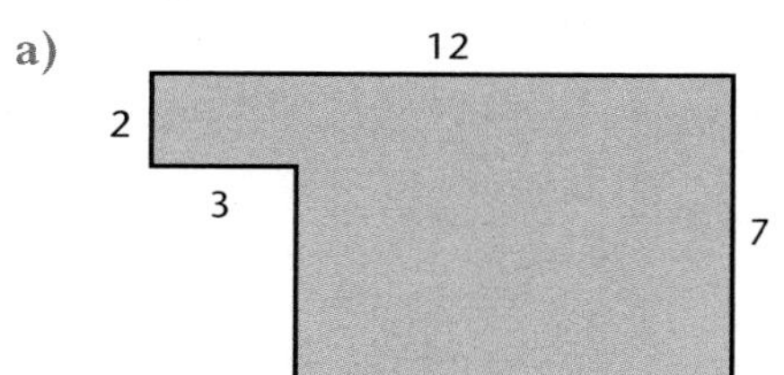

b)

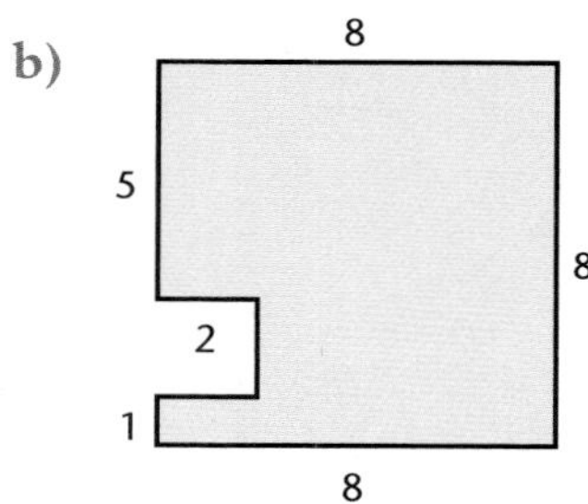

c)

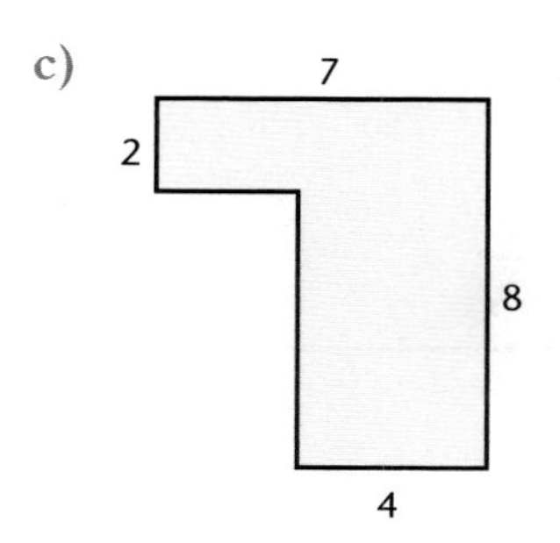

d)

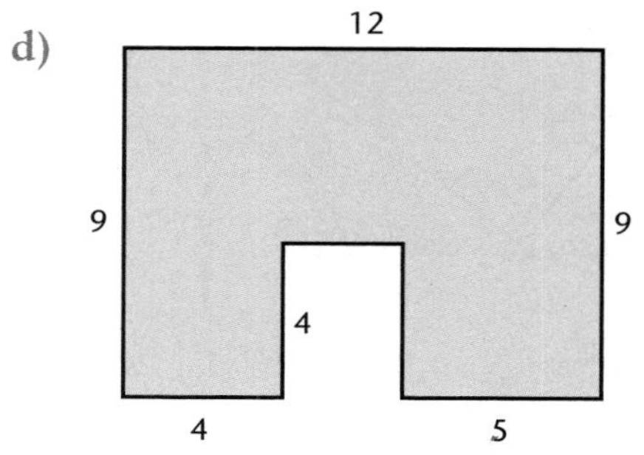

e)

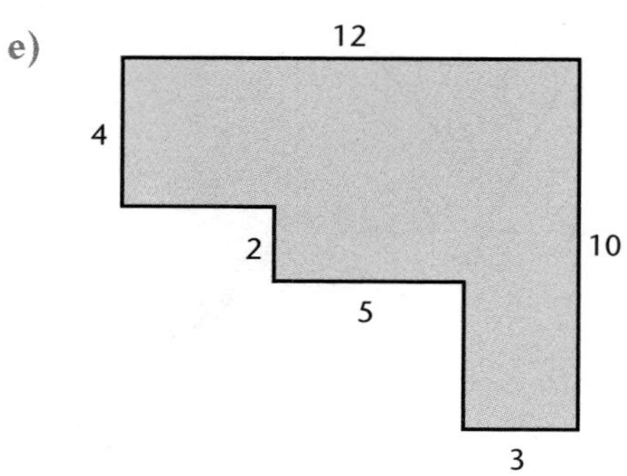

f)

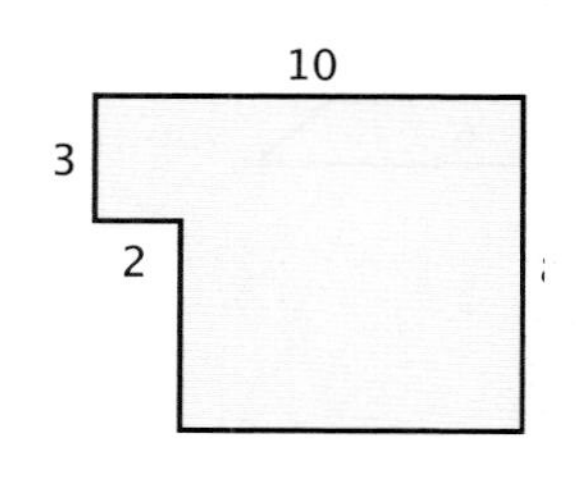

g)

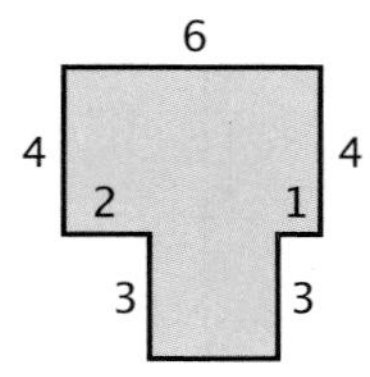

h)

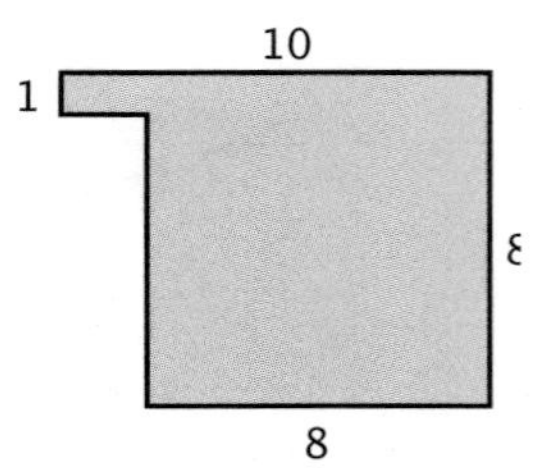

i)

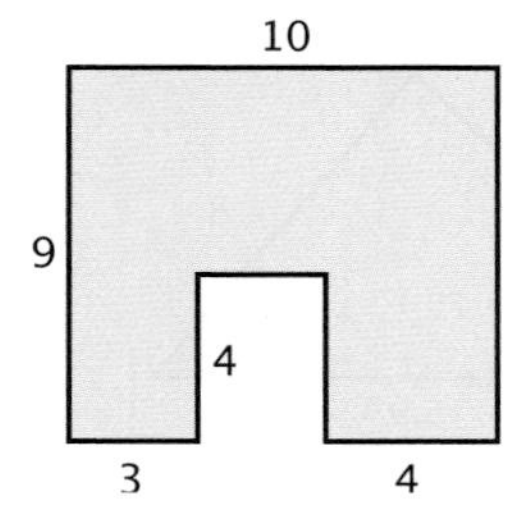

j)

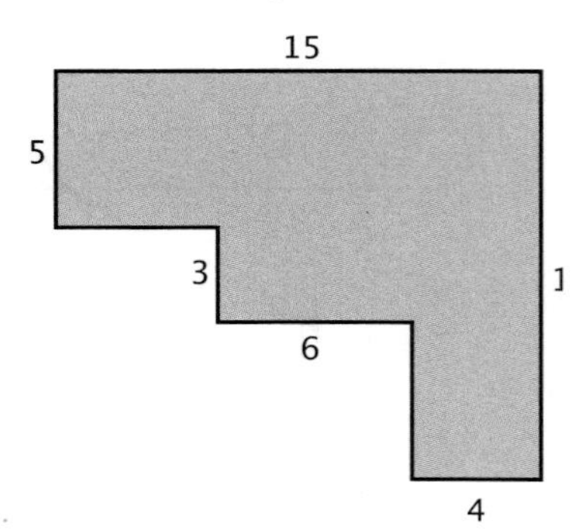

3 Work out the area of each of these shapes. All lengths are in centimetres.

a)

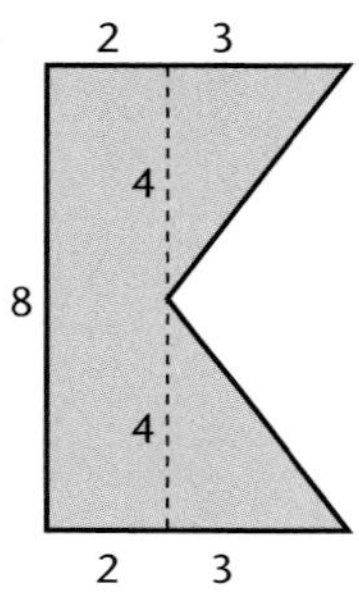

b)

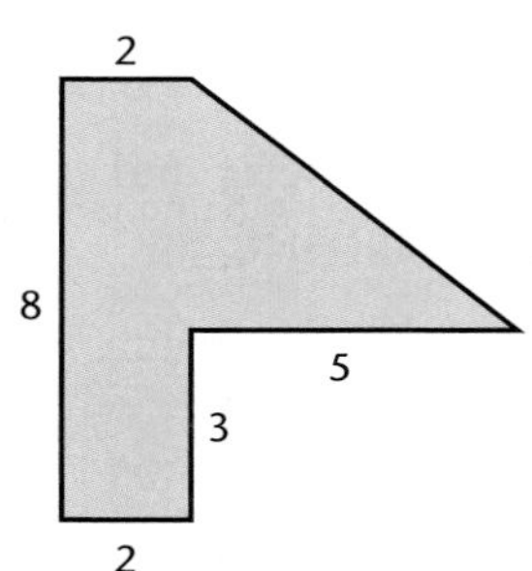

c)

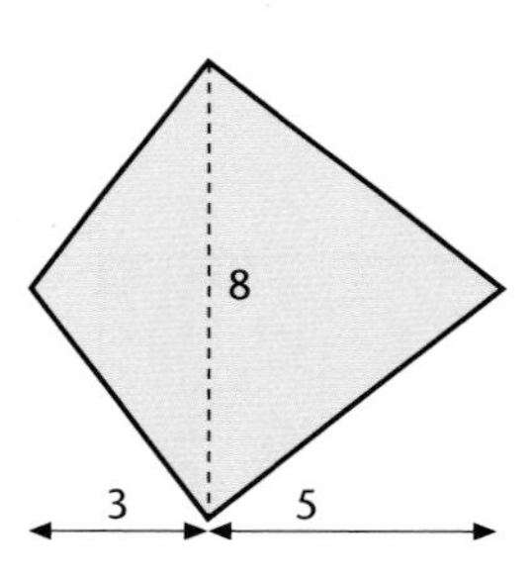

d)

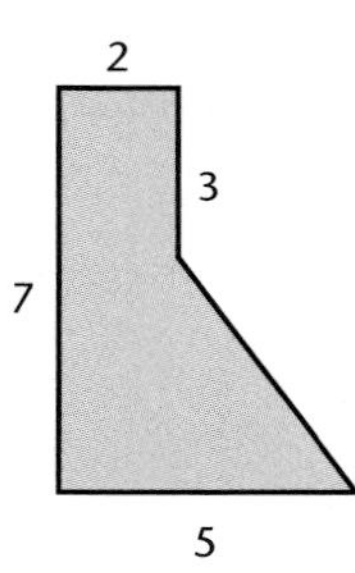

e)

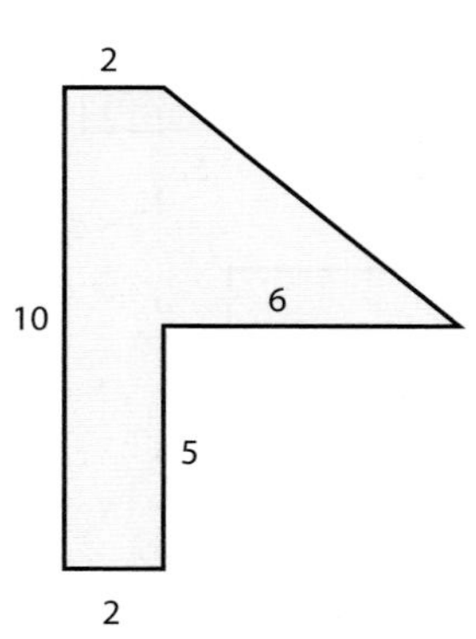

f)

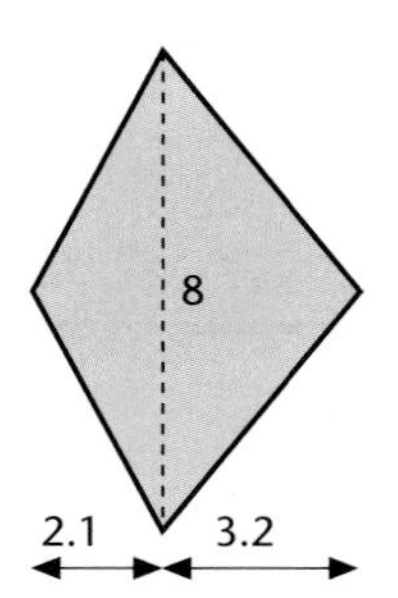

4 a) On squared paper, plot the points A(4, 2), B(2, 6), C(4, 8) and D(6, 6). Join them to make the quadrilateral ABCD.
b) Find the area of the shape.

5 a) On squared paper, plot the points A(1, 2), B(2, 6), C(4, 8) and D(6, 6). Join them to make the quadrilateral ABCD.
b) Find the area of the shape.

6 a) On squared paper, plot the points A(1, 1), B(4, 5), C(7, 5) and D(5, 1). Join them to make the quadrilateral ABCD.
b) Find the area of the shape.

7 a) On squared paper, plot the points A(1, 2), B(5, 4), C(6, 2) and D(5, 1). Join them to make the quadrilateral ABCD.
b) Find the area of the shape.

8 a) On squared paper, plot the points A(1, 1), B(1, 4), C(3, 6), D(5, 4) and E(5, 1). Join them to make the pentagon ABCDE.
b) Find the area of the shape.

Surface area

The surface area of a solid is the total area of all its faces. To find the surface area of a shape, you find the area of each of its faces and add them together. The faces can be rectangles, triangles or other shapes.

Example 8

Here is a cardboard box with no lid.

Calculate its surface area.

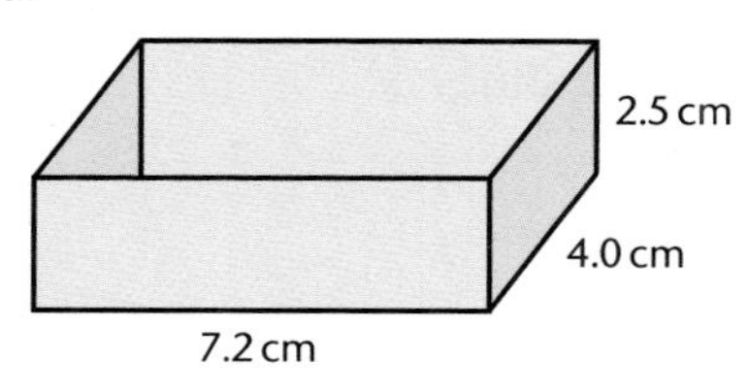

Solution

To find the surface area, you can make a list of the size of the faces and then add up their areas.

Base	$7.2 \times 4.0 = 28.8$
Front	$7.2 \times 2.5 = 18.0$
Back	$7.2 \times 2.5 = 18.0$
Right end	$4.0 \times 2.5 = 10.0$
Left end	$4.0 \times 2.5 = 10.0$
Total surface area	$= 84.8\,\text{cm}^2$

Alternatively, you can sketch its net. You can then write the area of each face on the net before adding them to get the total surface area.

7.2 cm
$18.0\,\text{cm}^2$ 2.5 cm
$10.0\,\text{cm}^2$ $28.8\,\text{cm}^2$ $10.0\,\text{cm}^2$ 4.0 cm
$18.0\,\text{cm}^2$

Activity 2

Using a ruler and compasses, draw an accurate net for this triangular prism on centimetre-squared paper.

Calculate the surface area of the prism.

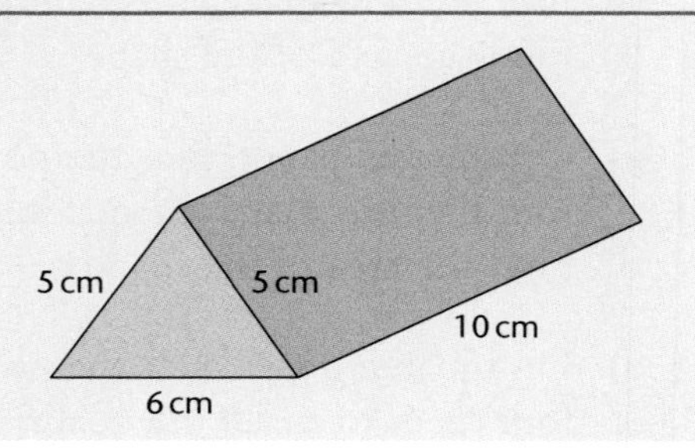

Exercise 10.5

1 Find the surface area of these cuboids.

a)

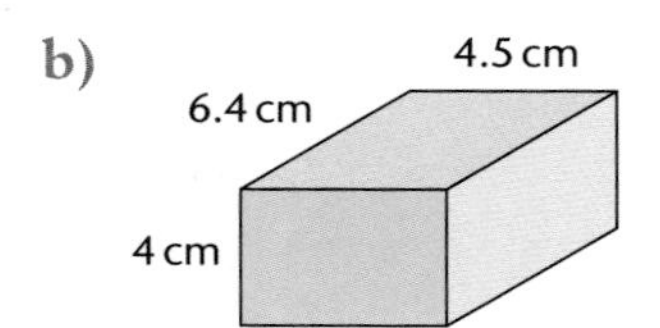

b)

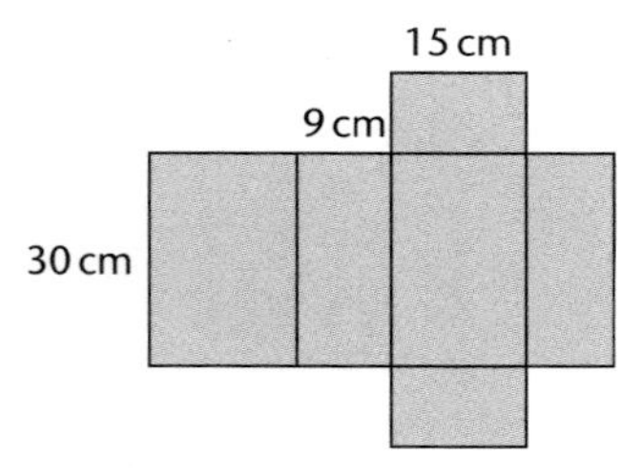

2 Find the surface area of a cube of side 20 cm.

3 The diagram shows the net of a shoe box. Calculate its surface area.

4 A fish tank is 80 cm long, 45 cm wide and 40 cm high. Draw a net for the fish tank. It does not have a lid. Use a scale of 1 cm to 10 cm. Calculate the surface area of the glass used to make the fish tank.

5 Calculate the surface area of this triangular prism.

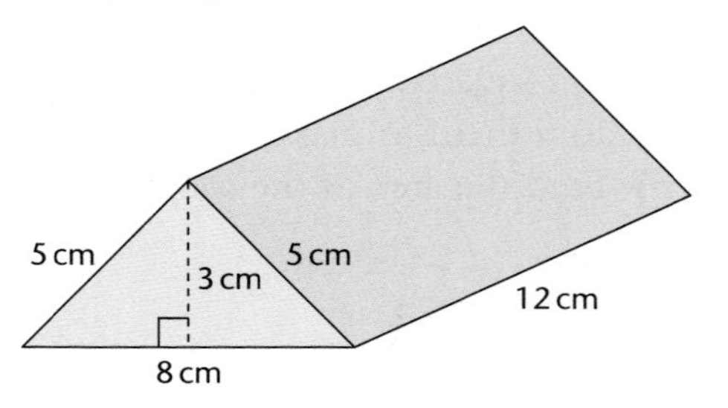

6 This is a sketch of the net for a square-based pyramid.
Calculate the surface area of the pyramid.

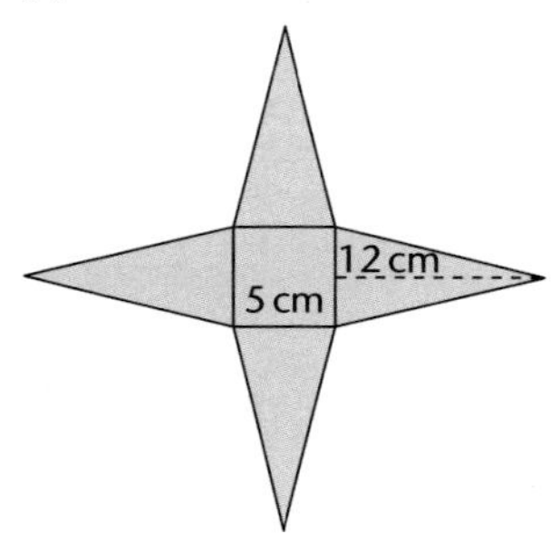

7 These two cuboids have the same volume.

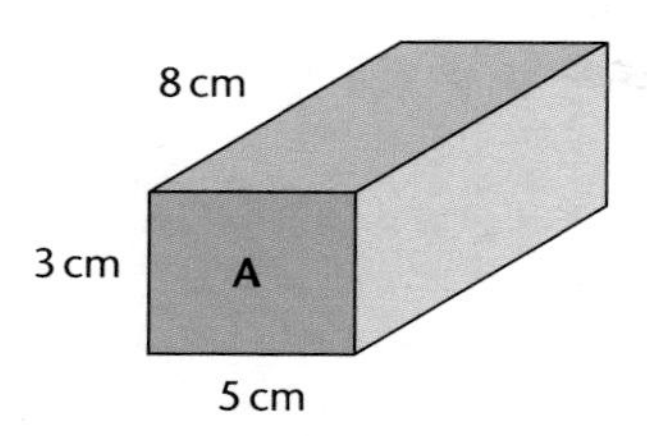

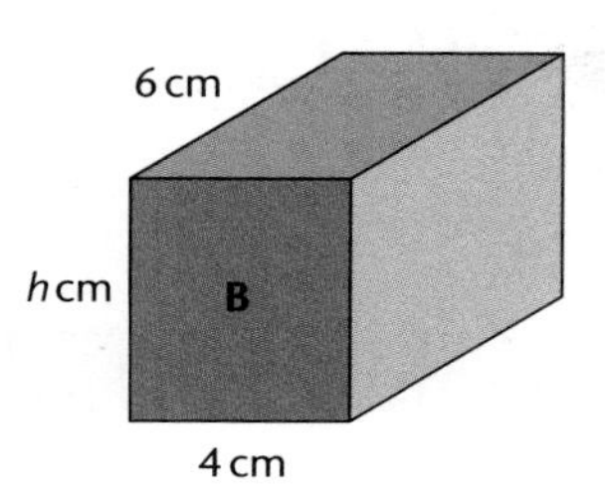

This question uses a topic from an earlier unit. You need to remember how to find the volume of a cuboid.

a) Find h.

b) Which cuboid has the smaller surface area? Show your calculations.

Key Ideas

- The area of a triangle is $\frac{1}{2} \times$ base $\times$ perpendicular height or $\frac{1}{2}bh$.
- The area of a parallelogram is base $\times$ height.
- To work out the area of a trapezium you can use the formula $\frac{1}{2}(a + b)h$ or, in words, the area of a trapezium is half the sum of the parallel sides $\times$ the height.
- You can find the area of many compound shapes by splitting them up into rectangles and triangles.
- The surface area of a solid is the total area of all its faces.

CHAPTER 11

Surveys and two-way tables

YOU WILL LEARN ABOUT	YOU SHOULD ALREADY KNOW
○ Collecting data ○ Designing a questionnaire ○ Designing and using two-way tables	○ How to make and use tally charts

Surveys

When you are trying to find out some statistical information, it is often impossible to find this out from everyone concerned. Instead, you ask a smaller section of the population. This selection is called a **sample**.

Activity 1

Are boys taller than girls?

Discuss how you could begin to answer this question.

- What information would you need to collect?
- How would you collect it?

Statistical questions such as 'are boys taller than girls' are often rewritten as a statement, called a **hypothesis**. For instance, the hypothesis here is 'boys are taller than girls'. In trying to answer the question you are investigating to find evidence for and against the statement.

In this chapter, we are looking at the first stage of answering a statistical question. In Chapter 19, you will learn about some more tools to help you analyse your findings.

Choosing a sample

The size of your sample is important. If your sample is too small, the results may not be very reliable. In general, the sample size needs to be at least 30. If your sample is too large, the data may take a long time to collect and analyse. You need to decide what is a reasonable sample size for the hypothesis you are investigating.

You also need to eliminate **bias**. A biased sample is unreliable because it means that some results are more likely than others.

It is often a good idea to use a **random** sample, where every person or piece of data has an equal chance of being selected. You may want, however, to make sure that your sample has certain characteristics.

For example, in investigating the hypothesis 'boys are taller than girls', random sampling within the whole school could mean that all the boys selected happen to be in Year 7 and all the girls in Year 11: this would be a biased sample, as older children tend to be taller. So you may instead want to use random sampling to select five girls and five boys from within each year group.

Example 1	Solution
Candace is doing a survey about school meals. She asks every tenth person going into lunch. Why may this not be a good method of sampling?	She will not get the opinions of those who dislike school meals and have stopped having them.

Activity 2

A borough council wants to survey public opinion about its library facilities.

How should it choose a sample of people to ask?

Discuss the advantages and disadvantages of each method you suggest.

Data collection sheets

When you collect large amounts of data, you may need to group it in order to analyse it or to present it clearly. It is usually best to use equal class widths for this. Tally charts are a good way of obtaining a frequency table, or you can use a spreadsheet or other statistics program to help you. Before you collect your data, make sure you design a suitable data collection sheet or spreadsheet.

Think first how you will show the results as this may influence the way you collect the data.

You could quite easily collect data from every student in your class on the number of children in their family. One way to collect this data would be to ask each person individually and make a list like this.

Class 10G														
1	2	1	1	2	3	2	1	2	1	1	2	4	2	1
5	2	3	1	1	4	10	3	2	5	1	2	1	1	2

Using the same method to collect data for all the students in your year could get very messy and one way to make the collection of data easier is to use a data collection sheet. Designing a table like the one on the left-hand side of the next page can make collecting the data easy and quick.

The data for Class 10G is shown in the completed frequency table on the right-hand side at the top of the next page.

Number of children	Tally
1	
2	
3	
4	
5	
6	
7	
8	
9	
10	

Number of children	Tally	Total (Frequency)
1	~~||||~~ ~~||||~~ ||	12
2	~~||||~~ ~~||||~~	10
3	|||	3
4	||	2
5	||	2
6		0
7		0
8		0
9		0
10	|	1

Before you design a data collection sheet it is useful to know what the answers might be but this is not always possible. For example, what would happen if you were using the tally chart above and someone said 13?

One way to deal with this problem is to have an extra line at the end of the table to record all other responses. The table below shows how this might look.

Number of children	Tally	Frequency
1	~~||||~~ ~~||||~~ ||	12
2	~~||||~~ ~~||||~~	10
3	|||	3
4	||	2
5	||	2
More than 5	|	1

Adding the 'More than 5' line allows all possible responses to be recorded and gets rid of some of the lines where there are no responses. It could just as easily have been 'More than 6' with zero shown as the frequency for 6. It all depends on what you think the answers to your question might be.

Designing a questionnaire

A **questionnaire** is often a good way of collecting data.

You need to think carefully about what information you need and how you will analyse the answers to each question. This will help you get the data in the form you need.

For instance, if you are investigating the hypothesis 'boys are taller than girls', you need to know a person's sex as well as their height. If you know their age as well you can find out whether your hypothesis is true for all ages of boys and girls. However, asking people their height would probably not be the best way of finding this information – you would be more likely to get reliable results if you could measure their height.

Here are some points to bear in mind when you design a questionnaire.

- Make the questions short, clear and relevant to your task.
- Only ask one thing at a time.
- Make sure your questions are not 'leading' questions. Leading questions show bias. They 'lead' the person answering them towards a particular answer. For example: Because of global warming, don't you think that families should be banned from having more than one car?'
- If you give a choice of answers, make sure there are neither too few nor too many.

Example 2

Suggest a sensible way of asking an adult their age.

Solution

Please tick your age group:

☐ 18–25 years ☐ 26–30 years ☐ 31–40 years
☐ 41–50 years ☐ 51–60 years ☐ Over 60 years

This means that the person does not have to tell you their exact age.

Exam tip

When you are using groups, make sure that all the possibilities are covered and that there are no overlaps.

When you have written your questionnaire, test it out on a few people. This is called doing a **pilot survey**. Try also to analyse the data from this pilot, so that you can check whether it is possible. You may then wish to reword one or two questions, regroup your data, or change your method of sampling, before you do the proper survey.

If you encounter practical problems in collecting your data, describe them in your report.

Activity 3

- Think of a survey topic about school lunches. Make sure that it is relevant to your own school or college. For instance, you may wish to test the hypothesis 'The favourite meal is Chicken Tikka.'
- Write some suitable questions for a survey to test your hypothesis.
- Try them out in a pilot survey. Discuss the results and how you could improve your questions.

Two-way tables

Sometimes the data collected involve more than one factor. Look at this example.

Example 3

Peter has collected data about cars in a car park. He has recorded the colour of each car and where it was made.

He can show both of these factors in a two-way table.

He has only completed some of the entries. Complete the table.

	Made in Europe	Made in Asia	Made in the USA	Total
Red	15	4	2	
Not red	83			154
Total		73		

Solution

	Made in Europe	Made in Asia	Made in the USA	Total
Red	15	4	2	21
Not red	83	69	2	154
Total	98	73	4	175

73 cars are made in Asia so 73 – 4 are not red.

154 cars are not red so 154 – 83 – 69 are not red and are made in the USA.

All the totals can now be completed by adding across the rows or down the columns.

Exam tip

A useful check is to calculate the grand total (the number in the bottom right corner of the table) twice. The number you get by adding down the last column should be the same as the number you get by adding across the bottom row.

Exercise 11.1

1 A borough council wants to survey public opinion about the local swimming pool. Give one disadvantage of each of the following sampling situations.

a) Selecting people to ring at random from the local phone directory

b) Asking people who are shopping on Saturday morning

2 Paul plans to ask 50 people at random how long they spent doing homework yesterday evening.

Here is the first draft of his data collection sheet.

Time spent	Tally	Frequency
Up to 1 hour		
1–2 hours		
2–3 hours		

Give two ways in which Paul could improve his collection sheet.

3 State what is wrong with each of these survey questions. Then write a better version for each of them.
 a) What is your favourite sport: cricket, tennis or athletics?
 b) Do you do lots of exercise each week?
 c) Don't you think this government should encourage more people to recycle waste?

4 Janine is doing a survey about how often people have a meal out in a restaurant. Here are two questions she has written.
 Q1. How often do you eat out?
 ☐ A lot ☐ Sometimes ☐ Never
 Q2. What food did you eat the last time you ate out?
 a) Give a reason why each of these questions is unsuitable.
 b) Write a better version of Q1.

5 Design a questionnaire to investigate use of the school library or resource centre. You need to know
- which year group the student is in.
- how often they use the library.
- how many books they usually borrow on each visit.

6 Here is a two-way table showing the results of a car survey.
 a) Copy and complete the table.

	Japanese	Not Japanese	Total
Red	35	65	
Not red	72	438	
Total			

 b) How many cars were surveyed?
 c) How many Japanese cars were in the survey?
 d) How many of the Japanese cars were not red?
 e) How many red cars were in the survey?

7 A drugs company compared a new type of drug for hay fever with an existing drug. The two-way table shows the results of the trial.
 a) Copy and complete the table.

	Existing drug	New drug	Total
Symptoms eased	700	550	
No change in symptoms	350	250	
Total			

 b) How many people took part in the trial?
 c) How many people using the new drug had their symptoms eased?

8 A group of students voted on what they wanted to do for an activity day.
Copy and complete the table.

	Riding	Sport	Total
Boys		18	
Girls	15		
Total		25	48

9 A group of students were surveyed about which sports they play.
a) Copy and complete the table.

	Hockey	Not hockey	Total
Badminton	33		
Not badminton			39
Total	57		85

b) How many students do not play either hockey or badminton?

10 At the indoor athletics championships, the USA, Germany and China won most medals.
a) Copy and complete the table.

	Gold	Silver	Bronze	Total
USA	31		10	
Germany	18	16		43
China		9	11	42
Total		43		

b) Which country won the most gold medals?
c) Which country won the most bronze medals?

11 Katy collected information about TV watching at the weekend.
This table shows the results for the people she asked.
a) Copy and complete the table.

Time watching TV (to the nearest hour)	Male	Female	Total
0–4	6	14	
5–8	11	6	
9–12	7		
over 12			13
Totals	30		60

b) How many males watched (to the nearest hour) five or more hours?
c) What fraction of those Katy asked watched TV for 0–4 hours, to the nearest hour? Give your answer in its lowest terms.
d) What percentage of the females watched TV for 5–8 hours, to the nearest hour?

12 Students in years 10 and 11 who spent time babysitting were asked how many hours they had worked in the last month.
This table shows the results.

Number of hours	Year 10	Year 11	Total
1–10	4	2	
11–20	11	10	
21–30	18	14	
over 30	3	1	
Totals	36		

a) Copy and complete the table.
b) What fraction of the year 10 babysitters worked over 20 hours in the month?
c) What percentage of the babysitters were in year 10? Give your answer correct to 1 decimal place.

Key Ideas

- Avoid bias when sampling or asking questions.
- In a random sample, every member of the population being considered has an equal chance of being selected.
- Make sure the size of the sample is sensible.
- In a questionnaire, questions should be short, clear and relevant to your task.
- You can do a pilot survey to test out a questionnaire or data collection sheet.
- When designing questions or collection sheets using groups, make sure there are no gaps or overlaps. Each response should have just one place it can go.

CHAPTER 12

Equations

YOU WILL LEARN ABOUT	YOU SHOULD ALREADY KNOW
○ Solving equations with brackets ○ Solving equations with the unknown on both sides ○ Forming and solving linear equations to solve problems	○ How to calculate with negative numbers ○ How to solve simple equations

Solving equations

When you solve an equation, you always do the same thing to both sides of the equation.

Here is a reminder about solving simple equations.

Example 1

Solve these equations.

a) $5x + 2 = 12$

b) $3x - 2 = 13$

Solution

a)
$5x + 2 = 12$
$[5x + 2 - 2 = 12 - 2]$ Subtract 2 from each side.
$5x = 10$
$[5x \div 5 = 10 \div 5]$ Divide each side by 5.
$x = 2$

b)
$3x - 2 = 13$
$[3x - 2 + 2 = 13 + 2]$ Add 2 to each side.
$3x = 15$
$[3x \div 3 = 15 \div 3]$ Divide both sides by 3.
$x = 5$

Exam tip

You don't need to show the working in square brackets.

Exercise 12.1

Solve these equations.

1 $3x + 4 = 16$

2 $5x - 4 = 16$

3 $7x - 3 = 18$

4 $9a + 4 = 40$

5 $8y + 3 = 27$

Equations with brackets

Sometimes the equation formed to solve a problem will involve brackets.

Example 2

Solve these equations.

a) $2(x + 3) = 6$

b) $4(x - 5) = 18$

Solution

a) There are two methods you can use.

Either:

$2(x + 3) = 6$

$[2 \times x + 2 \times 3 = 6]$ Expand the brackets.

$2x + 6 = 6$

$[2x + 6 - 6 = 6 - 6]$ Subtract 6 from each side.

$2x = 0$

$[2x \div 2 = 0 \div 2]$ Divide each side by 2.

$x = 0$

Or:

$2(x + 3) = 6$

$[2(x + 3) \div 2 = 6 \div 2]$ Divide each side by 2.

$x + 3 = 3$

$[x + 3 - 3 = 3 - 3]$ Subtract 3 from each side.

$x = 0$

b) Either:

$4(x - 5) = 18$

$[4 \times x + 4 \times {}^{-}5 = 18]$ Expand the brackets.

$4x - 20 = 18$

$[4x - 20 + 20 = 18 + 20]$ Add 20 to each side.

$4x = 38$

$[4x \div 4 = 38 \div 4]$ Divide each side by 4.

$x = 9\frac{1}{2}$

Or:

$4(x - 5) = 18$

$[4(x - 5) \div 4 = 18 \div 4]$ Divide each side by 4.

$x - 5 = 4\frac{1}{2}$

$[x - 5 + 5 = 4\frac{1}{2} + 5]$ Add 5 to each side.

$x = 9\frac{1}{2}$

Exam tip

You may find it easier to use this method. By expanding the brackets, you will have an equation like those in Example 1.

Exam tip

This method is usually shorter.

Exercise 12.2

Solve these equations.

1 $2(x + 1) = 10$

2 $3(x + 2) = 9$

3 $4(x - 1) = 12$

4 $5(x + 6) = 20$

5 $2(x - 3) = 7$

6 $2(x + 4) = 8$

7 $2(x - 4) = 8$

8 $5(x + 1) = 35$

9 $3(x + 7) = 9$

10 $2(x - 7) = 3$

11 $3(2x - 1) = 15$

12 $2(2x + 3) = 18$

13 $5(x - 1) = 12$

14 $3(4x - 7) = 24$

15 $2(5 + 2x) = 17$

16 $4(3x - 1) = 20$

17 $7(x + 4) = 21$

18 $3(5x - 13) = 21$

19 $2(4x + 7) = 12$

20 $2(2x - 5) = 11$

21 $5(x - 6) = 20$

22 $7(x + 3) = 28$

23 $8(2x + 3) = 40$

24 $5(3x - 1) = 40$

25 $2(5x - 3) = 14$

26 $4(3x - 2) = 28$

27 $7(x - 4) = 28$

28 $3(5x - 12) = 24$

29 $2(4x + 2) = 20$

30 $2(2x - 5) = 12$

Equations with the unknown on both sides

Sometimes the equation formed to solve a problem will have the unknown on both sides.

Example 3	Solution	
Solve $2x + 1 = x + 5$.	The first step is the same as before.	
	$2x + 1 = x + 5$	
	$[2x + 1 - 1 = x + 5 - 1]$	Subtract 1 from each side.
	$2x = x + 4$	
	Now use the same idea for the *x*-term on the right-hand side.	
	$[2x - x = x - x + 4]$	Subtract *x* from each side.
	$x = 4$	

Example 4	Solution	
Solve $2(3x - 1) = 3(x - 2)$.	$2(3x - 1) = 3(x - 2)$	
	$[2 \times 3x + 2 \times {}^{-}1 = 3 \times x + 3 \times {}^{-}2]$	Expand the brackets.
	$6x - 2 = 3x - 6$	
	$[6x - 2 + 2 = 3x - 6 + 2]$	Add 2 to each side.
	$6x = 3x - 4$	
	$[6x - 3x = 3x - 3x - 4]$	Subtract 3*x* from each side.
	$3x = {}^{-}4$	
	$[3x \div 3 = {}^{-}4 \div 3]$	Divide each side by 3.
	$x = {}^{-}1\frac{1}{3}$	

Example 5

Solve these equations.

a) $3x + 7 = 8x - 3$

b) $18 - 6x = 3x + 9$

Solution

a)

$$3x + 7 = 8x - 3$$
$$[3x + 7 - 3x = 8x - 3 - 3x]$$

You could start by subtracting $8x$ from each side, but this would give a negative x-term. So subtract $3x$ from each side.

$$7 = 5x - 3$$
$$10 = 5x$$ Add 3 to each side.
$$2 = x$$ Divide each side by 5.
$$x = 2$$ Swap the sides of the equation.

b)

$$18 - 6x = 3x + 9$$
$$[18 - 6x + 6x = 3x + 9 + 6x]$$ Add $6x$ to each side.
$$18 = 9x + 9$$
$$9 = 9x$$ Subtract 9 from each side.
$$1 = x$$ Divide each side by 9.
$$x = 1$$ Swap the sides of the equation.

Exercise 12.3

Solve these equations.

1. $2x - 1 = x + 3$
2. $3x + 4 = x + 10$
3. $5x - 6 = 3x$
4. $4x + 1 = x - 8$
5. $2x + 3 = x + 6$
6. $4x - 1 = 3x + 7$
7. $4x - 3 = x$
8. $5x + 7 = 2x + 16$
9. $2(x + 3) = x + 7$
10. $5(2x - 1) = 3x + 9$
11. $2(5x + 3) = 5x - 1$
12. $2(x - 1) = x + 2$
13. $2(2x + 3) = 3x - 7$
14. $5(3x + 2) = 10x$
15. $2x - 3 = 7 - 3x$
16. $4x - 1 = 2 + x$
17. $2x - 7 = x - 4$
18. $3x - 2 = x + 7$
19. $x - 5 = 2x - 9$
20. $x + 9 = 3x - 3$
21. $3x - 4 = 2 - 3x$
22. $5x - 6 = 16 - 6x$
23. $3(x + 1) = 2x$
24. $49 - 3x = x + 21$
25. $3(x - 1) = 2(x + 1)$
26. $3(3x + 2) = 2(2x + 3)$
27. $3(4x - 3) = 10x - 1$
28. $3(4x - 1) = 5(2x + 3)$
29. $3(3x + 1) = 5(x - 7)$
30. $7(x - 2) = 3(2x - 7)$

Challenge 1

A rectangle measures $(2x + 1)$ cm by $(x + 9)$ cm.

Find the value of x for which the rectangle is a square.

Forming equations

Some everyday problems can be solved by forming equations and solving them.

Example 6

The length of a rectangle is 4 cm greater than its width, which is x cm.

a) Write down an expression in terms of x for the perimeter of the rectangle.

b) The perimeter is 32 cm. Write down an equation in x and solve it.

c) What are the length and the width of the rectangle?

Solution

a) The length is 4 cm greater than the width, so the length is $(x + 4)$ cm.
Perimeter $= x + x + 4 + x + x + 4 = 4x + 8$

b) The perimeter is 32 cm, so

$$4x + 8 = 32$$
$$4x = 24$$
$$x = 6$$

$x + 4$

x

c) The width is x cm $= 6$ cm.
The length is $(x + 4)$ cm $= 10$ cm.

Example 7

The length of a rectangle is a cm, and the width is 15 cm shorter.
The length is three times the width.

Write down an equation in a and solve it to find the length and width of the rectangle.

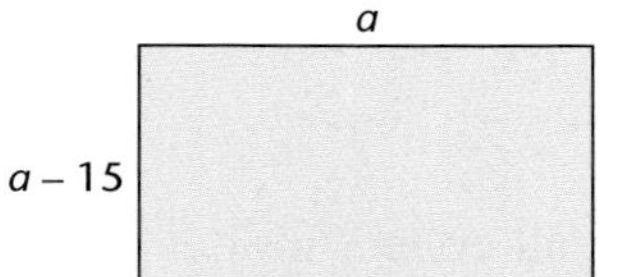

Solution

If the length $= a$, the width $= a - 15$ and the length $= 3 \times$ width $= 3(a - 15)$.

The equation is $a = 3(a - 15)$.

$a = 3(a - 15)$	
$a = 3a - 45$	Multiply out the brackets.
$a + 45 = 3a$	
$45 = 2a$	Subtract a from each side.
$22.5 = a$	Divide each side by 2.
$a = 22.5$	Rewrite with the subject on the left-hand side.

So the length $= 22.5$ cm and the width $= 7.5$ cm.

Exam tip

When you are asked to set up an equation and solve it, you will not get any marks if you just give the answer without the equation.

Exercise 12.4

1 Erica is x years old and Jayne is three years older than Erica. Their ages add up to 23.
Write down an equation in x and solve it to find their ages.

2 Two angles of a triangle are the same and the other is 15° bigger. Call the two equal angles a.
Write down an equation and solve it to find all the angles.

3 It costs £x to hire a bike for an adult, and it is £2 cheaper for a child's bike. Mr Newton hires bikes for two adults and three children.

a) Write down an expression in x for the cost of the bikes.
b) The cost is £19. Write down an equation and solve it to find x.
c) How much did each bike cost?

4 Mark, Patrick and Iain all collect model cars. Mark has four more than Patrick, and Iain has three more than Mark. They have 41 cars altogether.
Set up an equation and solve it to find how many model cars each boy has.

5 Chris and Alex go shopping. Chris spends £x and Alex spends twice as much. They spend £45 altogether.
Set up an equation and solve it to find how much each spends.

6 A pentagon has two angles of 150°, two of $x°$ and one of $(x + 30)°$. The sum of the angles in a pentagon is 540°.
a) Write down an equation in x and solve it.
b) State the size of each of the angles.

7 On a school trip to France, there are 15 more girls than boys. Altogether 53 students go on the trip.
a) If the number of boys is x, write down an equation in x and solve it.
b) How many boys and how many girls go on the trip?

8 At a café a cup of tea costs x pence, and a cup of coffee costs 10p more than a cup of tea. David spends £1.20 on three teas and two coffees.
a) Write down an equation in x and solve it.
b) What do tea and coffee cost at the café?

9 A firm employs 140 people, of whom x are men. There are ten fewer women than men.
Use algebra to find how many men and how many women work for the firm.

Exam tip

In these questions you can sometimes work out the answer without writing down the equation. But you must write it down when asked to in an examination, otherwise you will lose marks.

10 It costs £5 for each person to go skating. Skates can be hired for £2. Ten friends went skating and n of them hired skates.

a) Write down an expression in pounds for the total amount they spent.

b) They spent £62. Write down an equation in n and solve it to find how many hired skates.

11 Two angles in a triangle are x and $2x - 30°$. The first angle is twice the size of the second.
Set up an equation and solve it to find the size of the two angles.

12 The width of a rectangle is 3 cm and the length is $x + 4$ cm. The area is 27 cm^2.
Set up an equation and solve it to find x.

13 In a class of 32 students, x are girls. There are three times as many girls as boys.
Set up an equation and solve it to find how many boys and how many girls there are.

14 Stephen thinks of a number. If he doubles the number and then subtracts 5, he gets the same answer as if he subtracts 2 from the number and then multiplies by 3.
Let the number be n. Set up an equation and solve it to find n.

15 On a bus trip, each child pays £p and each adult pays £12 more than this. There are 28 children and four adults on the bus. The same amount of money is collected from all the children as from all the adults.
Set up an equation and solve it to find how much each child and each adult pays.

16 At Joe's Diner, one-course meals cost £x. Two-course meals cost £2 more. A group of eight people bought three one-course meals and five two-course meals. They paid £38 altogether.
Set up an equation and solve it to find the cost of a one-course meal.

17 Two angles of a pentagon are $x°$ and the other three are each $(2x - 20)°$. The total of all the angles is 540°.
Write down an equation and solve it to find the size of each angle.

18 The cost per person of a flight from Sheffield Airport is the charge made by the airline plus £40 tax. Four people flew from Sheffield to Cairo and the total they had to pay was £1640.
Let the charge made by the airline be £x. Write down an equation in x and solve it to find the charge made by the airline.

19 A 32-year-old man has three children who are x, $2x$ and $(2x + 4)$ years old. The man is four times as old as his eldest child.
Set up an equation and solve it to find the ages of the children.

20 At Deno's Pizza Place, a basic pizza costs £x and extra toppings are 50p each. Bernard and four of his friends each have pizzas with two extra toppings. They pay £25.50.
Set up an equation and find the cost of a basic pizza.

Key Ideas

- The two sides of an equation must always be kept equal. Operations carried out to simplify or solve an equation must always be the same for each side.
- To solve equations involving brackets, deal with the brackets first.
- To solve equations with x on both sides, first collect the x terms together on one side of the equation.
- Some problems can be solved by setting up and solving an equation.

CHAPTER 13

Angles with parallel lines

YOU WILL LEARN ABOUT	YOU SHOULD ALREADY KNOW
○ The angle properties associated with parallel lines	○ Basic angle facts about ○ straight lines ○ angles round a point ○ triangles ○ quadrilaterals

Angles with parallel lines

When a line crosses a pair of parallel lines, sets of equal angles are formed. The line cutting the parallels is called a **transversal**.

Learn to recognise each type of pair so that you are able to give reasons for angles being equal.

The diagrams show pairs of equal **alternate** angles. These are on opposite sides of the transversal.

> **Exam tip**
>
> Thinking of a Z-shape may help you to remember that alternate angles are equal. Alternatively, turn the page round and look at the diagrams upside down and you will see the same shapes!

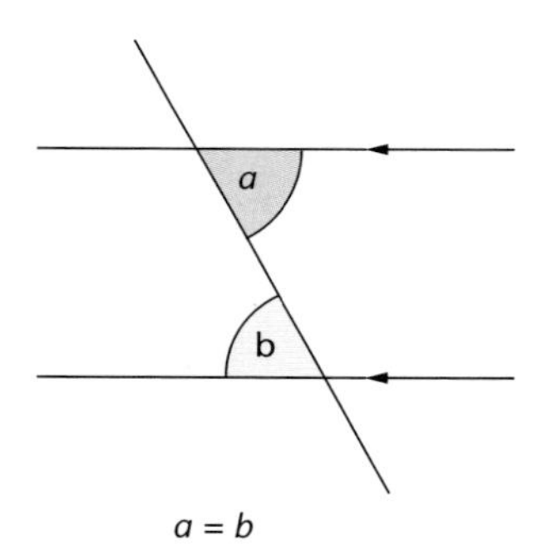

$a = b$

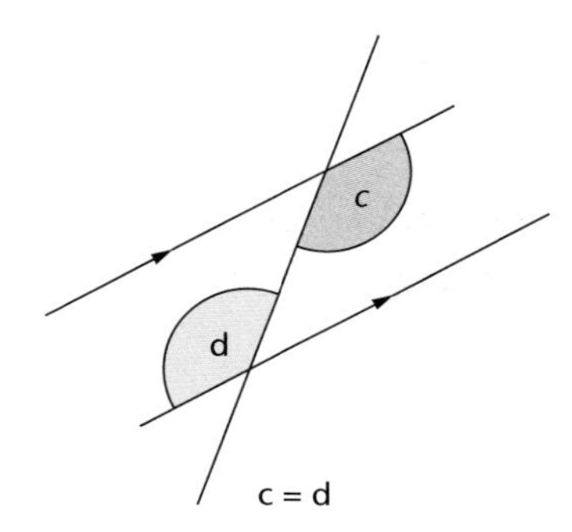

c = d

These diagrams show pairs of equal **corresponding** angles. These are in the same position between the transversal and one of the parallel lines.

> **Exam tip**
>
> Thinking of an F-shape or a translation may help you to remember that corresponding angles are equal.

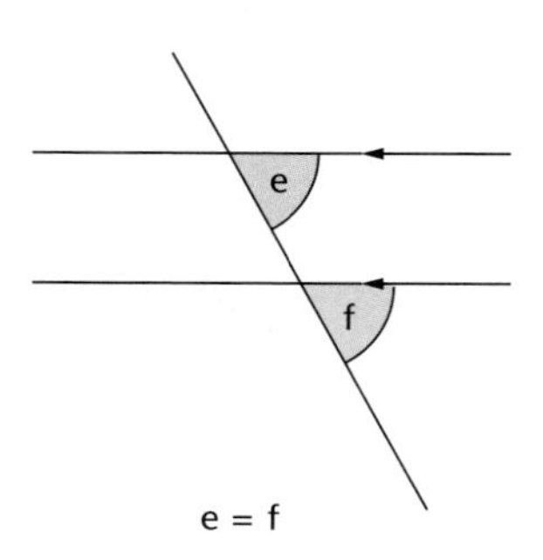

e = f

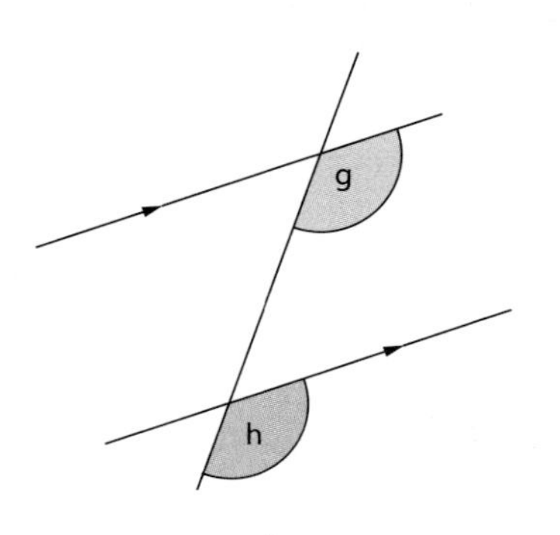

g = h

The pairs of angles in these diagrams are not equal. Instead, they add up to 180°. They are called **allied** angles or **co-interior** angles.

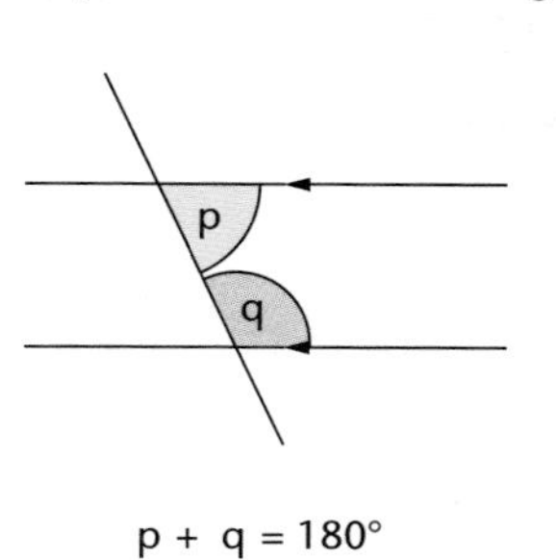

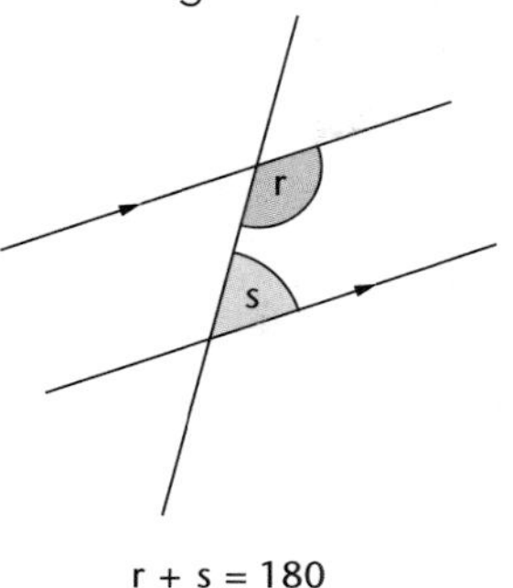

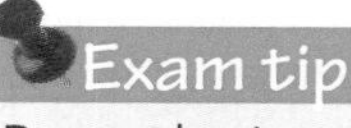

Exam tip

Remembering a C-shape or using facts about angles on a straight line may help you to remember that allied angles add up to 180°.

Example 1

Find the size of each of the lettered angles in these diagrams, giving your reasons.

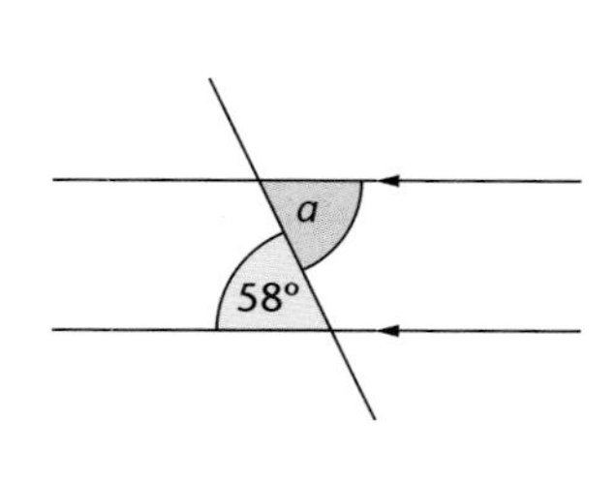

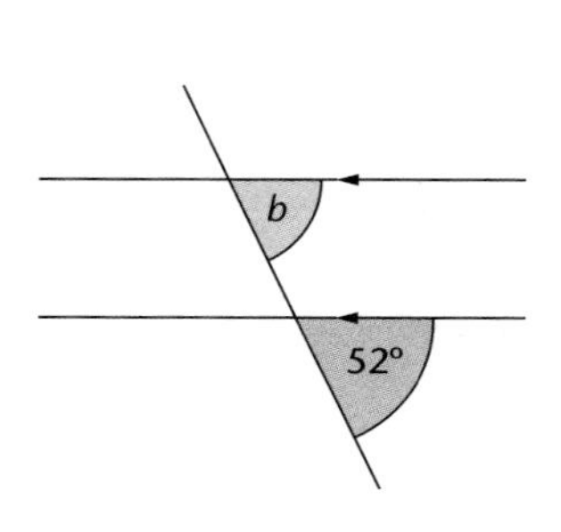

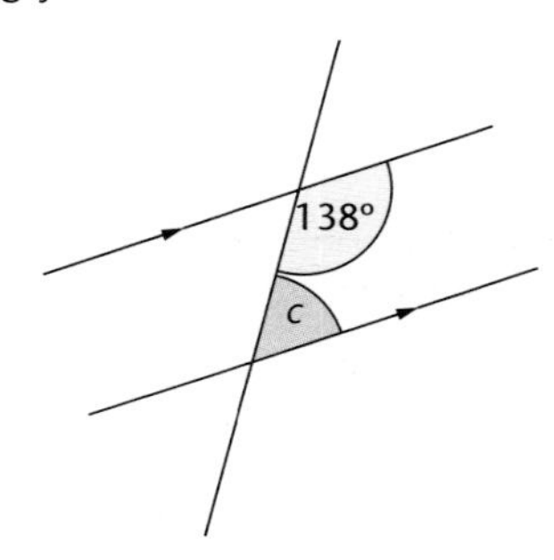

Solution

$a = 58°$	Alternate angles are equal.
$b = 52°$	Corresponding angles are equal.
$c = 42°$	Allied angles add up to 180°.

Notation for lines and angles

You often use single letters for angles, as in this chapter so far. Sometimes, the ends of a line are labelled with letters. Then you can use these letters to name the angles.

In this diagram, the line AB is the line joining A and B.

The angle at B is the angle made between the lines AB and BC. It may be written as angle ABC. In the diagram, angle ABC = 50°.

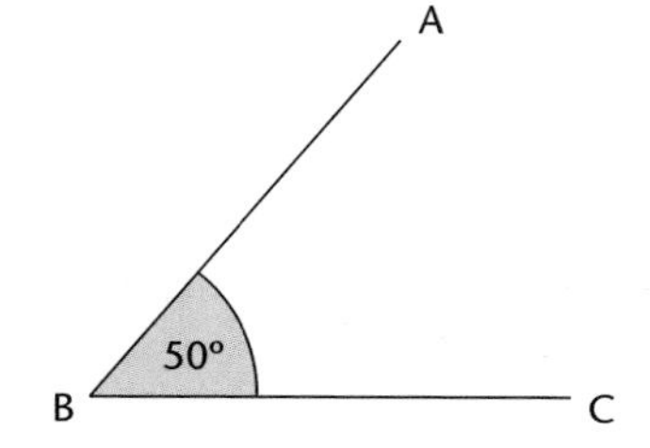

It is essential to use three letters when there is more than one angle at a point.

In this diagram, there are four angles at O and they are two different sizes.

Angle AOC = angle BOD = 50°

Angle BOC = angle AOD = 130°

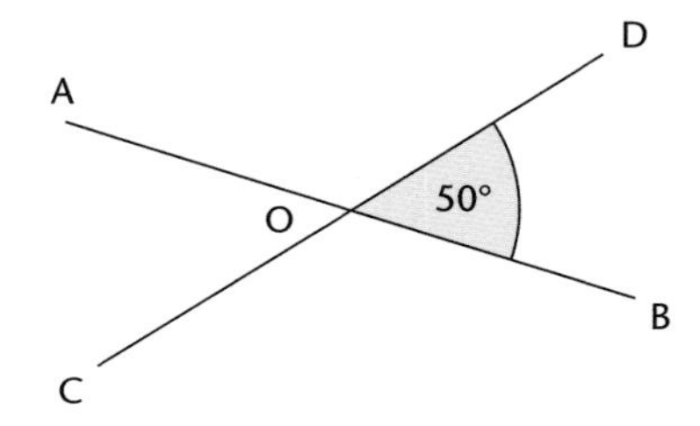

We can use the angle facts for parallel lines to prove that the sum of the angles in a triangle is 180°.

ABC is a triangle.

Line AX is parallel to line BC.

Angle x + angle y + angle z = 180° because the sum of the angles on a straight line is 180°.

Angle x = angle p because they are alternate angles.

For the same reason, angle z = angle q.

Therefore angle p + angle y + angle q = 180°.

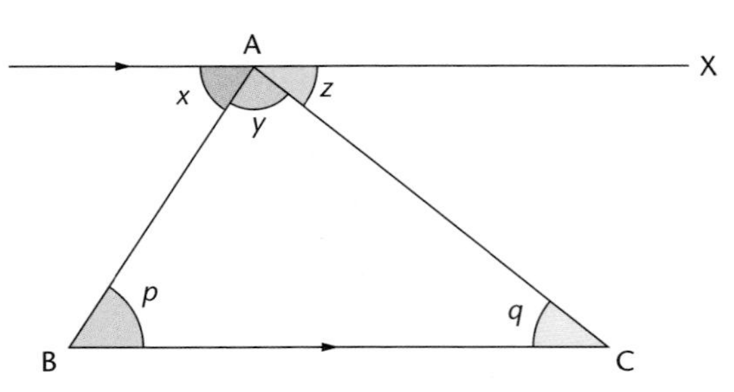

Exercise 13.1

1 Find the value of x, y and z in these diagrams, giving your reasons.

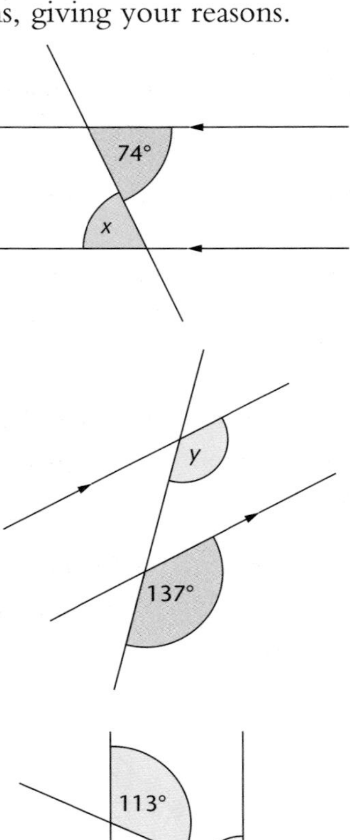

2 Draw accurately a pair of parallel lines and a line crossing them.
Mark on your drawing a pair of acute corresponding angles.
Measure them and check that they are equal.

3 Draw accurately a pair of parallel lines and a line crossing them.
Mark on your drawing a pair of obtuse alternate angles.
Measure them and check that they are equal.

4 Draw accurately a pair of parallel lines and a line crossing them.
Mark on your drawing a pair of acute alternate angles.
Measure them and check that they are equal.

5 Draw accurately a pair of parallel lines and a line crossing them.
Mark on your drawing a pair of obtuse corresponding angles.
Measure them and check that they are equal.

6 Draw accurately a pair of parallel lines and a line crossing them.
Mark on your drawing a pair of allied angles.
Measure them and check that they add up to 180°.

7 Find the size of each of the lettered angles in these diagrams.

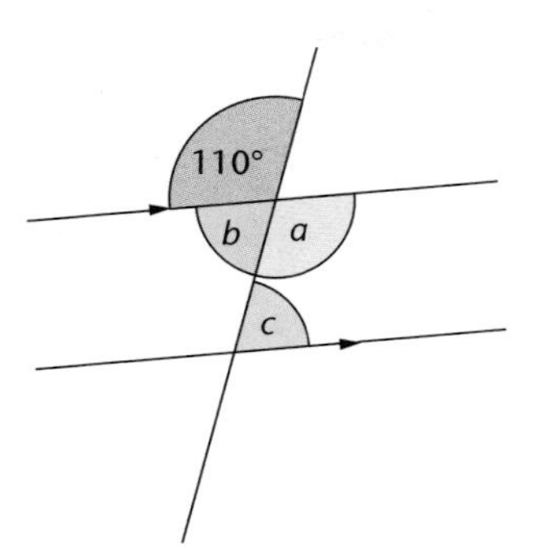

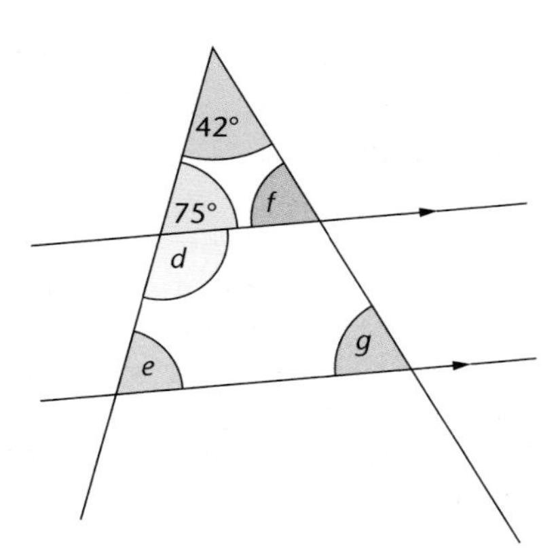

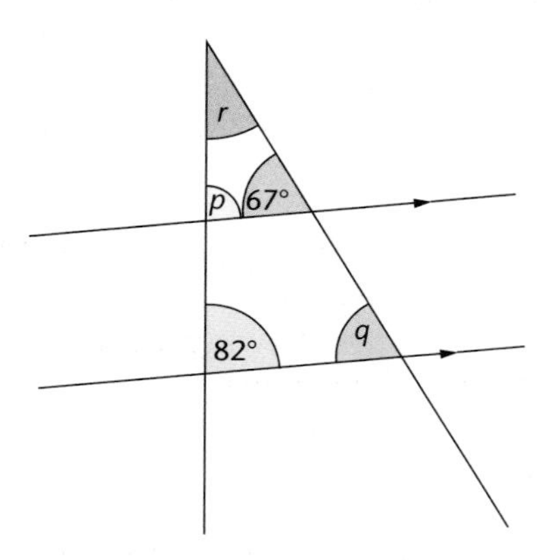

8 In a parallelogram, one angle is 126°.
Make a sketch of the parallelogram and mark this angle.
Calculate the other three angles and label them on your sketch.

9 In a parallelogram ABCD, angle ABC = 75°.
Make a sketch and mark this angle.
Calculate the other three angles and label them on your sketch.

10 In an isosceles trapezium ABCD, angle BCD is 127°.
Make a sketch of the trapezium and mark this angle.
Calculate the other three angles and label them on your sketch.

11 One of the angles of an isosceles trapezium is 64°.
Draw a sketch and write the sizes of all the angles in suitable positions on your sketch.

12 Sketch a parallelogram and write down all you can about its angles and its symmetry.

13 Sketch an isosceles trapezium and write down all you can about its angles and its symmetry.

14 Find the size of each of the lettered angles in these diagrams and give a reason for your answers.

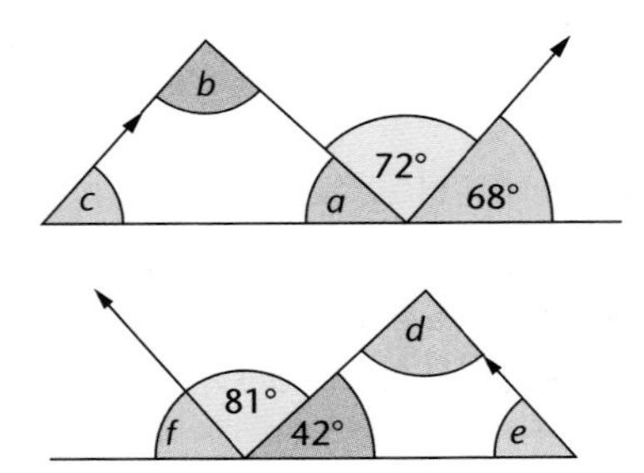

15 Use what you know about triangles to find the sizes of angles ABE and CDE in this diagram.
Show why BE and CD are parallel.

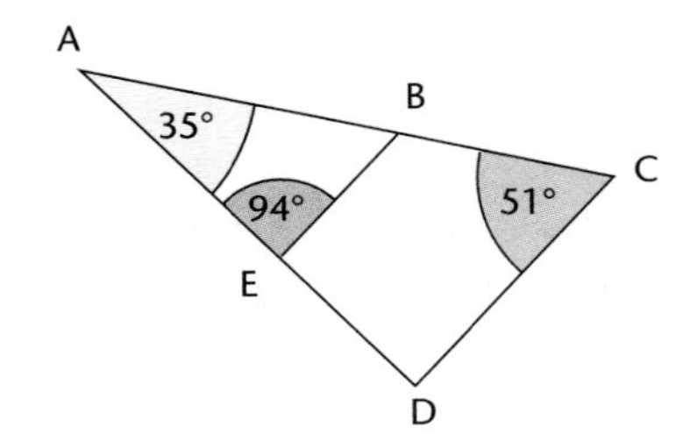

Key Ideas

When a line crosses a pair of parallel lines,
- alternate angles are equal.
- corresponding angles are equal.
- allied angles add up to 180°.

14 Circles

Foundation Silver/ Higher Initial

YOU WILL LEARN ABOUT	YOU SHOULD ALREADY KNOW
○ The language associated with circles ○ The circumference and area of a circle	○ The terms *area, perimeter, circumference, radius, diameter* ○ The units for area: cm^2, m^2

You need to be able to identify the parts of a circle.

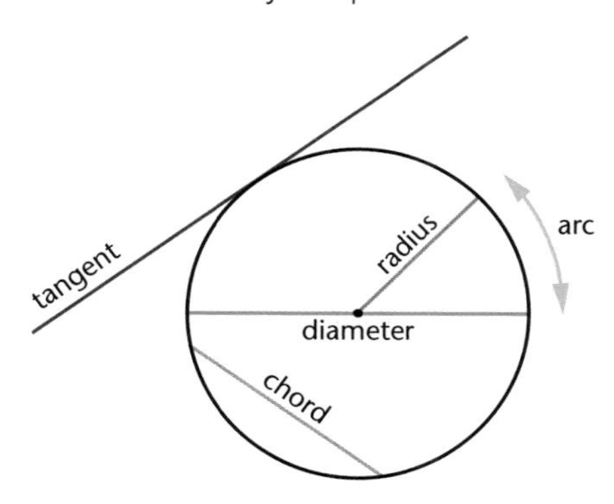

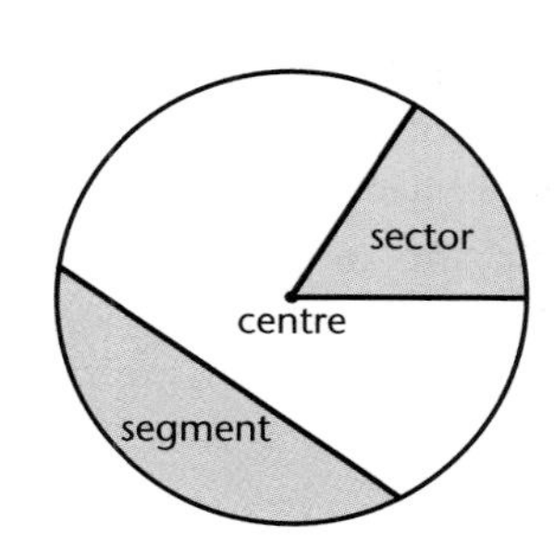

The circumference of a circle

The circumference of a circle is the distance all the way round – the perimeter of the circle.

The diameter of a circle is the distance across it, through the centre.

Activity 1

Using a tape measure or a piece of string and a ruler, measure the circumference and diameter of eight circular objects. Make sure you use the same units for both measurements.

Copy and complete this table. For the fourth column, use your calculator to work out the value of the circumference, C, divided by the diameter, d.

Name of object	Circumference, C	Diameter, d	$C \div d$

The numbers in the fourth column of the table should all be about 3.

The approximate relationship $C \approx 3 \times d$ has been known for thousands of years. In fact, very accurate calculations have shown that, instead of 3, the multiplier should be approximately 3.142. Even more accurate calculations have found this number to hundreds of decimal places. The number is a never-ending decimal, and is denoted by the Greek letter π.

Your calculator has the number which π represents stored in its memory.

Find the π button on your calculator and write down all the digits it shows. On some calculators you may have to press = after pressing the π button.

The relationship between circumference and diameter can now be written as $C = \pi \times d$.

Remember, when two letters are written next to each other in algebra, it means multiply. So the formula for the circumference is written in this way.

$$C = \pi d$$

If you know the radius of the circle instead of the diameter, use the fact that the diameter is double the radius, $d = 2r$.

The formula then becomes $C = \pi \times 2r$.

Since multiplication can be done in any order, this can be written as $C = 2 \times \pi \times r$ and so you have this formula for the circumference of a circle.

$$C = 2\pi r$$

Example 1	Solution
A circle has a diameter of 5 cm. Find its circumference.	Circumference $= \pi d$ $= \pi \times 5$ $= 15.707\ldots$ $= 15.7$ cm (to 1 d.p.)

Exam tip

If you are using your calculator, always use the π button rather than another approximation.

Example 2	Solution
A circle has a radius of 8 cm. Find its circumference.	Circumference $= 2\pi r$ $= 2 \times \pi \times 8$ $= 50.265\ldots$ $= 50.3$ cm (to 1 d.p.)

Example 3	Solution
A circle has a circumference of 20 m. Find its diameter.	Circumference $= \pi d$ $20 = \pi \times d$ Solving the equation gives $d = 20 \div \pi$ $= 6.366\ldots$ $= 6.37$ m (to 2 d.p.)

Exercise 14.1

1 Calculate the circumference of the circles with these diameters, giving your answers correct to 1 decimal place.

a) 12 cm b) 9 cm
c) 20 m d) 16.3 cm
e) 15.2 m f) 25 m
g) 0.3 cm h) 17 m
i) 5.07 m j) 6.5 cm

2 Find the circumference of circles with these radii, giving your answers correct to 1 decimal place.

a) 5 cm b) 7 cm
c) 16 m d) 18.1 m
e) 5.3 m f) 28 cm
g) 3.2 cm h) 60 m
i) 1.9 m j) 73 cm

3 The centre circle on a football pitch has a radius of 9.15 metres.
Calculate the circumference of the circle.

4 The radius of the Earth at the equator is 6378 km.
Calculate the circumference of the Earth at the equator.

5 The diagram shows a wastepaper bin in the shape of a cylinder.

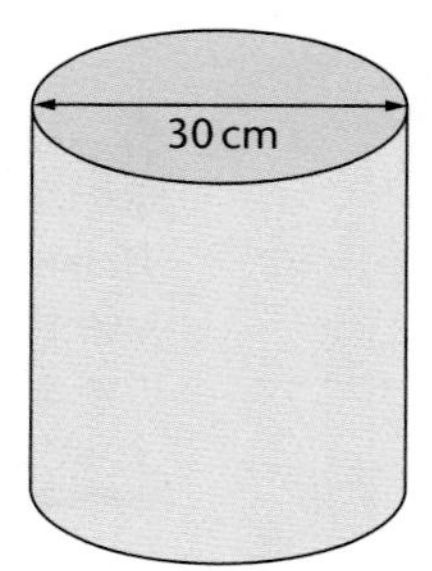

Ali is going to decorate it with braid around the top rim.
Calculate the length of braid she needs.

6 Bijan and Claire have a ride on the roundabout at a fair.
Bijan rides on a motorcycle which is 2.5 metres from the centre of the roundabout.
Claire rides on a horse which is 3 metres from the centre.
On each ride, the roundabout goes round ten times.
How much further does Claire travel than Bijan?

7 Calculate the diameter of the circles with these circumferences, giving your answers correct to 1 decimal place.

a) 75 cm b) 18 cm

8 The circumference of a circle is 50 cm.
Calculate the diameter of the circle.

9 A circular racetrack is 300 metres in circumference.
Calculate the diameter of the racetrack.

10 Mr Jones has a circular pond of radius 3 m. He wants to surround the pond with a gravel path 1 metre wide. An edging strip is needed along each side of the gravel path.
What length of edging strip is needed altogether?

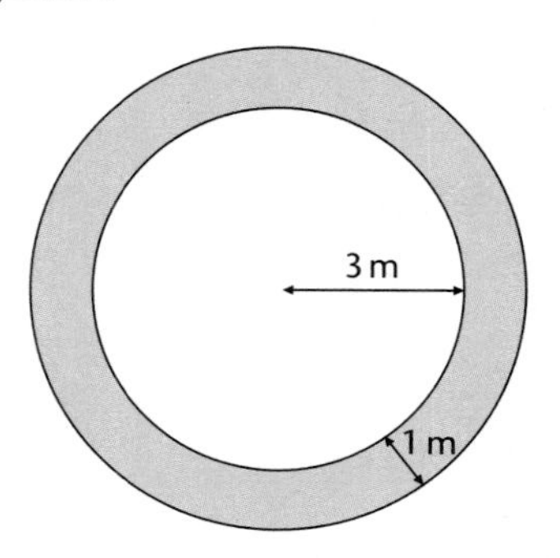

Challenge 1

Calculate the radius of the circles with these circumferences, giving your answers correct to 1 decimal place.

a) 30 cm b) 16 cm

The area of a circle

Activity 2

Take a disc of paper and cut it into 12 narrow sectors, all the same size.

Arrange them, reversing every other piece, as in the diagram.

This is nearly a rectangle.

If you had cut the disc into 100 sectors, it would be more accurate.

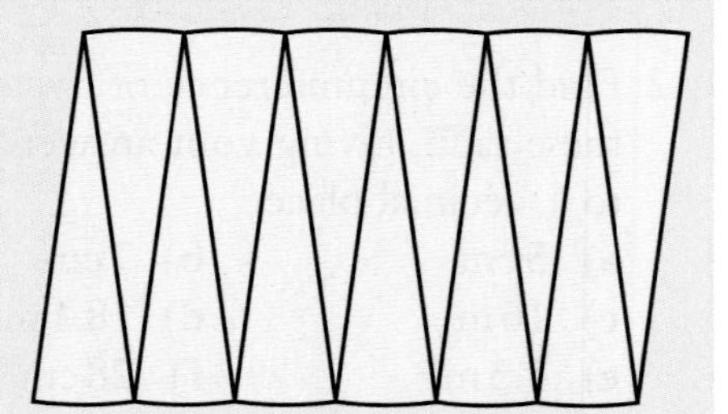

a) What are the dimensions of the rectangle?

b) What is its area?

The height of the rectangle in Activity 2 is the radius of the circle, r.

The width is half the circumference of the circle, $\frac{1}{2}\pi d$ or πr.

This gives a formula to calculate the area, A, of a circle of radius r.

$$A = \pi r^2$$

Exam tip

One of the most common errors is to mix up diameter and radius. Every time you do a calculation, make sure you have used the right one.

Example 4	Solution
The radius of a circle is 6 cm. Calculate the area of the circle.	$A = \pi r^2$ $= \pi \times 6^2$ $= \pi \times 36$ $= 113\text{ cm}^2$ (to the nearest cm^2)

Exam tip

Make sure you can use the π button and the square button on your calculator.

Example 5	Solution
The radius of a circle is 4.3 m. Calculate the area of the circle.	$A = \pi r^2$ $= \pi \times 4.3^2$ $= 58.1\text{ m}^2$ (to 1 d.p.)

Exam tip

Always round your answer to a sensible degree of accuracy.

Example 6	Solution
The diameter of a circle is 18.4 cm. Calculate the area of the circle.	$r = 18.4 \div 2$ $= 9.2\text{ cm}$ $A = \pi r^2$ $= \pi \times 9.2^2$ $= 266\text{ cm}^2$ (to the nearest cm^2)

Exercise 14.2

In all of these questions, make sure you state the units of your answer.

1 Find the area of the circles with these radii.

a) 4 cm   b) 16 m
c) 11.3 m   d) 13.6 m
e) 8.9 cm

2 Find the area of each of these circles.

a)

3 cm

b)

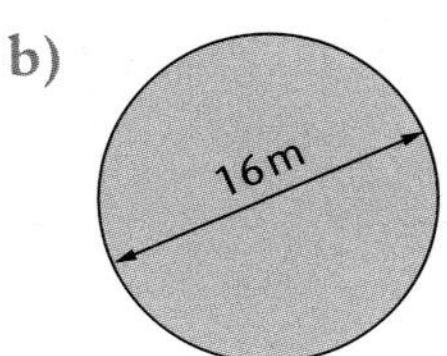

c)

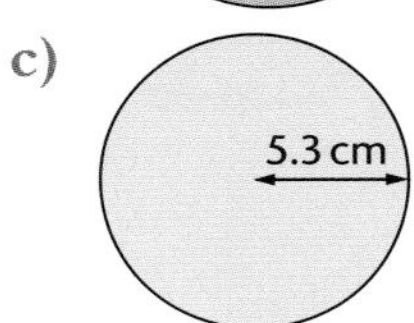

d)

e)

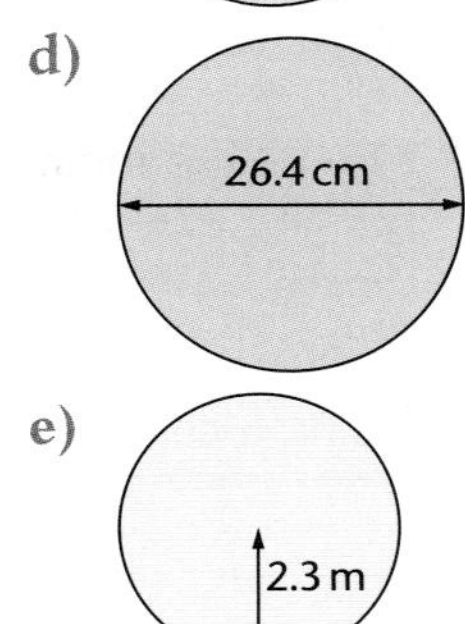

3 The radius of a circular fish pond is 1.5 m.
Find the area of the surface of the water.

4 A circular mouse mat has a radius of 9 cm.
Find the area of the mouse mat.

5 The diameter of a circular table is 0.8 m.
Find the area of the table.

6 According to the *Guiness Book of Records*, the largest pizza ever made was 37.4 m in diameter.
What was the area of the pizza?

7 To make a table mat, a circle of radius 12 cm is cut from a square of side 24 cm, as shown in the diagram.

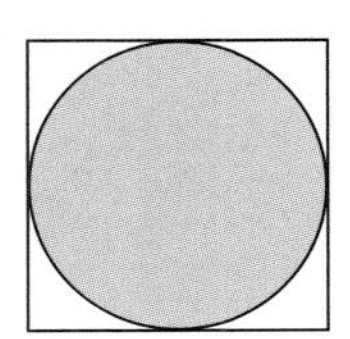

Calculate the area of the material that is wasted.

8 A square has a side of 3.5 cm and a circle has a radius of 2 cm.
Which has the bigger area?
Show your calculations.

9 The radius of the circular face of a church clock is 1.2 m.
Calculate the area of the clock face.

10 Charlie is making a circular lawn with a radius of 15 m.
The packets of grass seed say 'sufficient to cover 50 m^2'.
How many packets will she need?

11 Use your calculator to find the area of a circle with radius 6.8 cm.
Without using your calculator, do an approximate calculation to check your answer.

Exam tip

In an examination you will have to choose which formula to use. Remember r^2 must give units such as cm^2 or m^2 and so πr^2 must be the area formula.

12 The diagram shows a child's plastic ring.

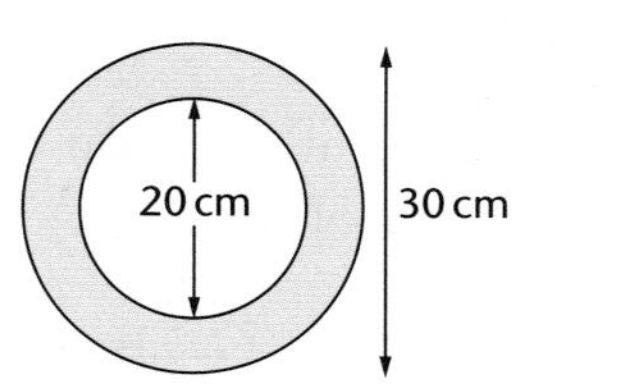

The diameter of the large circle is 30 cm.
The diameter of the small circle is 20 cm.
Calculate the area of the ring (green).

13 Mr Jones has a circular pond of radius 3 m. He decides to surround the pond with a gravel path 1 metre wide and wants to calculate how much gravel he needs. He uses this formula:
Tonnes of gravel needed = area of path in square metres × 0·034
Calculate how much gravel he needs.

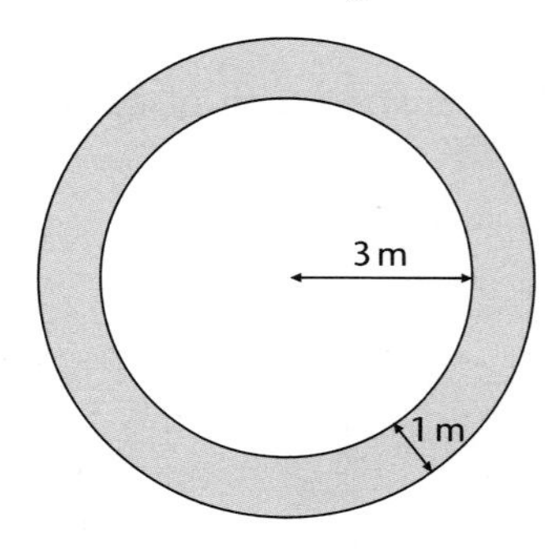

Challenge 2

The distance round a circular flower bed is 36 metres.

a) Find the radius of the flower bed.

b) Find the area of the flower bed.

Challenge 3

A circle has radius 4 m.

Calculate an estimate of the circumference of the circle without using a calculator.

Compare the answers you obtain with others, then compare with the answer using a calculator.

Using π without a calculator

When finding the area and circumference of a circle, you need to use π.

Since $\pi = 3.141\,592\ldots$, you often round it to 1 significant figure when working without a calculator.

An alternative is to give an exact answer by leaving π in the answer.

Example 7	Solution
Find the area of a circle of radius 5 cm, leaving π in your answer.	Area $= \pi r^2$ $= \pi \times 5^2$ $= \pi \times 25$ $= 25\pi$ cm^2

Example 8

A circular pond of radius 3 m is surrounded by a path 2 m wide.

Find the area of the path.

Give your answer as a multiple of π.

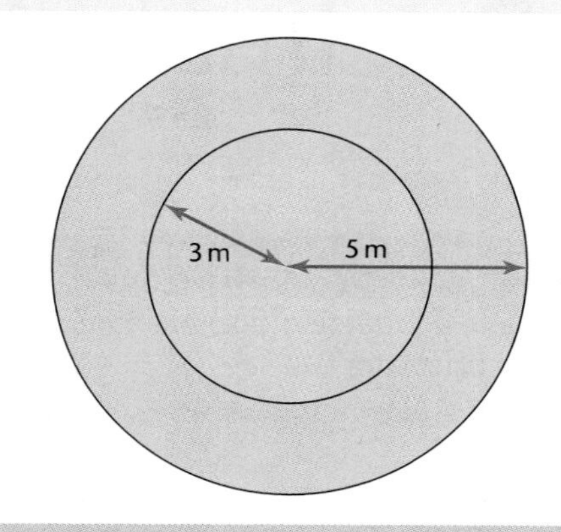

Solution

Area of path = area of large circle − area of small circle

$= \pi \times 5^2 - \pi \times 3^2$

$= 25\pi - 9\pi$ You already know that you can **collect like terms**.

$= 16\pi\ \text{m}^2$ You can treat π in the same way.

Exercise 14.3

Give your answers to these questions as simply as possible.
Leave π in your answers where appropriate.

1 a) $2 \times 4 \times \pi$ b) $\pi \times 8^2$ c) $\pi \times 6^2$
d) $2 \times 13 \times \pi$ e) $\pi \times 9^2$ f) $2 \times \pi \times 3.5$

2 a) $4\pi + 10\pi$ b) $\pi \times 8^2 + \pi \times 4^2$ c) $\pi \times 6^2 - \pi \times 2^2$
d) $2 \times 25\pi$ e) $\dfrac{24\pi}{6\pi}$ f) $2 \times \pi \times 5 + 2 \times \pi \times 3$

3 The circumferences of two circles are in the ratio $10\pi : 4\pi$. Simplify this ratio.

4 Find the area of a circle with radius 15 cm.

5 A circular hole of radius 2 cm is drilled in a square of side 8 cm.
Find the area that is left.

Key Ideas

- The circumference of a circle is given by $2\pi r$ or πd.
- The area of a circle is given by πr^2.
- If you do not have a calculator, you can give an exact answer by leaving π in the answer, simplifying the other numbers.

CHAPTER 15

Decimals 1

YOU WILL LEARN ABOUT	YOU SHOULD ALREADY KNOW
○ Adding and subtracting decimal numbers without a calculator ○ Multiplying and dividing decimals without a calculator	○ How to add, subtract, multiply and divide integers ○ How to multiply and divide decimals by powers of 10

Place value

You need to remember

- the links between decimals and place value.
- that decimals are really fractions with denominators of 10 or 100 or 1000, etc.

hundreds	tens	ones	.	tenths	hundredths
4	3	5	.	2	7

4 hundreds + 3 tens + 5 ones + 2 tenths + 7 hundredths

$= 400 + 30 + 5 + \frac{2}{10} + \frac{7}{100}$
$= 435.27$

Addition and subtraction of decimals

When adding or subtracting decimals without using a calculator, take care to line up the decimal points.

Example 1

Work out these.

a) 16.45 + 2.62

b) 13.78 – 1.24

Solution

a) 16.45 + 2.62 = 16.45
+ 2.62
19.07

b) 13.78 – 1.24 = 13.78
– 1.24
12.54

Remember, every time you add and subtract amounts of money you are dealing with decimals; remember too that you should write three pounds twenty pence as £3.20 and not £3.2 and, for example, £3.22 + £2.18 = £5.40 not £5.4.

Exercise 15.1

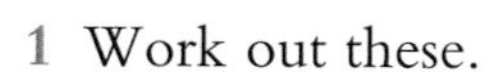

1 Work out these.
   a) £2.10 + £3.45
   b) £5.78 + £2.82
   c) £7.15 – £6.13
   d) £34.02 + £14.89
   e) £123.67 – £65.77

2 Work out these.
   a) £4.37 + £6.53
   b) £15.78 + £9.89
   c) £6.18 + £6.32
   d) £15.42 – £9.34
   e) £19.21 – £17.04

3 Work out these.
   a) 1.6 + 3.4
   b) 4.9 + 5.21
   c) 17.77 + 19.54
   d) 10.78 – 4.9
   e) 21.99 – 11.9
   f) 5.53 – 3.09

4 Work out these.
   a) 4.9 + 5.6
   b) 34.94 + 7.62
   c) 24.24 + 16.16
   d) 8.6 – 3.4
   e) 9.42 – 8.57
   f) 24.2 – 4.63

5 Anna buys a T-shirt costing £8.99 and a umper costing £18.99.
How much change does she get from £30?

6 Helen buys a packet of sandwiches costing £1.60, a bar of chocolate costing 45p and a drink costing 65p.
How much change does she get from £5?

7 Class 7 are collecting money each week for a charity.
The totals collected in the first four weeks are £12.56, £14.66, £18.13 and £11.82.
How much more do they need to raise to reach their target of £100?

8 Kate buys a 4-metre length of material.
She makes two curtains, each of which is 1.8 m long.
How much material does she have left?

9 In the long jump, Jim jumps 13.42 m and Dai jumps 15.18 m.
Find the difference between the lengths of their jumps.

10 The times for the first and last places in a 200-metre race were 24.42 seconds and 27.38 seconds.
Find the difference between these times.

Multiplying decimals by integers or decimals

When multiplying two decimal numbers the method is as follows.

- First 'ignore' the decimal points.
- Then do the multiplication.
- Finally, count up the number of decimal places in the numbers you are multiplying – the total will give the number of decimal places in the answer.

You can use the same method when multiplying a decimal by an integer.
Alternatively, you can set the calculation out as a formal multiplication with the decimal point in the answer underneath the decimal point in the number you are multiplying.

Example 2

Work out
5.3×4.

Solution

Either:

5.3×4

First work out $53 \times 4 = 212$.

Then put in the decimal point.

There is 1 decimal place in 5.3 and no decimal places in 4, so there will be 1 decimal place in the answer.

$5.3 \times 4 = 21.2$

Or:

$$\begin{array}{r} 5.3 \\ \times\ 4 \\ \hline 21.2 \end{array}$$

Example 3

Work out these.

a) 0.4×0.5

b) 1.23×1.2

Solution

a) 0.4×0.5

First work out $4 \times 5 = 20$.
Then put in the decimal point.
There is 1 decimal place in 0.4 and 1 decimal place in 0.5, so there will be 2 decimal places in the answer.

$0.4 \times 0.5 = 0.20 = 0.2$

b) 1.23×1.2

First work out $123 \times 12 = 1476$.
Then put in the decimal point.
There are 2 decimal places in 1.23 and 1 decimal place in 1.2 so there will be 3 decimal places in the answer.

$1.23 \times 1.2 = 1.476$.

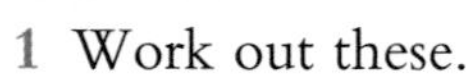

Exercise 15.2

1 Work out these.

a) 2×4.6 **b)** 4×7.9
c) 3×4.5 **d)** 12×3.2
e) 6×7.9 **f)** 5×45.2

2 Work out these.

a) 14×6.4 **b)** 9×7.3
c) 8.4×6 **d)** 13.3×5

3 Work out these.

a) 200×0.56 **b)** 8.6×40
c) 5.9×200 **d)** 0.07×300
e) 0.01×4000 **f)** 600×4.5

4 Work out these.

a) 100×0.49 **b)** 33.3×40
c) 200×0.6 **d)** 400×3.8

5 Given that $63 \times 231 = 14\,553$ write down the answers to these.

a) 6.3×2.31 **b)** 63×23.1
c) 0.63×23.1 **d)** 63×0.231
e) $6.3 \times 23\,100$

6 Given that $12.4 \times 8.5 = 105.4$ write down the answers to these.

a) 124×8.5 **b)** 12.4×0.85
c) 0.124×8.5 **d)** 1.24×8.5
e) 0.124×850

7 Work out these.

a) 4×0.3 **b)** 0.5×7
c) 3×0.6 **d)** 0.8×9
e) 0.6×0.4 **f)** 0.8×0.6
g) 40×0.3 **h)** 0.5×70
i) 0.3×0.2 **j)** 0.8×0.1
k) $(0.7)^2$ **l)** $(0.3)^2$

8 Work out these.

a) 4.2 × 1.5
b) 6.2 × 2.3
c) 5.9 × 6.1
d) 7.2 × 2.7
e) 63 × 1.8
f) 72 × 5.4
g) 5.6 × 8.9
h) 10.9 × 2.4
i) 12.7 × 0.4
j) 2.34 × 0.8
k) 5.46 × 0.7
l) 6.23 × 1.6

9 Kate buys three of these packs of meat.

a) What is the total weight?
b) What is the total cost?

10 Find the cost of 5 kg of new potatoes at £1.18 per kilogram.

11 Pali buys two shirts at £8.95 each and a pair of trousers at £17.99.
How much change does he get from £50?

12 Gemma buys two cucumbers at 68p each and three cauliflowers at £1.25 each.
How much change does she get from £10?

Dividing decimals by integers

When dividing decimals by a whole number, line up the decimal point in the answer with the decimal point in the question.

Example 4	Solution
Work out 75.5 ÷ 5.	$75.5 \div 5 = 5\overline{)75.5}$ with answer 15.1 above

Exercise 15.3

1 Work out these.

a) 40.5 ÷ 5
b) 210.3 ÷ 3
c) 2.644 ÷ 4
d) 36.33 ÷ 3

2 Work out these.

a) 81.9 ÷ 9
b) 64.8 ÷ 8
c) 124.4 ÷ 4
d) 64.4 ÷ 7

3 Work out these.

a) 3.2 ÷ 4
b) 39.6 ÷ 6
c) 12.5 ÷ 5
d) 49.63 ÷ 7

4 Work out these.

a) 39.3 ÷ 3
b) 56.8 ÷ 8
c) 13.6 ÷ 4
d) 15.6 ÷ 4

5 Sam cut a 4.72 m length of string into eight equal pieces. How long was each piece?

6 Grandma left £25 643.10 to be shared equally between her six grandchildren. How much did they each receive?

Dividing decimals by decimals

When dividing a decimal number by another decimal number, first write the division as a fraction, then multiply the top number and the bottom number by 10 or 100 or a bigger power of 10 to get a whole number on the bottom and finally cancel, if possible, and divide.

Example 5

Work out 0.9 ÷ 1.5.

Solution

$0.9 \div 1.5 = \frac{0.9}{1.5}$ Write the division as a fraction.

$= \frac{9}{15}$ Multiply the top number and the bottom number by 10 to get a whole number on the bottom.

$= \frac{3}{5}$ Simplify.

$\frac{3}{5} = 5\overline{)3.0}$ = 0.6 Divide.

0.9 ÷ 1.5 = 0.6

Exercise 15.4

Work out these.

1 a) 0.42 ÷ 1.5 b) 460 ÷ 0.2
c) 1.43 ÷ 1.1 d) 1.2 ÷ 1.5
e) 12.8 ÷ 3.2

2 a) 2.4 ÷ 0.6 b) 17.5 ÷ 3.5
c) 10.8 ÷ 1.2 d) 12.4 ÷ 3.1

3 a) 21.6 ÷ 2.4 b) 16 ÷ 3.2
c) 4 ÷ 0.8 d) 7.4 ÷ 3.7

4 a) 1.8 ÷ 0.3 b) 2.8 ÷ 0.7
c) 4.2 ÷ 0.6 d) 6.5 ÷ 1.3

5 a) 8 ÷ 0.2 b) 1.2 ÷ 0.3
c) 5.6 ÷ 0.7 d) 3.6 ÷ 0.4
e) 24 ÷ 1.2 f) 50 ÷ 2.5
g) 9 ÷ 0.3 h) 15 ÷ 0.3
i) 16 ÷ 0.2 j) 24 ÷ 0.8
k) 1.55 ÷ 0.5 l) 48.8 ÷ 0.4

6 a) 14.7 ÷ 0.3 b) 13.6 ÷ 0.8
c) 14.4 ÷ 0.6 d) 22.4 ÷ 0.7
e) 47.7 ÷ 0.9 f) 85.8 ÷ 1.1
g) 3.42 ÷ 0.6 h) 1.96 ÷ 0.4
i) 1.45 ÷ 0.5 j) 3.51 ÷ 1.3
k) 5.55 ÷ 1.5 l) 6.3 ÷ 1.4

7 A jug holds 0.4 litres of water.
How many jugfuls are needed to fill a 10 litre bucket with water?

8 Jim buys 1.8 kg of carrots at 75p per kg and some potatoes at 1.20 per kg.
He gives the assistant a £10 note and gets £5.41 change. Find how many kg of potatoes he buys.

Key Ideas

- When adding or subtracting decimals, line up the decimal points.
- To multiply a decimal by an integer or by a decimal, first 'ignore' the decimal points, then do the multiplication and, finally, count up the number of decimal places in the numbers you are multiplying – the total will give the number of decimal places in the answer.
- To divide a decimal by an integer, divide as usual and line up the decimal points in your answer.
- To divide a decimal by a decimal, first write as a fraction, then multiply both numbers by a power of 10 so that you divide by an integer.

16 Rotations

YOU WILL LEARN ABOUT	YOU SHOULD ALREADY KNOW
o Rotating objects about a given point o Recognising rotations	o About reflection and rotation symmetry o How to plot and read coordinates o About congruent shapes

Rotations

You can **rotate** an object about a point, C. The point is called the **centre of rotation**.

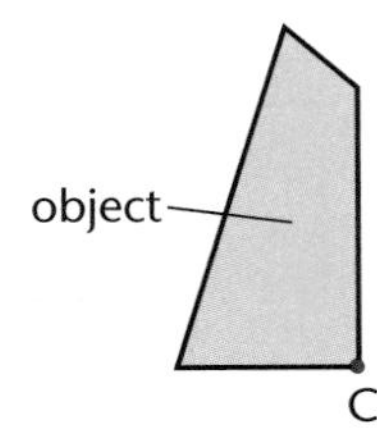

Rotating an object is like completing one step towards making a drawing with rotation symmetry.

Trace the object then use a pencil or compass point to keep C still. Turn the tracing paper through the angle of rotation required.

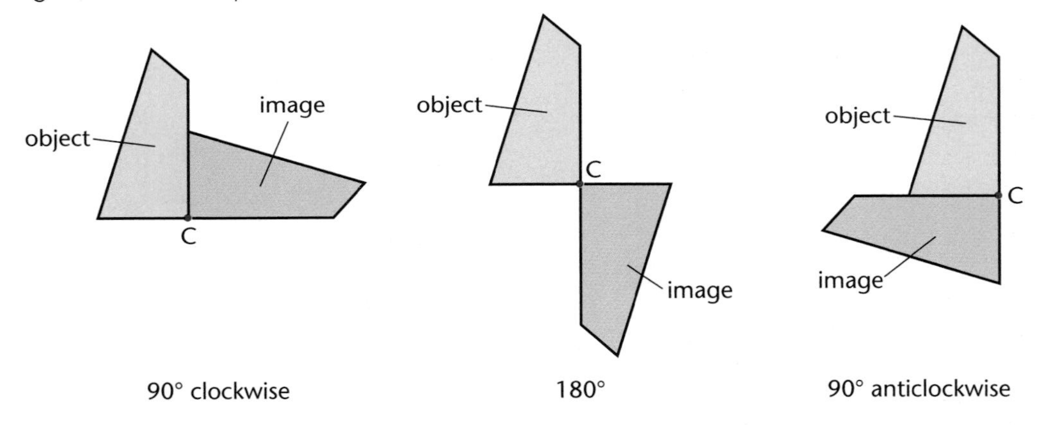

Since 180° is a half-turn, you can rotate the shape clockwise or anticlockwise.

90° anticlockwise is the same as 270° clockwise.

Each point on the **image** is the same distance away from the centre of rotation as the matching point on the **object**.

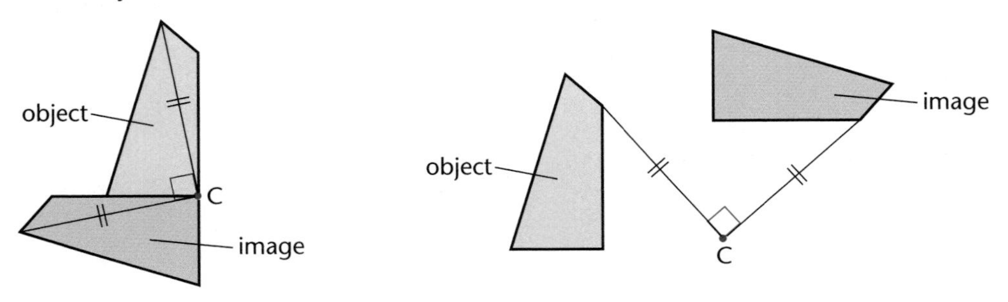

When an object is rotated, its image is **congruent** to it.

Unlike reflections, when an object is rotated, the object and the image are the same way round. (You turn the tracing paper round for a rotation but for a reflection you turn it over.)

You can also describe a rotation of 90° as a quarter-turn and a rotation of 180° as a half-turn. Rotations are anticlockwise, unless you are told otherwise.

In this diagram, the shape P has been rotated through a quarter-turn (90° anticlockwise) about point O, the origin.

You can do this by counting squares or by tracing the shape and turning it or by using a pair of compasses with the point at O and drawing quarter circles from each point.

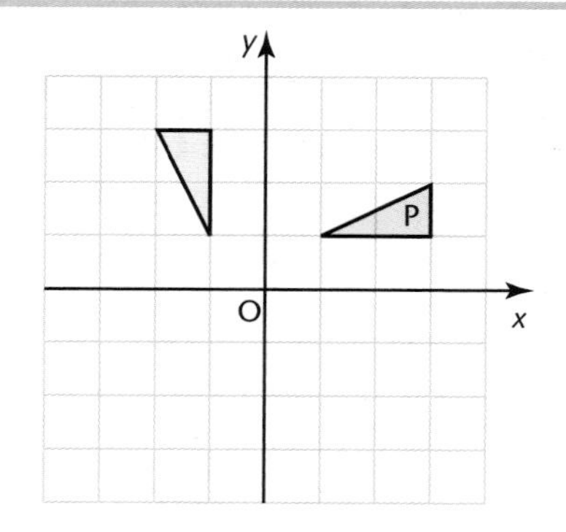

Exam tip

When you have drawn the rotation, turn the page through the correct angle to check it looks like the original.

Example 1

Rotate this shape through a three-quarter-turn about O.

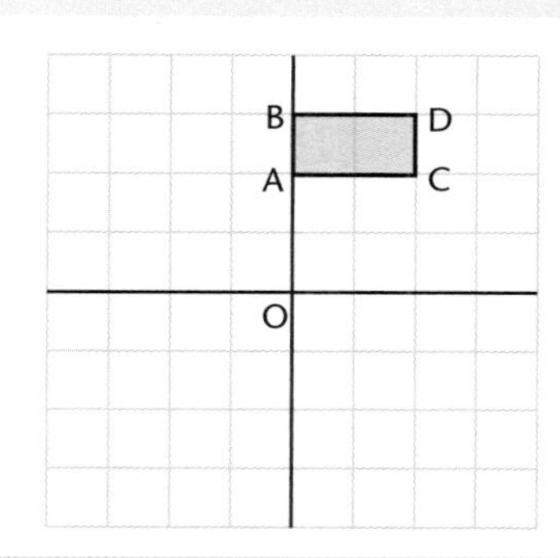

Solution

You can think of a three-quarter-turn (anticlockwise) as a quarter-turn clockwise.

You can draw the rotation by counting squares.

A is 2 squares above O so its image, A′, is 2 squares to the right.

B is 3 squares above O so its image, B′, is 3 squares to the right.

C is 2 squares above and 2 to the right of O so its image, C′, is 2 to the right and 2 below O.

D is 3 squares above and 2 to the right of O so its image, D′, is 3 to the right and 2 below O.

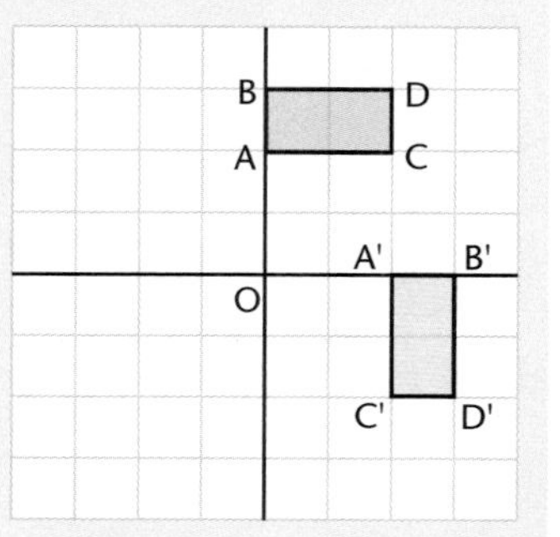

Alternatively, you could do the rotation using tracing paper.

Trace the shape.

Hold the centre of rotation still with the point of a pin or a pencil and rotate the paper through a quarter-turn clockwise.

Use another pin or the point of your compasses to prick through the corners.

Join the pin holes to form the image.

Exam tip

To rotate a shape you need to know three things.

- The angle of rotation
- The direction of rotation
- The centre of rotation

Exercise 16.1

1 Through what angle is the blue object rotated to fit the green image in each of these diagrams?

a)

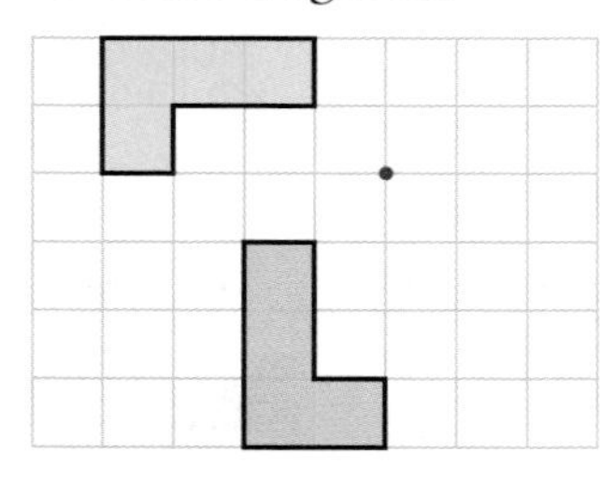

b)

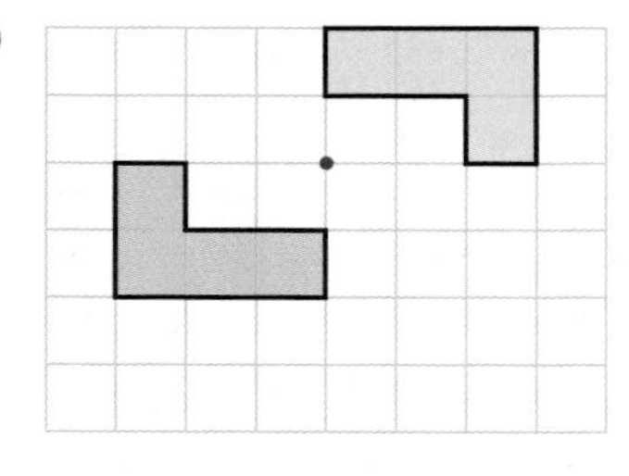

2 This shape is made using flags A, B and C.
The shape has rotation symmetry of order 3.

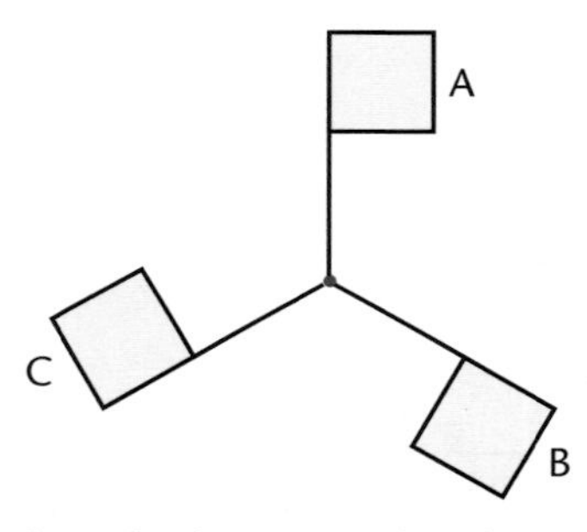

What clockwise angle of rotation maps
a) A on to B?
b) A on to C?

3 What fraction of a turn are each of these rotations.

a)

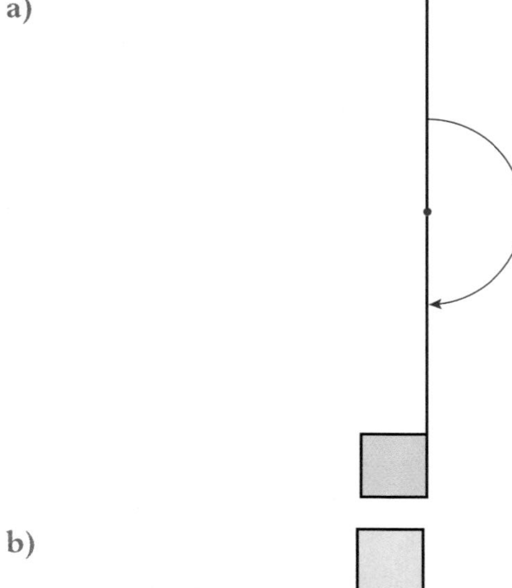

b)

4 Make three copies of this diagram and answer each part of the question on a separate diagram.

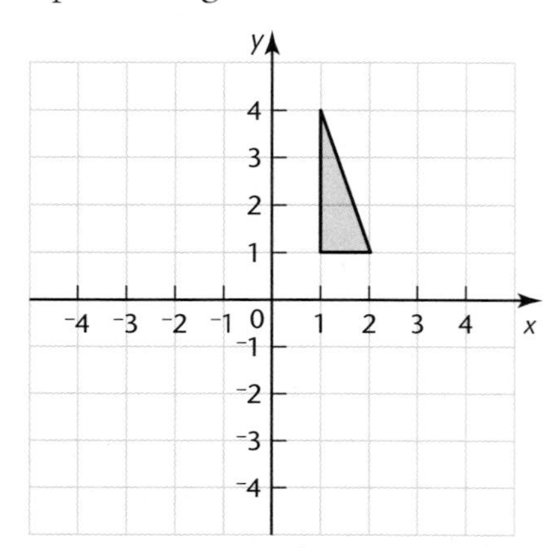

a) Rotate the shape through a half-turn about the origin.
b) Rotate the shape through 90° clockwise about the origin.
c) Rotate the shape through 90° anticlockwise about the origin.

5 Make three copies of this diagram and answer each part of the question on a separate diagram.

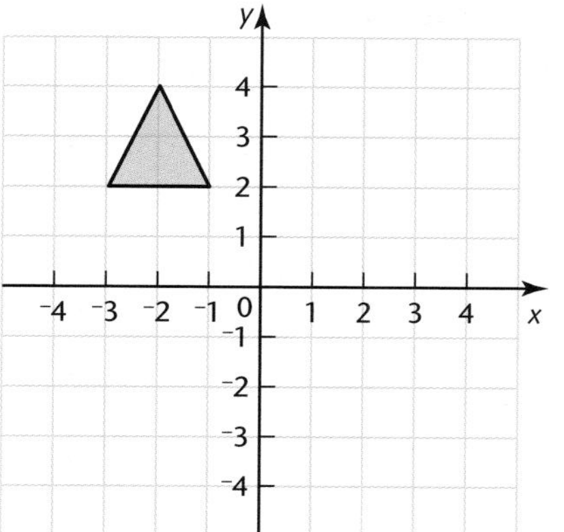

a) Rotate the shape through a half-turn about the origin.
b) Rotate the shape through 90° clockwise about the origin.
c) Rotate the shape through 90° anticlockwise about the origin.

6 Make three copies of this diagram and answer each part of the question on a separate diagram.

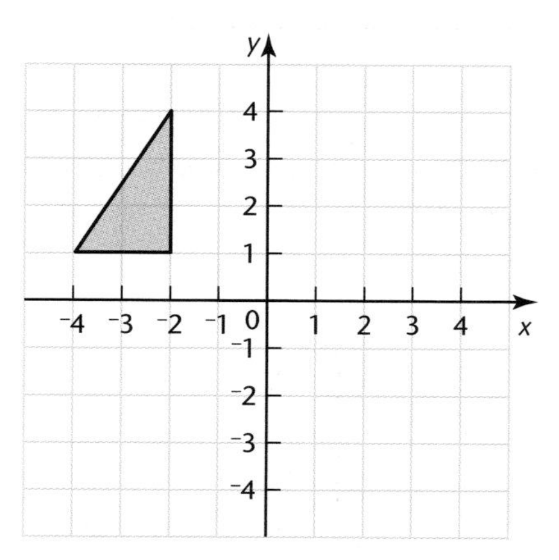

a) Rotate the shape through 90° (anticlockwise) about the origin.
b) Rotate the shape through 90° clockwise about the origin.
c) Rotate the shape through 180° about the origin.

7 Make three copies of this diagram and answer each part of the question on a separate diagram.

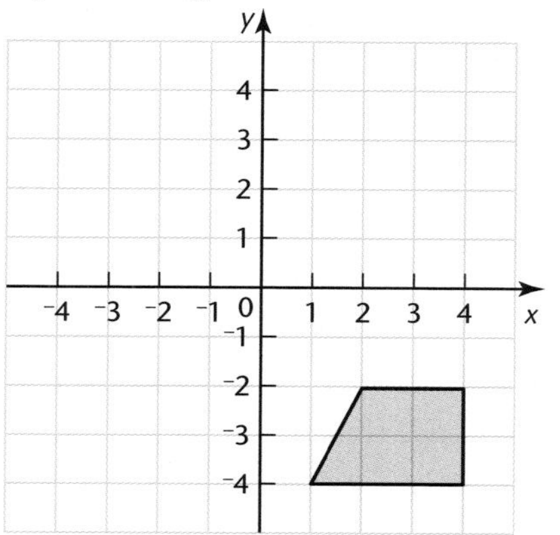

a) Rotate the shape through 90° anticlockwise about the origin.
b) Rotate the shape through 90° clockwise about the origin.
c) Rotate the shape through 180° about the origin.

8 Rotate this shape through a half-turn about the point (4, 3).

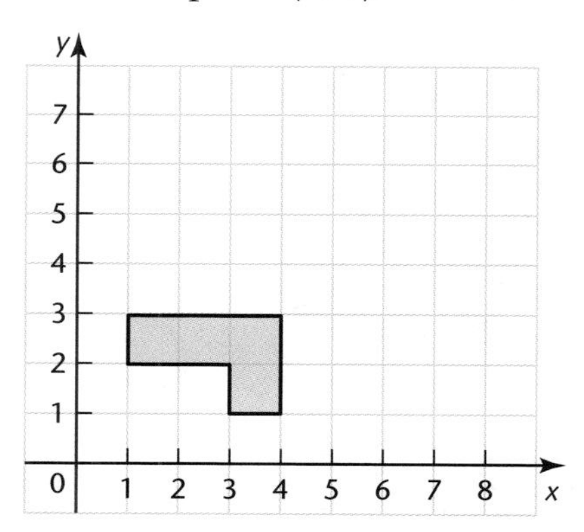

9 Rotate this shape through a half-turn about the point (5, 3).

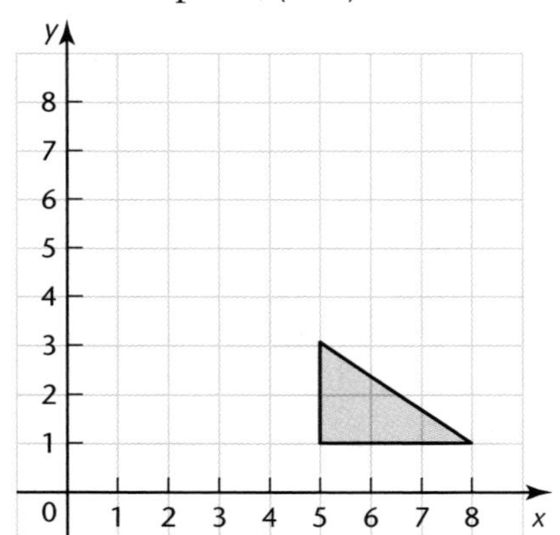

10 Rotate this shape through 90° anticlockwise about the point (5, 2).

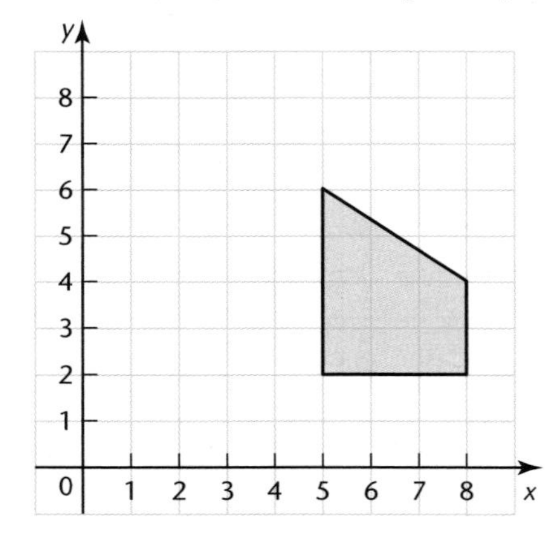

Exam tip

Tracing paper is always stated as optional extra material in examinations. When doing transformation questions, always ask for it.

Challenge 1

Work in a group.

- Each draw the same object on a piece of squared paper.
- Trace over your object.
- Each mark a different centre of rotation.
- Use a pencil or compass point to keep the centre of rotation still. Turn your tracing paper though 90° clockwise.
- Compare the diagrams in your group.
- See how the position of the image changes when the centre of rotation changes.

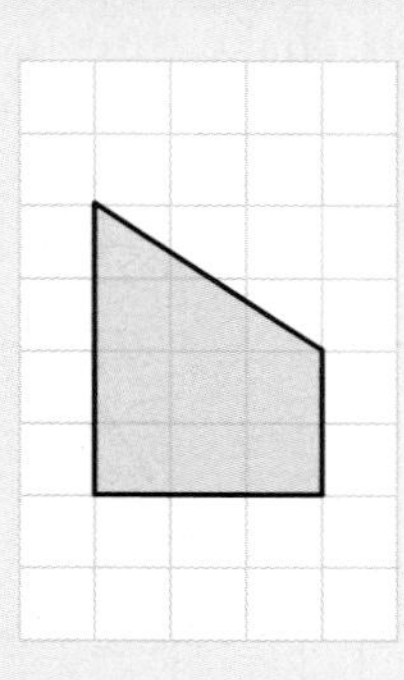

Challenge 2

Repeat Challenge 1 for a rotation of 90° anticlockwise or 180°.

Drawing rotations on plain paper

When drawing rotations on plain paper, you have to measure the angle of rotation using a set square or protractor.

Example 2

Rotate triangle ABC through 90° clockwise about C.

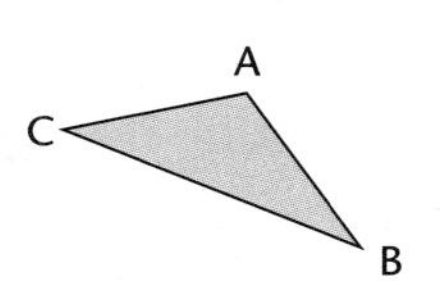

Solution

Measure an angle of 90° clockwise from the line AC.

Trace the shape ABC.

Put a pencil or pin at C to hold the tracing still at that point.

Rotate the tracing paper until AC coincides with the new line you have drawn.

Use another pin or the point of your compasses to prick through the other corners (A and B).

Join up the new points to form the image (A′ B′ C).

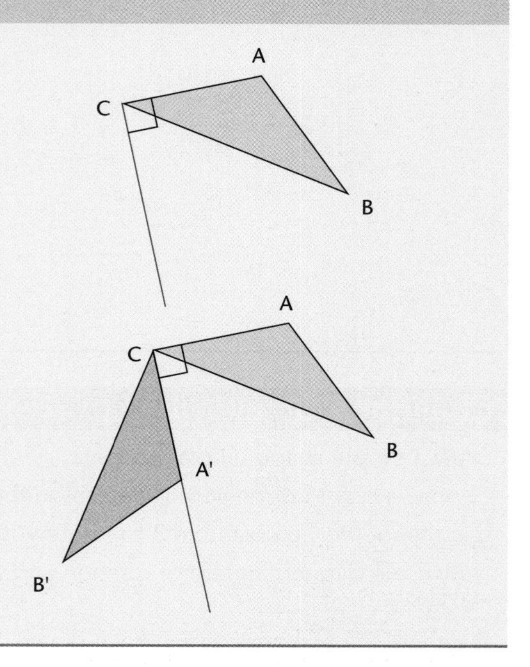

If the centre of rotation is not on the object, then the method is slightly more complicated.

Example 3

Rotate the triangle ABC through 90° clockwise about the point O.

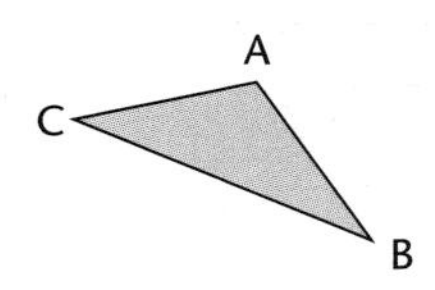

O•

Solution

Join O to a point on the object (C).

Measure an angle of 90° clockwise from OC and draw a line.

Trace the triangle ABC and the line OC.

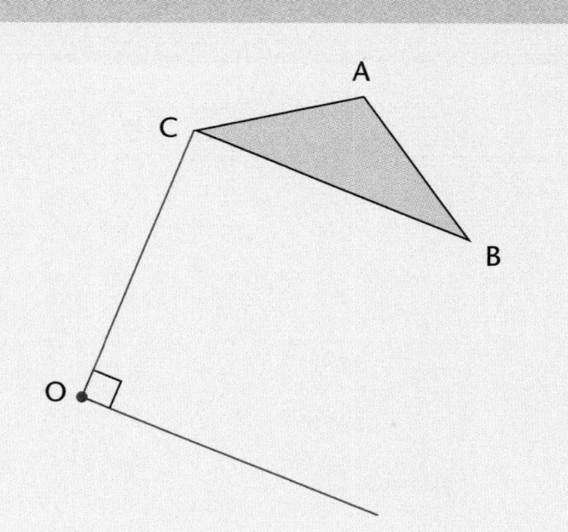

Rotate the tracing about O until the line OC coincides with the new line you have drawn.

Use a pin or the point of your compasses to prick through the other corners (A, B and C).

Join up the pin holes to form the image (A' B' C').

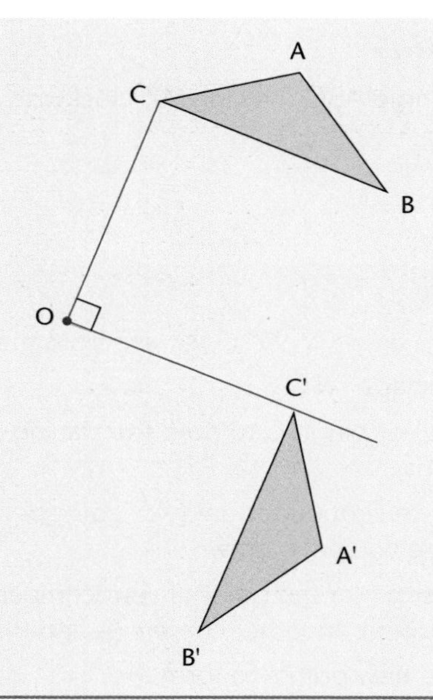

Challenge 3

Draw a simple shape on plain paper.

Choose a point not on your shape to be the centre of rotation.

Use the method of Example 3 to rotate your shape through 90° clockwise.

Now use a different centre of rotation and, again, rotate your original shape through 90° clockwise.

Compare the two images you have drawn. What is the same about them and what is different?

For other angles of rotation, such as 120° clockwise, the first angle is measured as 120° rather than 90° but otherwise the method is identical.

Challenge 4

a) Draw a simple shape on plain paper.
Choose a point not on your shape to be the centre of rotation.
Use the method of Example 3 to rotate your shape through 60° clockwise.

b) Draw another simple shape on plain paper.
Choose a point not on your shape to be the centre of rotation.
Use the method of Example 3 to rotate your shape through 45° anticlockwise.

Recognising rotations

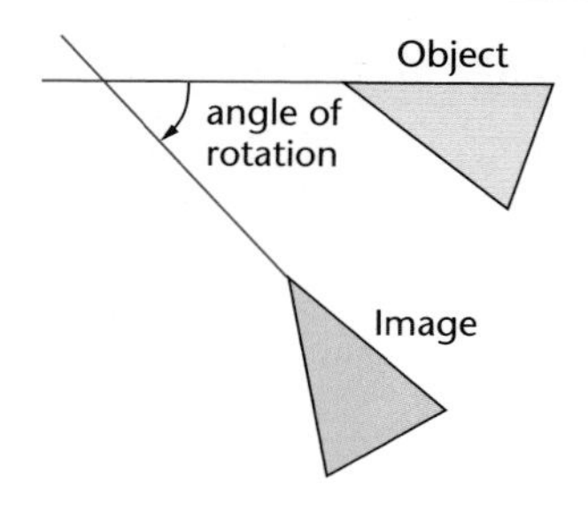

It is usually easy to recognise when a transformation is a rotation, as it should be possible to place a tracing of the object over the image without turning the tracing paper over.

To find the angle of rotation, find a pair of sides that correspond in the object and the image. Measure the angle between them. You may need to extend both of these sides to do this.

If the centre of rotation is not on the object, its position may not be obvious. The easiest method to use is trial and error, either by counting squares or using tracing paper.

Exam tip

Always remember to state whether the rotation is clockwise or anticlockwise.

Example 4

Describe fully the single transformation that maps flag A on to flag B.

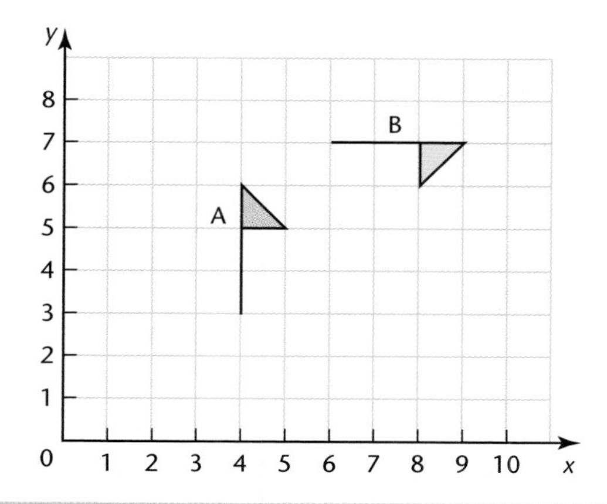

Solution

It appears that the transformation is a rotation and that the angle is 90° clockwise. You may need to make a few trials, using tracing paper and a compass point centred on different points, to find that the centre of rotation is (7, 4).

If you did not spot it, try it now with tracing paper.

Exercise 16.2

1 a) Copy the diagram.

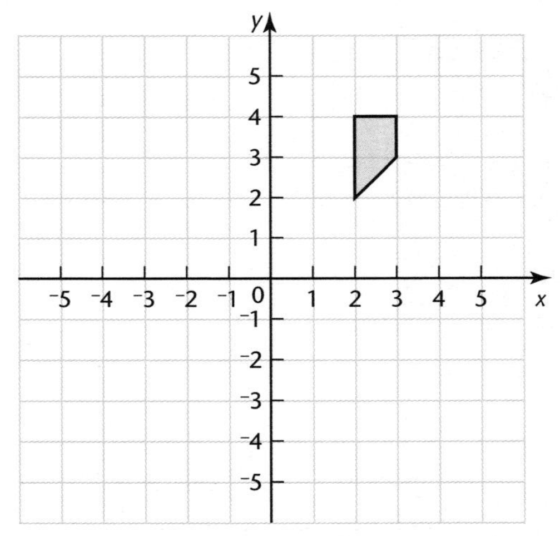

Rotate the shape through 90° clockwise about (2, 2).
Label the image A.

b) On the same diagram, rotate the original shape through 90° anticlockwise about (1, 1).
Label the image B.

2 a) Copy the diagram.

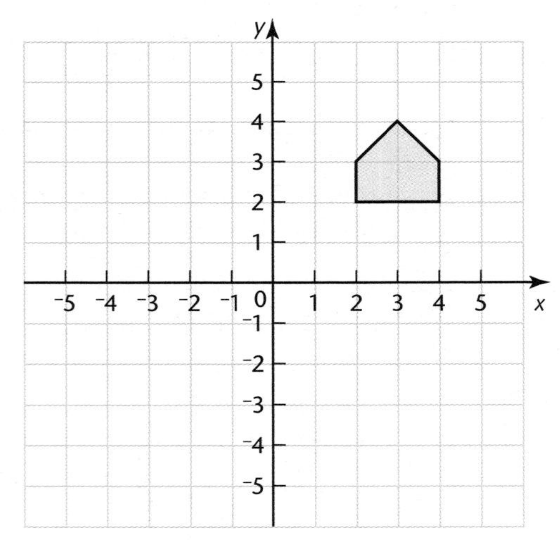

Rotate the shape through 90° clockwise about (2, 2).
Label the image A.

b) On the same diagram, rotate the original shape through 90° anticlockwise about (1, 1).
Label the image B.

3 a) Copy the diagram.

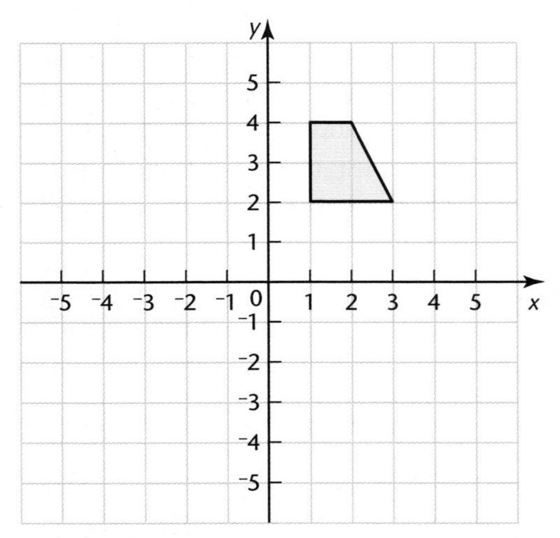

Rotate the shape through 180° about (3, 2).
Label the image A.

b) On the same diagram, rotate the original shape through 90° anticlockwise about (1, 1).
Label the image B.

4 Copy the diagram.

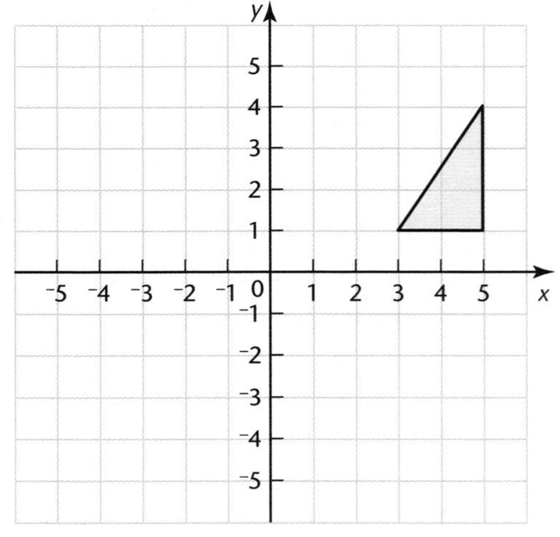

Rotate the triangle by 90° anticlockwise about the point (2, 0).

5 a) Copy the diagram.

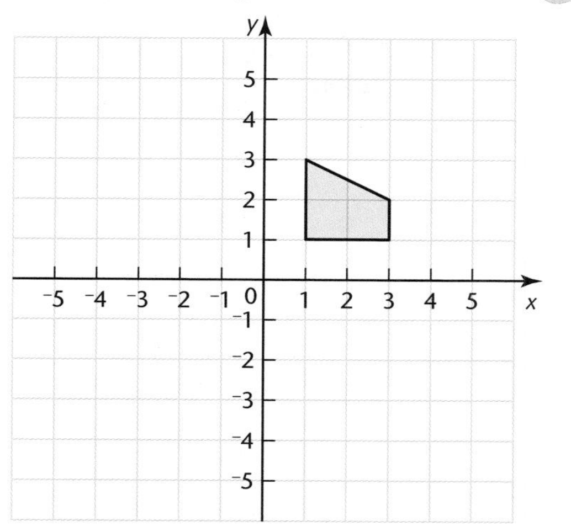

Rotate the shape through 90° anticlockwise about (1, 3).
Label the image A.

b) On the same diagram, rotate the original shape through 180° about (1, −2).
Label the image B.

6 a) Copy the diagram.

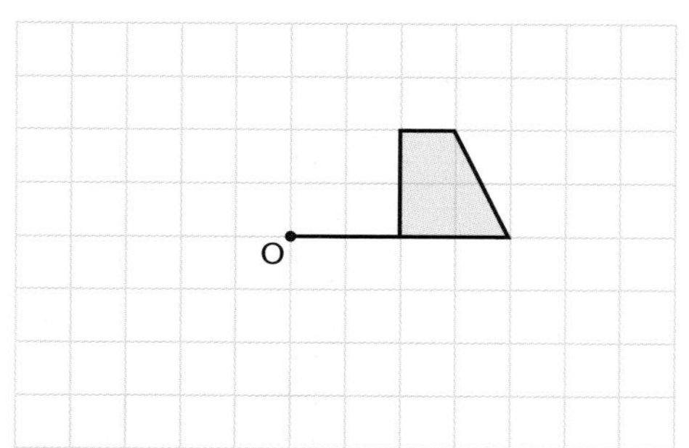

Rotate the shape through 90° clockwise about O.
Label the image A.

b) Rotate A through 90° clockwise about O.
Label the image B.

c) Rotate B to complete a symmetrical pattern.
Label the image C.

d) Describe fully the single transformation that maps A on to C.

7 a) Copy the diagram.

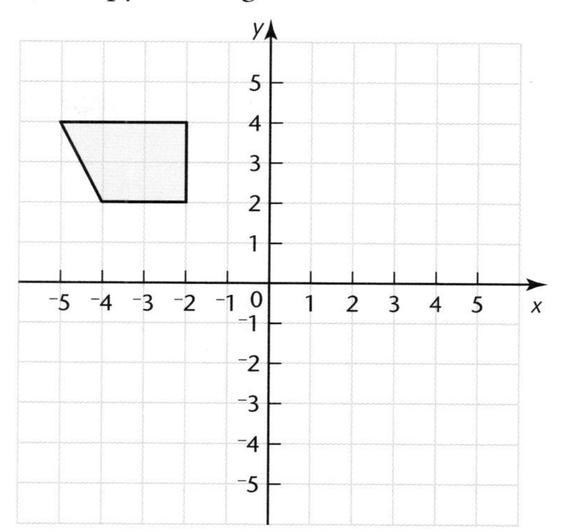

Rotate the shape through 90° clockwise about (0, 0).
Label the image A.

b) On the same diagram, rotate the original shape through 90° anticlockwise about (−2, 1).
Label the image B.

c) Describe fully the single transformation that maps A on to B.

8 a) On squared paper, draw *x*- and *y*-axes from –5 to 5.
Plot the points (−2, 1), (−2, 4) and (−4, 3).
Join the points to form a triangle and label it T.

b) Rotate T through 90° anticlockwise about (−1, 0).
Label the image A.

c) Rotate T through 90° clockwise about (0, −1).
Label the image B.

d) Describe fully the single transformation that maps A on to B.

9 For this diagram, describe fully the single transformation that maps

a) A on to B. **b)** A on to C.

c) A on to D. **d)** B on to E.

e) D on to F.

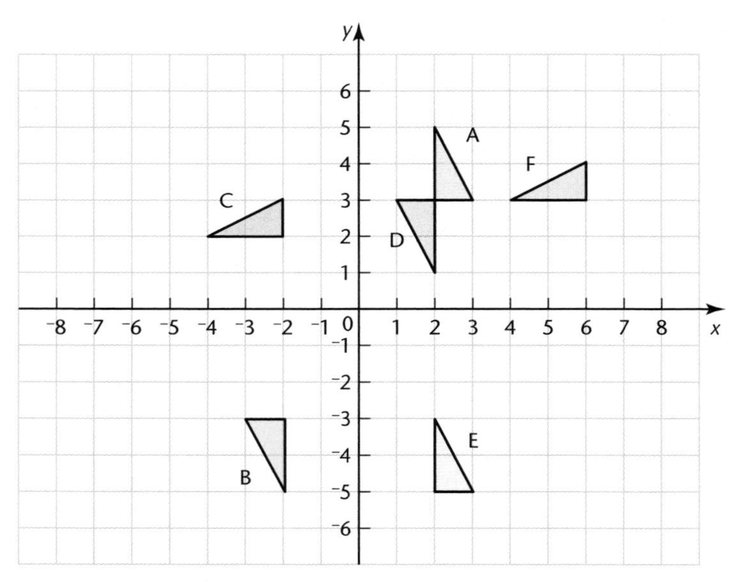

10 For this diagram, describe fully the single transformation that maps

a) A on to B. **b)** B on to C.

c) B on to D. **d)** E on to F.

e) A on to G.

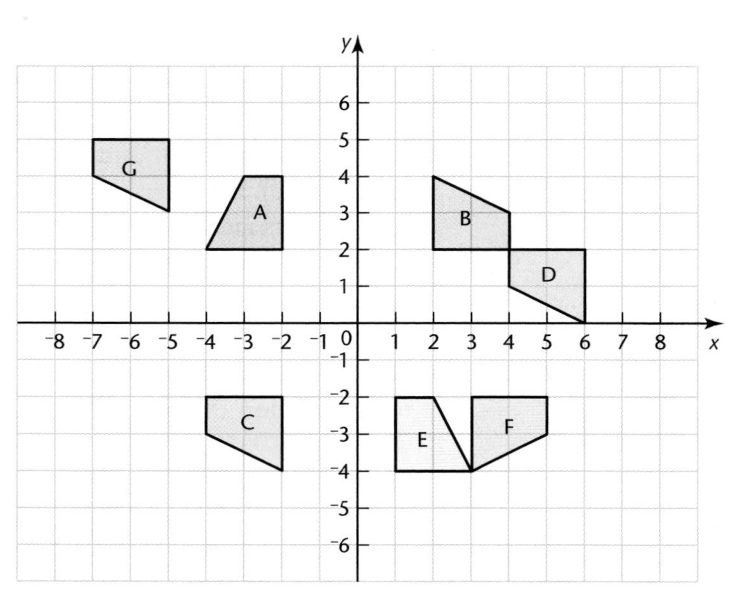

Exam tip

When describing transformations, always state the type of transformation first and then give all the necessary extra information. For rotations this is the angle, the direction and the centre of rotation.

Key Ideas

- In a rotation, all the points move through the same angle in the same direction.
- In a rotation, all the points remain at the same distance from the centre of rotation.
- In a rotation, the original shape and the image are congruent.
- When describing a rotation, you need to give the angle, the direction (anticlockwise is positive) and the centre of rotation.

CHAPTER 17 Drawing linear graphs

YOU WILL LEARN ABOUT	YOU SHOULD ALREADY KNOW
o Using tables for linear equations o Drawing graphs of linear equations	o How to plot points in all four quadrants o How to substitute in equations

Lines parallel to the axes

The simplest equations that give straight-line graphs are of the type $y = 3$ or $x = 2$.

Every point on the line $y = 3$ has 3 as its y-coordinate.

Every point on the line $x = 2$ has 2 as its x-coordinate.

The x-axis has the equation $y = 0$ and the y-axis has the equation $x = 0$.

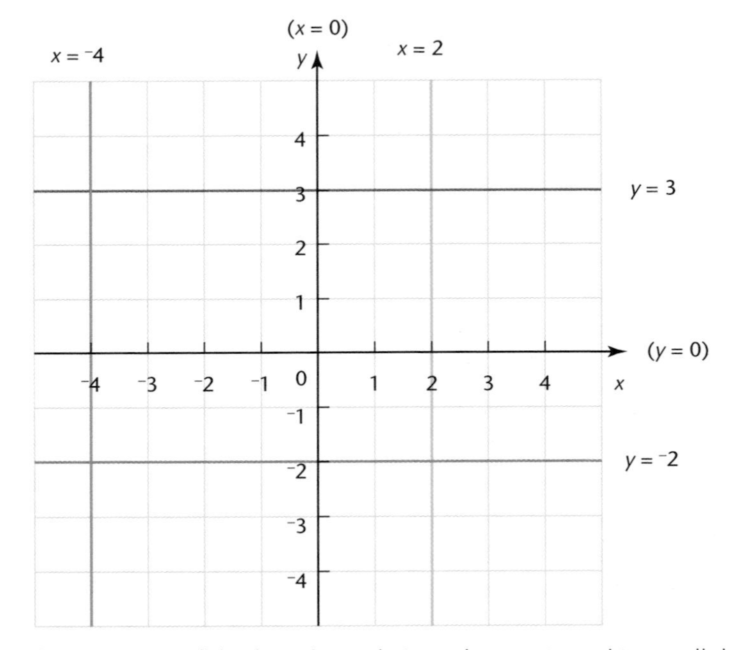

So $y = 3$ is the equation of the line through 3 on the y-axis and is parallel to $y = 0$, and $x = 2$ is the equation of the line through 2 on the x-axis and is parallel to $x = 0$.

These are both drawn on the graph. The lines $y = {}^{-}2$ and $x = {}^{-}4$ are also drawn on the graph.

Tables of values

$y = 2x$, $y = 3x + 1$, $y = 4x - 2$ and $y = {}^{-}3x + 4$ are also equations of straight lines.

To plot the line you can choose whatever values of x you like and make a table of values.

Exam tip

Always label the axes clearly.

Example 1

Copy and complete this table for the equation $y = 2x$.

x	⁻3	⁻2	⁻1	0	1	2	3
y = 2x							

Solution

To find the y-value you multiply the x-value by 2.

For example, when $x = ^-3$, $y = 2 \times ^-3 = ^-6$.

x	⁻3	⁻2	⁻1	0	1	2	3
y = 2x	⁻6	⁻4	⁻2	0	2	4	6

As there is only one term for y, the table is very simple.

Example 2

Copy and complete this table for the equation $y = 3x - 2$.

x	⁻3	⁻2	⁻1	0	1	2	3
3x							
– 2							
y = 3x – 2							

Solution

x	⁻3	⁻2	⁻1	0	1	2	3
3x	⁻9	⁻6	⁻3	0	3	6	9
– 2	⁻2	⁻2	⁻2	⁻2	⁻2	⁻2	⁻2
y = 3x – 2	⁻11	⁻8	⁻5	⁻2	1	4	7

Here you have separate rows for $3x$, $^-2$ and $y = 3x - 2$.

Complete each row in turn. Remember that $3x = 3 \times ^-3 = ^-9$, $3 \times ^-2 = ^-6$ and so on, and notice that $^-2$ does not change for different values of x.

To complete the bottom row and so find y, you add the two rows above.

Example 3

Draw a table for the equation $y = 5 - 4x$ for values of x from $^-2$ to 4.

Solution

x	⁻2	⁻1	0	1	2	3	4
5	5	5	5	5	5	5	5
– 4x	8	4	0	⁻4	⁻8	⁻12	⁻16
y = 5 – 4x	13	9	5	1	⁻3	⁻7	⁻11

$y = 5 - 4x$ could also be written as $y = ^-4x + 5$.

Having worked out the rows for 5 and ^-4x, you still add the two rows.

Challenge 1

a) Can you spot a pattern in the results for Examples 1 to 3?

b) Explain the pattern.

c) What pattern would you get for these equations?

(i) $y = 4x$ (ii) $y = 6x - 9$ (iii) $y = {}^{-}7x + 16$

Particularly if you have a calculator, you may use tables of values where the working lines are not shown. The next example shows this.

Example 4

Complete this table of values for $y = 5x - 4$.

x	$^{-}2$	$^{-}1$	0	1	2	3
y				1		

Exam tip

Start with positive values of x when you complete a table – they are easier!

Solution

Choose an easy x-value to start with, such as $x = 2$.

In your head, or on the calculator, work out $5 \times 2 - 4$. The answer is 6, so this is the y-value.

Continue similarly for all the other values of x.

Here is the completed table:

$y = 5x - 4$

x	$^{-}2$	$^{-}1$	0	1	2	3
y	$^{-}14$	$^{-}9$	$^{-}4$	1	6	11

Exam tip

You can use the pattern in the y numbers as a check on the values you have calculated.

Exercise 17.1

1 On a grid, draw and label axes from $^{-}6$ to 6 for both x and y. Then draw the line for each of these equations.
$x = 3$, $x = {}^{-}3$, $y = 4$, $y = {}^{-}5$

2 On a grid, draw and label axes from $^{-}6$ to 6 for both x and y. Then draw the line for each of these equations.
$x = 5$, $x = {}^{-}2$, $y = 2$, $y = {}^{-}3$

3 Copy and complete this table for the equation $y = 3x$.

x	$^{-}3$	$^{-}2$	$^{-}1$	0	1	2	3
y = 3x		$^{-}6$					9

4 Copy and complete this table for the equation $y = 5x - 3$.

x	⁻3	⁻2	⁻1	0	1	2	3
5x		⁻10					
– 3		⁻3					
y = 5x – 3		⁻13					

5 Copy and complete this table for the equation $y = 4x + 2$.

x	⁻3	⁻2	⁻1	0	1	2	3
4x							
+ 2							
y = 4x + 2							

6 Copy and complete this table for the equation $y = x + 2$.

x	⁻4	⁻3	⁻2	⁻1	0	1	2
y = x + 2							

7 Copy and complete this table for the equation $y = 2x - 5$.

x	⁻1	0	1	2	3	4	5
y = 2x – 5							

8 Copy and complete this table for the equation $y = 5 - 2x$.

x	⁻1	0	1	2	3	4	5
y = 5 – 2x							

9 Draw a table for the equation $y = 2x - 3$ for values of x from $^-2$ to 4.

10 Draw a table for the equation $y = 3x + 6$ for values of x from $^-4$ to 2.

11 Draw a table for the equation $y = 1 - 3x$ for values of x from $^-3$ to 3.

12 Draw a table for the equation $y = {}^-4x + 7$ for values of x from $^-2$ to 4.

Drawing straight-line graphs

You only need two points to draw a straight line, but you should always work out at least three points as a check. It is best to choose two points as far apart as possible. You would often choose the point with an x-coordinate of zero as your third point.

Always use a ruler to draw a straight-line graph.

Example 5

a) Copy and complete this table for the equation $y = 3x - 1$.

x	⁻3	0	3
y = 3x – 1			

b) Draw the graph of $y = 3x - 1$ for values of x from ⁻3 to 3.

c) From the graph, find the value of x when $y = 4$. Give your answer correct to 1 decimal place.

Solution

a)

x	⁻3	0	3
y = 3x – 1	⁻10	⁻1	8

b) A grid is needed with x from ⁻3 to 3 and y from ⁻10 to 8.

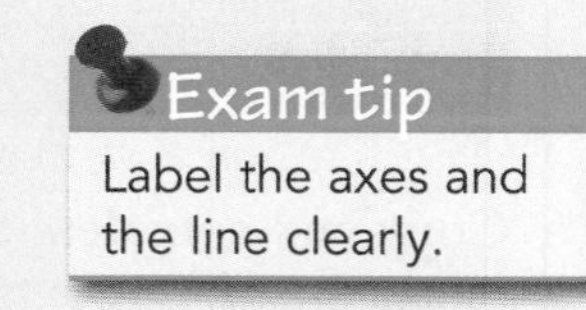

Exam tip

Label the axes and the line clearly.

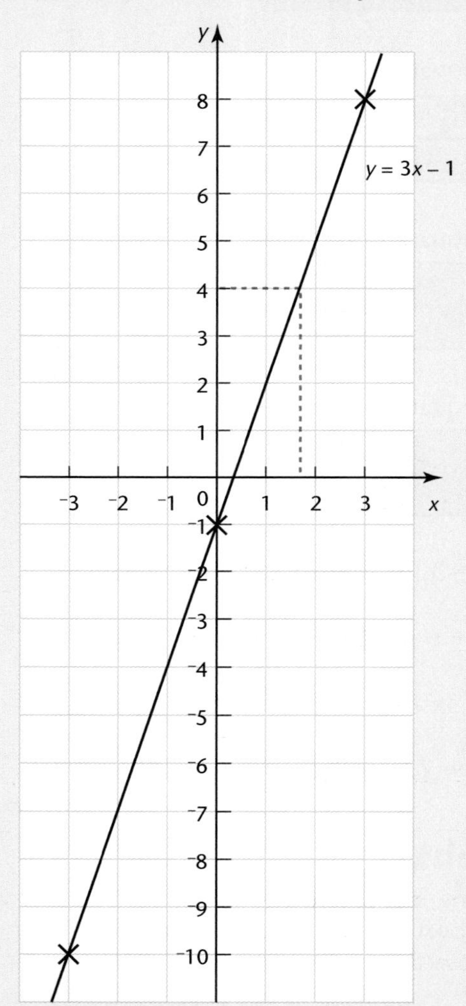

c) Draw a horizontal line through $y = 4$ across to the line. Then draw a vertical line down from that point to the x-axis. It meets the x-axis at $x = 1.7$.

So the answer is $x = 1.7$.

Example 6

a) Draw the graph of $y = 3 - 2x$ for values of x from $^-2$ to 4.

b) From the graph, find the value of x when $y = 0$.
Give your answer correct to 1 decimal place.

Solution

a) First make a table. For the values of x, choose the two end points and zero.

x	$^-2$	0	4
$y = 3 - 2x$	7	3	$^-5$

A grid is needed with x from $^-2$ to 4 and y from $^-5$ to 7.

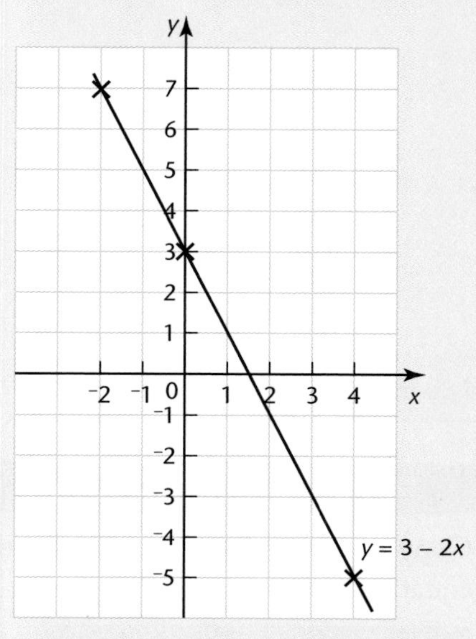

b) From the graph, $x = 1.5$ when $y = 0$.

> **Exam tip**
> When the number in front of the x is positive, the line slopes up from left to right. When it is negative, the line slopes down from left to right.

All the equations studied so far have been for y and x. They can be for other letters and for real situations.

Example 7

To hire a coach it costs £20 plus £2 a mile. This can be written as

$$C = 20 + 2m$$

where C is the cost in pounds and m is the number of miles travelled.

a) Draw the graph of $C = 20 + 2m$ for values of m from 0 to 100.

b) Millhouses sports club hired the coach and paid £180.
How far did they travel?

Solution

a) First make a table.

m	0	50	100
$C = 20 + 2m$	20	120	220

A grid is needed with m from 0 to 100 and C from 0 to 220.

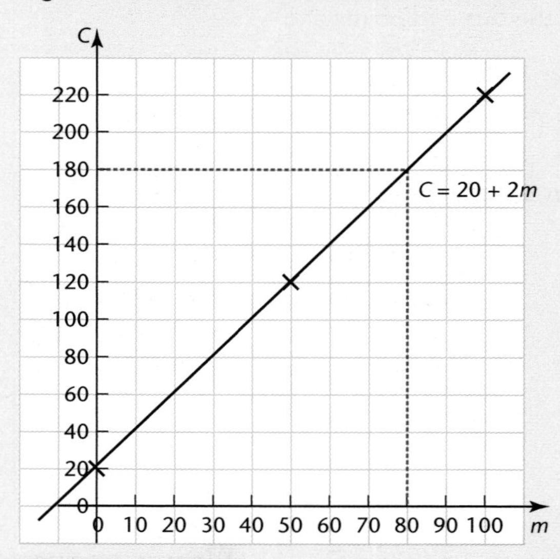

b) From the graph, when $C = 180$, $m = 80$. So they travelled 80 miles.

Exercise 17.2

1 a) Copy and complete this table for the equation $y = 4x$.

x	$^-3$	0	3
$y = 4x$			

b) Draw the graph of $y = 4x$ for values of x from $^-3$ to 3.

2 a) Copy and complete this table for the equation $y = 2x + 5$.

x	$^-4$	0	2
$y = 2x + 5$			

b) Draw the graph of $y = 2x + 5$ for values of x from $^-4$ to 2.

3 a) Copy and complete this table for the equation $y = 2 - x$.

x	$^-3$	0	3
$y = 2 - x$			

b) Draw the graph of $y = 2 - x$ for values of x from $^-3$ to 3.

4 a) Copy and complete this table for the equation $y = 3x - 4$.

x	$^-2$	0	4
$y = 3x - 4$			

b) Draw the graph of the equation $y = 3x - 4$ for values of x from $^-2$ to 4.

c) From the graph find the value of x when $y = 4$. Give your answer correct to 1 decimal place.

5 Draw the graph of the equation $y = x - 3$ for values of x from $^-1$ to 5.

6 a) Draw the graph of the equation $y = 5x + 2$ for values of x from $^-3$ to 3.

b) From the graph find the value of x when $y = 10$. Give your answer correct to 1 decimal place.

7 a) Draw the graph of the equation $y = 4 - 2x$ for values of x from $^-2$ to 4.

b) From the graph find the value of x when $y = {^-1}$. Give your answer correct to 1 decimal place.

8 a) Draw the graph of the equation $C = 5n + 15$ for values of n from 0 to 20.

b) From the graph find the value of n when $C = 80$. Give your answer correct to the nearest whole number.

9 The cost £C of printing a leaflet is given by the formula $C = 5 + 0.2n$, where n is the number of copies.

a) Draw the graph of $C = 5 + 0.2n$ for values of n from 0 to 200.

b) From the graph find the value of n when $C = 30$. Give your answer correct to the nearest whole number.

10 A plasterer works out his daily charge (£C) using the equation $C = 12n + 40$, where n is the number of hours he works on a job.

a) Draw the graph of C against n for values of n from 0 to 10.

b) From your graph find how many hours he works when the charge is £130.

Challenge 2

To hire a minibus, Delaney's Cabs charge £20 plus £2 a mile and Tracey's Cars charge £50 plus £1.50 a mile.

C is the charge and n is the number of miles.

For Delaney's Cabs the equation is $C = 2n + 20$.

For Tracey's Cars the equation is $C = 1.5n + 50$.

a) On the same graph, draw a line for each equation for values of n up to 100.

b) Use your graph to find the number of miles for which the two firms charge the same amount.

Key Ideas

- To draw a straight-line graph, first make a table of values.
- Always work out at least three points on the line.
- Label axes clearly and check you understand the scales of graphs which have axes drawn for you.

Chapter 18 Ratio and proportion

YOU WILL LEARN ABOUT	YOU SHOULD ALREADY KNOW
○ Ratio notation ○ Using ratios in proportion calculations ○ Dividing a quantity in a given ratio ○ Comparing proportions	○ How to multiply and divide without a calculator ○ How to find common factors ○ How to simplify fractions ○ What is meant by an enlargement ○ How to change between metric units

Ratios

When something is divided into a number of parts, ratios are used to compare the sizes of those parts.

For example, if ten sweets are divided between two people so that one person gets six sweets and the other four sweets, the sweets have been divided in the ratio 6 : 4.

- As with fractions, ratios can be simplified and should be if possible.
  For example, the fraction $\frac{6}{4} = \frac{3}{2}$; similarly, the ratio 6 : 4 = 3 : 2.

When simplifying ratios, you need to look for common factors. So 40 : 60 is simplified to 2 : 3 by dividing both numbers by 20. You could have arrived at 2 : 3 by dividing by different factors. For example: 40 : 60 = 4 : 6 = 2 : 3 (dividing by 10 and then 2); or 40 : 60 = 20 : 30 = 10 : 15 = 2 : 3 (dividing by 2, then 2 again and then 5).

Activity 1

Find as many ways as you can to simplify these ratios.

a) 20 : 50 **b)** 16 : 20 **c)** 14 : 56 **d)** 9 : 27 **e)** 6 : 48

When the ratio you are using involves measurements, the units must be the same for both measurements when simplifying.

Here are some examples where the units need to be changed first.

Example 1

Write each of these ratios in its lowest terms.

a) 1 millilitre : 1 litre **b)** 1 kilogram : 200 grams

Solution

a) 1 millilitre : 1 litre = 1 millilitre : 1000 millilitres
= 1 : 1000

Write each part in the same units.
When the units are the same, you do not need to include them in the ratio.

b) 1 kilogram : 200 grams = 1000 grams : 200 grams
= 5 : 1

Write each part in the same units.
Divide each part by 200.

Example 2

Write each of these ratios in its lowest terms.

a) 50p : £2 **b)** 2 cm : 6 mm **c)** 600 g : 2 kg : 750 g

Solution

a) 50p : £2 = 50p : 200p
= 1 : 4

Write each part in the same units.
Divide each part by 50.

b) 2 cm : 6 mm = 20 mm : 6 mm
= 10 : 3

Write each part in the same units.
Divide each part by 2.

c) 600 g : 2 kg : 750 g = 600 g : 2000 g : 750 g
= 12 : 40 : 15

Write each part in the same units.
Divide each part by 50.

Exercise 18.1

1 Write each of these ratios in its simplest form.

a) 6 : 3
b) 25 : 75
c) 30 : 6
d) 10 : 15
e) 7 : 35
f) 15 : 12
g) 24 : 8
h) 4 : 48
i) 3 : 27
j) 9 : 81

2 Write each of these ratios in its simplest form.

a) 20 minutes : 1 hour
b) 50 g : 1 kg
c) 300 ml : 1 litre
d) 2 kg : 600 g
e) 2 minutes : 30 seconds

3 Write each of these ratios in its simplest form.

a) 30p : £1
b) 2 m : 50 cm
c) 45 seconds : 2 minutes
d) 5 kg : 750 g
e) 500 m : 3 km

4 Write each of these ratios in its simplest form.

a) 5 : 15 : 25
b) 6 : 12 : 8
c) 8 : 32 : 40

5 Write each ratio in its simplest form.

a) 50p : £2.50 : £5
b) 20 cm : 80 cm : 1.20 m
c) 600 g : 750 g : 1.5 kg

6 Write each of these ratios in its simplest form.
a) 50 g : 1000 g
b) 30p : £2
c) 2 minutes : 30 seconds
d) 4 m : 75 cm
e) 300 ml : 2 litres

7 At a concert there are 350 men and 420 women.
Write the ratio of men to women in its simplest form.

8 Alison, Peter and Dave invest £500, £800 and £1000 respectively in a business.
Write the ratio of their investments in its simplest form.

9 A recipe for vegetable soup uses 1 kg of potatoes, 500 g of leeks and 750 g of celery.
Write the ratio of the ingredients in its simplest form.

Challenge 1

Write each of these ratios in its simplest form.
a) 20 miles : 20 kilometres
b) 30 kilograms : 30 pounds
c) 40 litres : 40 gallons

Writing a ratio in the form 1 : *n*

It is sometimes useful to have a ratio with 1 on the left.

A common scale for a scale model is 1 : 24.

The scale of a map or enlargement is often given as 1 : *n*.

To change a ratio to this form, divide both numbers by the number on the left.

Example 3

Write these ratios in the form 1 : *n*.

a) 2 : 5 **b)** 8 mm : 3 cm **c)** 25 mm : 1.25 km

Solution

a) 2 : 5 = 1 : 2.5	Divide each side by 2.
b) 8 mm : 3 cm = 8 mm : 30 mm	Write each side in the same units.
= 1 : 3.75	Divide each side by 8.
c) 25 mm : 1.25 km = 25 : 1 250 000	Write each side in the same units.
= 1 : 50 000	Divide each side by 25.

Exam tip
Use a calculator, if necessary, to convert the ratio to the form 1 : *n*.

1 : 50 000 is a common map scale. It means that 1 cm on the map represents 50 000 cm, or 500 m, on the ground.

Exercise 18.2

1 Write each of these ratios in the form 1 : n.

a) 2 : 6 b) 3 : 15
c) 6 : 15 d) 4 : 7
e) 20p : £1.50 f) 4 cm : 5 m
g) 10 : 2 h) 2 mm : 1 km

2 Write each of these ratios in the form 1 : n.

a) 2 : 8 b) 5 : 12
c) 2 mm : 10 cm d) 2 cm : 5 km
e) 100 : 40

3 On a map a distance of 8 mm represents a distance of 2 km.
What is the scale of the map in the form 1 : n?

4 A passport photo is 35 mm long. An enlargement of the photo is 21 cm long. What is the ratio of the passport photo to the enlargement in the form 1 : n?

Using ratios

Sometimes you know one of the quantities in the ratio, but not the other.

If the ratio is in the form 1 : n, you can work out the second quantity by multiplying the first quantity by n. You can work out the first quantity by dividing the second quantity by n.

Example 4

a) A photo is enlarged in the ratio 1 : 20 to make a poster.
The photo measures 36 mm by 24 mm.
What size is the enlargement?

b) Another 1 : 20 enlargement measures 1000 mm × 1000 mm.
What size is the original photo?

Solution

a) The enlargement will be 20 times bigger than the photo, so multiply both dimensions by 20.

36 × 20 = 720
24 × 20 = 480

The poster measures 720 mm by 480 mm.

b) The original photo will be 20 times smaller than the enlargment, so divide the dimensions by 20.

1000 ÷ 20 = 50

The original photo measures 50 mm × 50 mm.

Example 5

A map is drawn to a scale of 1 cm : 2 km.

a) On the map, the distance between Amhope and Didburn is 5.4 cm. What is the real distance in kilometres?

b) The length of a straight railway track between two stations is 7.8 km. How long is this track on the map in centimetres?

Solution

a) The real distance, in kilometres, is twice as large as the map distance, in centimetres.

So multiply by 2.

$2 \times 5.4 = 10.8$

Real distance = 10.8 km

b) The map distance, in centimetres, is half of the real distance, in kilometres.

So divide by 2.

$7.8 \div 2 = 3.9$

Map distance = 3.9 cm

Challenge 2

What would the answer to part a) of Example 5 be in centimetres?

What ratio could you use to work this out?

Sometimes you have to work out quantities using a ratio that is not in the form 1 : *n*.

To work out an unknown quantity, you multiply each part of the ratio by the same number to get an equivalent ratio which contains the quantity you know. This number is called the **multiplier**.

For example, when you are mixing quantities, it is often important to keep the amount in the same proportion. If you are mixing black paint and white paint to make the same shade of grey each time, it will be important to keep the proportions of black and white paint the same.

If the colour you want is obtained by mixing 2 litres of black paint with 1 litre of white paint, then you will need 4 litres of black paint if you use 2 litres of white paint.

The ratio of black paint to white paint is 2 parts to 1 part.

Example 6

To mix the same shade of grey paint as above, how much white paint will you need to mix with 8 litres of black paint?

Solution

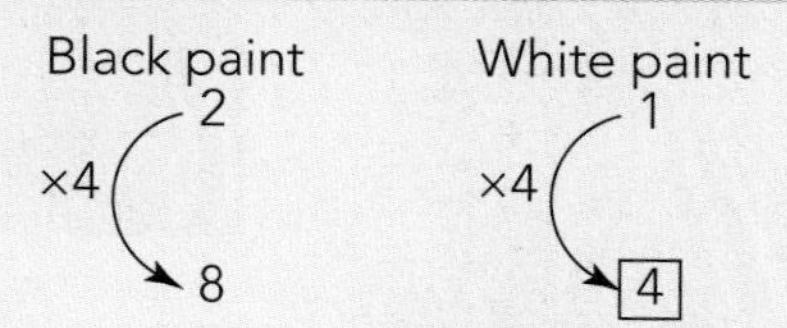

The **multiplier** for black paint is 4, so you must use the same multiplier for white paint: $1 \times 4 = 4$.

So you need 4 litres of white paint.

Example 7

To make pink paint, red paint is added to white paint in the ratio 3 parts red to 2 parts white.

a) How much white paint should be mixed with 6 litres of red paint?

b) How much red paint should be mixed with 10 litres of white paint?

Solution

a)

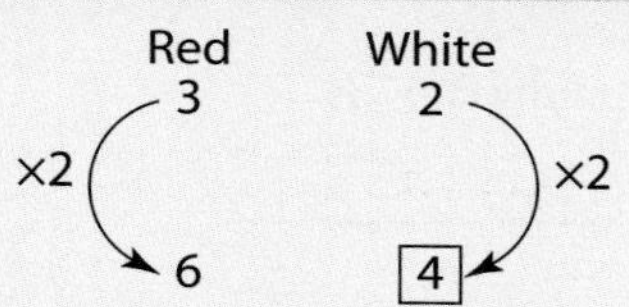

The multiplier is 2, so you need $2 \times 2 = 4$ litres of white paint.

b)

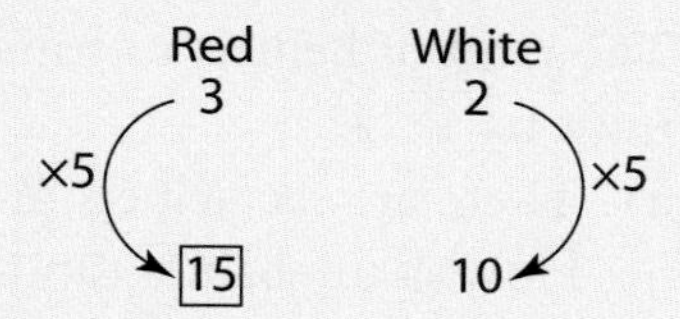

The multiplier is 5 so you need $3 \times 5 = 15$ litres of red paint.

If the multiplier is not immediately obvious you may need to use division to find it.

Example 8

To make light grey paint, Rosie mixes black and white paint in the ratio 1 part black to 5 parts white.

How much black paint does she need to mix with 7 litres of white paint?

Solution

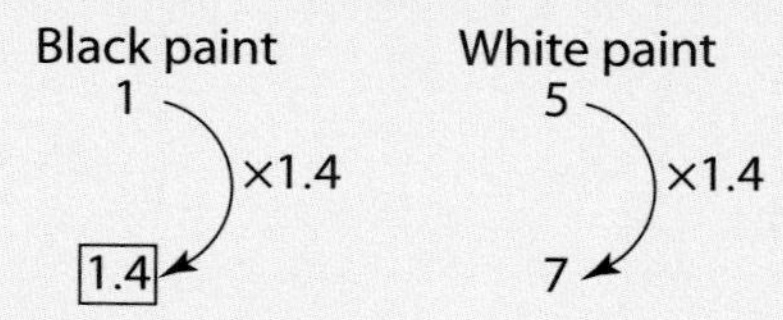

Here the multiplier is $7 \div 5 = 1.4$.

So she needs $1 \times 1.4 = 1.4$ litres of black paint.

Exam tip

The multiplier may not be a whole number. Work with the decimal or fraction and round the final answer if necessary.

Example 9

Two photos are in the ratio 2 : 5.

a) What is the height of the larger photo?

b) What is the width of the smaller photo?

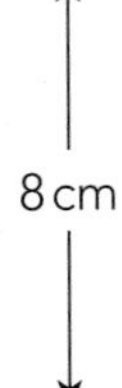

Solution

a) $8 \div 2 = 4$ — Divide the height of the smaller photo by the smaller part of the ratio to find the multiplier.

$2 : 5 = 8 : 20$ — Multiply each part of the ratio by the multiplier, 4.

Height of the larger photo = 20 cm

b) $15 \div 5 = 3$ — Divide the width of the larger photo by the larger part of the ratio to find the multiplier.

$2 : 5 = 6 : 15$ — Multiply each part of the ratio by the multiplier, 3.

Width of the smaller photo = 6 cm

Exercise 18.3

1 Michelle is making mortar. She mixes sand and cement in the ratio 5 parts sand to 1 part cement. She measures the quantities in bags.

a) How many bags of sand should she mix with 2 bags of cement?

b) How many bags of cement should she mix with 20 bags of sand?

2 The ratio of the lengths of the sides of two squares is 1 : 6.

a) The length of the side of a small square is 2 cm.
What is the length of the side of the large square?

b) The length of the side of a large square is 21 cm.
What is the length of the side of the small square?

3 The ratio of helpers to babies in a crèche must be 1 : 4.

a) There are six helpers on Tuesday.
How many babies can there be?

b) There are 36 babies on Thursday.
How many helpers must there be?

4 Sanjay is mixing pink paint. To get the shade he wants, he mixes red and white paint in the ratio 1 : 3.

a) How much white paint should he mix with 2 litres of red paint?

b) How much red paint should he mix with 12 litres of white paint?

5 A road atlas of Great Britain is to a scale of 1 inch to 4 miles.
  a) On the map, the distance between Forfar and Montrose is 7 inches. What is the real distance between the two towns in miles?
  b) It is 40 miles from Newcastle to Middlesbrough. How far is this on the map?

6 Graham is making pastry. He starts by mixing flour and fat in the ratio 8 parts flour to 3 parts fat.
  a) How much fat should he mix with 320 grams of flour?
  b) How much flour should he mix with 150 grams of fat?

7 For a recipe, Chelsy mixes water and lemon curd in the ratio 2 : 3.
  a) How much lemon curd should she mix with 20 ml of water?
  b) How much water should she mix with 15 teaspoons of lemon curd?

8 To make a solution of a chemical, a scientist mixes 3 parts of the chemical with 20 parts of water.
  a) How much water should he mix with 15 ml of the chemical?
  b) How much of the chemical should he mix with 240 ml of water?

9 An alloy is made by mixing 2 parts of silver with 5 parts of nickel.
  a) How much nickel must be mixed with 60 g of silver?
  b) How much silver must be mixed with 120 g of nickel?

10 Sachin and Rehan share a flat. They agree to share the rent in the same ratio as their wages. Sachin earns £600 a month and Rehan earns £800 a month. If Sachin pays £90, how much does Rehan pay?

11 A recipe for hotpot uses onions, carrots and stewing steak in the ratio 1 : 2 : 5 by mass.
  a) What quantity of steak is needed if 100 g of onion is used?
  b) What quantity of carrots is needed if 450 g of steak is used?

Dividing a quantity in a given ratio

Activity 2

Maya has an evening job making up party bags for a children's party organiser.
She shares out lemon sweets and raspberry sweets in the ratio 2 : 3.
Each bag contains five sweets.

a) On Monday Maya makes up 10 party bags.
  (i) How many sweets does she use in total?
  (ii) How many lemon sweets does she use?
  (iii) How many raspberry sweets does she use?

b) On Tuesday Maya makes up 15 party bags.
  (i) How many sweets does she use in total?
  (ii) How many lemon sweets does she use?
  (iii) How many raspberry sweets does she use?

What do you notice?

In the previous section you saw how to mix grey paint if you know how much black or white paint to use. In this section you will find out how much black and white paint to use to make up the amount of grey paint you want.

In Example 6, black and white paint were mixed in the ratio 2 parts black to 1 part white. How much of each colour paint will you need, to mix 12 litres of grey paint?

Two litres of black paint and one litre of white paint will make three litres of grey paint.

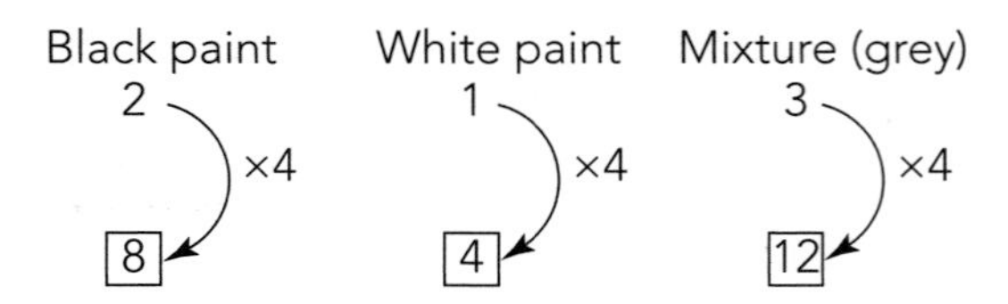

The multiplier is 4.

So you need $2 \times 4 = 8$ litres of black paint and $1 \times 4 = 4$ litres of white paint.

You can check your answer by making sure the quantities add up to 12.

To find the quantities shared in a ratio:

- Find the total number of shares.
- Divide the total quantity by the total number of shares to find the multiplier.
- Multiply each part of the ratio by the multiplier.

Exam tip

The multiplier may not be a whole number. Work with the decimal or fraction and round the final answer if necessary.

Example 10

Two families are going to Alton Towers for the day.

There are five people in the Smith family and four people in the Jones family.

They rent a minibus for the day. It costs £90.

How much does each family pay?

Solution

There are nine people altogether, so each person will pay £90 ÷ 9 = £10.

The Smith family pays 5 × £10 = £50.

The Jones family pays 4 × £10 = £40.

Check: 50 + 40 = 90

Example 11

To make fruit punch, orange juice and grapefruit juice are mixed in the ratio 5 : 3.

Jo wants to make 1 litre of punch.

a) How much orange juice does she need, in millilitres?

b) How much grapefruit juice does she need, in millilitres?

Solution

$5 + 3 = 8$ First work out the total number of shares.

$1000 \div 8 = 125$ Convert 1 litre to millilitres and divide by 8 to find the multiplier.

A table is often helpful for this sort of question.

	Orange	Grapefruit
Ratio	5	3
Amount	$5 \times 125 = 625$ ml	$3 \times 125 = 375$ ml

a) She needs 625 ml of orange juice.

b) She needs 375 ml of grapefruit juice.

Exam tip

To check your answers, add the parts together. Together they should equal the total quantity.

For example, 625 ml + 375 ml = 1000 ml ✓

Exercise 18.4

Do not use your calculator for questions **1** to **11**.

1 Share £20 between Dave and Sam in the ratio 2 : 3.

2 Paint is mixed in the ratio 3 parts red to 5 parts white to make 40 litres of pink paint.
  a) How much red paint is used?
  b) How much white paint is used?

3 Asif is making mortar by mixing sand and cement in the ratio 5 : 1. How much sand is needed to make 36 kg of mortar?

4 To make a solution of a chemical, a scientist mixes 1 part of the chemical with 5 parts of water. She makes 300 ml of the solution.
  a) How much of the chemical does she use?
  b) How much water does she use?

5 Amit, Bree and Chris share £1600 between them in the ratio 2 : 5 : 3. How much does each receive?

6 A 600 g bar of brass is made using the metals copper and zinc in the ratio 2 : 1. How much of each metal is used?

7 Fertiliser needs mixing with water in the ratio 20 ml to 1 litre. How much fertiliser would you need with 4 litres of water?

8 In a class of 36 children the ratio of boys to girls is 5 : 4. How many boys are there?

9 For a football match, tickets are allocated between the 'home' club and the 'away' club in the ratio 6 : 1. The total number of tickets available is 35 000. How many tickets does the home club receive?

10 To get home from work Michael runs and walks. The distances he runs and walks are in the ratio 2 : 3. He works 2 km from home. How far does he run?

11 Orange squash needs to be mixed in the ratio 1 part concentrate to 6 parts water. How much concentrate is needed to make 3.5 litres of orange squash?

You may use your calculator for questions **12** to **14**.

12 There are 572 senators in a national assembly. The numbers of senators in the Blue, Orange and Green parties are in the ratio 6 : 3 : 2.
How many senators are there in each of the parties?

13 St Anthony's College Summer Fayre raised £1750. The governors decided to share the money between the college and a local charity in the ratio 5 : 1.
How much did the local charity receive? Give your answer correct to the nearest pound.

14 Sally makes breakfast cereal by mixing bran, currants and wheat germ in the ratio 8 : 3 : 1 by mass.

a) How much bran does she use to make 600 g of the cereal?

b) One day, she has only 20 g of currants.
How much cereal can she make? She has plenty of bran and wheat germ.

The best value

Activity 3

Two packets of cornflakes are available at a supermarket.
Which is the better value for money?

To compare value, you need to compare *either*

- how much you get for a certain amount of money *or*
- how much a certain quantity (for example, volume or mass) costs.

In each case you are comparing **proportions**, either of size or of cost.

The better value item is the one with the **lower unit cost** or the **greater number of units per penny** (or per pound).

Example 12

Sunflower oil is sold in 700 ml bottles for 95p and in 2 litre bottles for £2.45.
Show which bottle is the better value.

Solution

Method 1

Work out the price per millilitre for each bottle.

	Small	Large
Capacity	700 ml	2 litres = 2000 ml
Price	95p	£2.45 = 245p
Price per ml	95 ÷ 700 = 0.14p	245 ÷ 2000 = 0.1225p = 0.12p

Use the same units for each bottle.

Round your answers to 2 decimal places if necessary.

The price per ml is lower for the 2 litre bottle.
It has the lower unit cost. In this case the unit is a millilitre.
The 2 litre bottle is the better value.

Method 2

Work out the amount per penny for each bottle.

	Small	Large
Capacity	700 ml	2 litres = 2000 ml
Price	95p	£2.45 = 245p
Amount per penny	700 ÷ 95 = 7.37 ml	2000 ÷ 245 = 8.16 ml

Again, use the same units for each bottle.

Round your answers to 2 decimal places if necessary.

The amount per penny is greater for the 2 litre bottle.
It has the greater number of units per penny.
The 2 litre bottle is the better value.

Exam tip

Make it clear whether you are working out the cost per unit or the amount per penny, and include the units in your answers. Always show your working.

Exercise 18.5

1 A 420 g bag of Choco bars costs £1.59 and a 325 g bag of Choco bars costs £1.09.
Which is the better value for money?

2 Spa water is sold in 2 litre bottles for 85p and in 5 litre bottles for £1.79.
Show which is the better value.

3 Wallace bought two packs of cheese, a 680 g pack for £3.20 and a 1.4 kg pack for £5.40.
Which was the better value?

4 One-inch nails are sold in packets of 50 for £1.25 and in packets of 144 for £3.80.
Which packet is the better value?

5 Toilet rolls are sold in packs of 12 for £1.79 and in packs of 50 for £7.20.
Show which is the better value.

6 Brillowhite toothpaste is sold in 80 ml tubes for £2.79 and in 150 ml tubes for £5.00.
Which size tube is the better value?

7 A supermarket sells cola in three different sized bottles: a 3 litre bottle costs £1.99, a 2 litre bottle costs £1.35 and a 1 litre bottle costs 57p.
Which bottle gives the best value?

8 Rice pops are sold in three sizes: 750 g for £1.79, 1.4 kg for £3.20 and 2 kg for £4.89.
Which packet gives the best value?

9 Here are some special offers at two discount music stores.

Explain which is the better offer if you want to buy
a) three CDs.
b) four CDs.

10 Shop A sells a 2 litre bottle of milk for £1.49.
Shop B sells a 4 pint bottle of milk for £1.74.
Find which bottle is better value for money. [1 pint = 0·568 litres]

Key Ideas

- Simplify ratios by dividing by common factors whenever you can.
- To simplify ratios involving measures, both quantities must be in the same units.
- If a ratio is in the form 1 : n, you can work out the second quantity by multiplying the first quantity by n. You can work out the first quantity by dividing the second quantity by n.
- To find an unknown quantity, each part of the ratio must be multiplied by the same number, called the multiplier.
- To divide a quantity in a given ratio, first find the total number of shares, then divide the total quantity by the total number of shares to find the multiplier. You can then multiply each part of the ratio by the multiplier.
- To compare value, work out either the cost per unit or the number of units per penny (or per pound). The better value item is the one with the lower cost per unit or the greater number of units per penny (or per pound).

Working with data

Foundation Silver/ Higher Initial

YOU WILL LEARN ABOUT	YOU SHOULD ALREADY KNOW
o Drawing and interpreting stem-and-leaf diagrams, statistical tables, bar graphs and frequency polygons o Finding the modal class o Calculating the mean from a frequency table	o How to find the median, mean, mode and range of a set of data o How to use them to compare data

Stem-and-leaf diagrams

Here are the marks gained by 30 students in an examination.

63	58	61	52	59	65	69	75	70	54
57	63	76	81	64	68	59	40	65	74
80	44	47	53	70	81	68	49	57	61

A different way of showing these data is to make a stem-and-leaf diagram.

1 Write the tens figures in the left-hand column of a diagram.
These are the 'stems'.

```
4 |
5 |
6 |
7 |
8 |
```

2 Go through the marks in turn and put in the units figures of each mark in the appropriate row.
These are the 'leaves'.

First 63.

```
4 |
5 |
6 | 3
7 |
8 |
```

Then 58.

```
4 |
5 | 8
6 | 3
7 |
8 |
```

Then 61.

```
4 |
5 | 8
6 | 3 1
7 |
8 |
```

3 When all the marks are entered the diagram will look like this.

```
4 | 0 4 7 9
5 | 8 2 9 4 7 9 3 7
6 | 3 1 5 9 3 4 8 5 8 1
7 | 5 0 6 4 0
8 | 1 0 1
```

4 Rewrite the diagram so the units figures in each row are in order of size, with the smallest first.

Finally, add a key.

4	0 4 7 9
5	2 3 4 7 7 8 9 9
6	1 1 3 3 4 5 5 8 8 9
7	0 0 4 5 6
8	0 1 1

Key: 5 | 2 represents 52 marks

This is a finished stem-and-leaf diagram. It is a sort of frequency chart and allows you to read off certain information, for example

- the **modal group**, (the one with the highest frequency), is the 60–69 group.
- there are 30 results so the median result is midway between the 15th and the 16th result. Starting at the first result, 40, and counting on 15 results gives 63, the 16th result is also 63 so the median result is 63.

Exercise 19.1

1 These are the weights, in kilograms, of 25 newborn babies.

2.6	2.9	3.2	2.5	3.1
1.9	3.5	3.9	4.0	2.8
4.1	1.7	3.8	2.6	3.1
2.4	4.1	2.6	4.2	3.6
2.9	2.8	2.7	3.3	3.8

a) Copy and complete this stem-and-leaf diagram.

1	
2	
3	
4	

Key: 1 | 7 represents 1.7 kg

b) Use your diagram to find the median weight.

2 The speeds of 30 cars, in mph, measured in a city street are given below.

41	15	26	14	28	22	27	18	21	32
43	37	30	25	18	25	29	34	28	30
25	52	36	9	21	25	16	29	32	19

a) Copy and complete the stem-and-leaf diagram to show these data.

0	
1	
2	
3	
4	
5	

Key:

b) Use your diagram to find the median speed.

c) The speed limit is 30 mph. What percentage of cars were breaking this limit?

3 A group of students took two mathematics tests.
The stem-and-leaf diagrams of the marks for the tests are given below.

Test 1

Stem	Leaf
2	3 4 5
3	1 3 6 8
4	0 2 5 6 7
5	2 3 5 7
6	1 3 5 5 6 9
7	0 1 3 4 4 6 8
8	0 2 3 3 6 6 7
9	1 2 2 4 5

Test 2

Stem	Leaf
2	0 1 2 6
3	1 2 2 4 5 7 8
4	2 2 5 5 6 8 8 9
5	0 2 2 6 7 8
6	1 3 4 4 9 9
7	0 2 3 5 7
8	3 7 9
9	1 5

Key: 3 | 3 represents 33 marks

a) Which test appeared to be harder?
b) Find the median mark for each test.

4 A TV repair company monitored the time, in hours, it took their engineers to visit 20 homes and repair or replace televisions and DVD players.

0.9	1.0	2.1	2.4	0.7	1.1	0.9	0.6	0.4	0.3
1.2	1.6	0.6	0.3	0.7	1.4	1.0	0.8	0.7	0.9

a) Copy and complete the stem-and-leaf diagram to show these data.

Stem	Leaf
0	
1	
2	

Key:

b) Use your diagram to find the median time.

5 A scientist measured the lengths of a type of fish caught in two different rivers.
Here are their lengths, in centimetres.

River A

38	47	43	51	45	33	62	57
36	40	49	66	55	49	45	31
40	44	57	58	35	52	73	39
38	69	46	55				

River B

48	52	54	37	42	65	70	49
61	50	54	45	61	72	74	64
56	38	65	69	71	67	71	70
68							

a) Make a stem-and-leaf diagram to show the lengths of fish in each river.
b) Find the median length for the fish caught in each river.

6 Rakhi's birthday is December 12th.
She decided to look in a newspaper and see how old other people were who shared the same birthday.
Her results were as follows.

56	74	75	56	67	71	48	30	45	58
60	62	21	24	36	38	51	56	43	22
18	32	44	40	21	18	50	40	30	50

a) Show these data in a stem-and-leaf diagram, choosing a suitable interval.
b) What is the median age?

7 Mr Smith recorded the marks his students got on a history test. These are the marks they scored.

34	56	54	76	84	48	32	18	43	66	50
67	52	43	44	76	88	35	44	60	39	43

a) Show these data in a stem-and-leaf diagram, choosing a suitable interval.
b) What is the median mark?

8 The data shows the ages of 25 employees in a business.

19	17	26	33	31	41	41	27	25	28
24	28	32	40	31	29	29	39	26	36
26	35	38	42	38					

a) Construct a stem-and-leaf diagram to show these ages.
b) What is the modal age?
c) What is the median age?

9 The data shows the times, in seconds, 45 students took to swim one length of a swimming pool.

14.6	15.2	15.7	19.4	18.5	14.7	16.1	18.5	15.9	19.5
16.2	15.6	21.7	20.0	18.4	17.1	17.8	16.2	17.7	15.1
17.0	15.8	16.1	18.1	16.2	14.9	15.5	18.6	17.4	18.3
16.3	19.2	16.4	15.5	18.7	14.9	18.3	14.1	17.4	19.8
16.4	16.3	21.2	15.8	16.8					

a) Construct a stem-and-leaf diagram to show these times. Use the key 14|6 represents 14.6 seconds.
b) Students who swam the length in less than 15 seconds qualified for the school team. How many students qualified?
c) What was the median time?

10 For their mock examinations, a group of 41 students took two maths papers. Here are their results.

Paper 1

25	46	33	76	55	40	86	83	83	92	36
38	69	45	87	65	53	52	95	94	80	
49	78	57	31	74	65	63	61	70	92	
24	23	66	42	86	74	73	71	82	91	

Paper 2

38	26	34	32	20	69	64	75	61	87	48
49	35	46	45	32	31	52	50	72	83	
26	48	58	57	45	42	52	63	89	95	
37	22	21	69	56	52	64	73	70	91	

a) Construct a stem-and-leaf diagram for each set of scores.
b) Which paper do you think was the harder? Give a reason for your answer.
c) Find the median mark for each paper.
d) The pass mark was 40 on each paper. How many students passed each paper?

Calculating the mean from a frequency table

When working with large amounts of data, it is often easier to see the pattern of the data if they are displayed in a table. For example, this is a list of goals scored in 20 matches.

1 1 3 2 0 0 1 4 0 2
2 0 6 3 4 1 1 3 2 1

This is the same data displayed in a frequency table.

Number of goals	Frequency
0	4
1	6
2	4
3	3
4	2
5	0
6	1

Mode

From the table, it is easy to identify the number of goals with the greatest frequency. That is the mode of the data. Here the mode is 1 goal.

Mean

The table can also be used to calculate the mean.

There are: four matches with 0 goals $4 \times 0 = 0$ goals
six matches with 1 goal $6 \times 1 = 6$ goals
four matches with 2 goals $4 \times 2 = 8$ goals

and so on.

To find the total number of goals scored altogether, multiply each number of goals by its frequency and then add the results. You can add an extra column to your table to help you work out the values multiplied by their frequencies.

Number of goals	Frequency	Number of goals × frequency
0	4	0
1	6	6
2	4	8
3	3	9
4	2	8
5	0	0
6	1	6
Totals	20	37

Then, dividing by the total number of matches (20) gives the mean.

The working for this is shown in the table.

Mean = 37 ÷ 20 = 1.85 goals.

In this example you have been given the original data. Check the answer by adding up the list of goals scored and dividing them by 20!

Example 1

Work out the mean, mode and range for the number of children in the houses in Berry Road, listed in this table.

Number of children (c)	Frequency (number of houses)	Total number of children (c × frequency)
0	6	0
1	4	4
2	5	10
3	7	21
4	1	4
5	2	10
Totals	25	49

Solution

Mean= 49 ÷ 25 = 1.96 children

Mode= 3 children

Range= 5 – 0 = 5 children

Using a spreadsheet to find the mean

You can also calculate the mean using a computer spreadsheet. Follow these steps to work out the mean for the data in Example 1.

1 Open a new spreadsheet.
2 In cell A1 type the title 'Number of children (c)'.
  In cell B1 type the title 'Frequency (number of houses)'.
  In cell C1 type the title 'Total number of children'.
3 In cell A2 type the number 0. Then type the numbers 1 to 5 in cells A3 to A7.
4 In cell B2 type the number 6. Then type the other frequencies in cells B3 to B7.
5 In cell C2 type **=A2*B2** and press the enter key.
  Click on cell C2, click on Edit in the toolbar and select Copy.
  Click on cell C3, hold down the mouse key and drag down to cell C7.
  Then click on Edit in the toolbar and select Paste.
6 In cell A8 type the word 'Total'.
7 In cell B8 type **=SUM(B2:B7)** and press the enter key.
  In cell C8 type **=SUM(C2:C7)** and press the enter key.
8 In cell A9 type the word 'Mean'.
9 In cell B9 type **=C8/B8** and press the enter key.
  Your spreadsheet should look like the one shown on the next page.

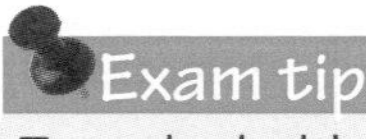

Type the bold text carefully: do not put in any spaces.

	A	B	C
1	Number of children (c)	Frequency (number of houses)	Total number of children
2	0	6	0
3	1	4	4
4	2	5	10
5	3	7	21
6	4	1	4
7	5	2	10
8	Total	25	49
9	Mean	1.96	

Exercise 19.2

1 Calculate the mean for each of these sets of data.

a)

Number of passengers in taxi	Frequency
1	84
2	63
3	34
4	15
5	4
Total	200

b)

Number of pets owned	Frequency
0	53
1	83
2	23
3	11
4	5
Total	175

c)

Number of books read in a month	Frequency
0	4
1	19
2	33
3	42
4	29
5	17
6	6
Total	150

d)

Number of drinks a day	Frequency
3	81
4	66
5	47
6	29
7	18
8	9
Total	250

2 Answer one of the questions in this exercise using a computer spreadsheet.

3 Tom is the manager of a garage. He records the numbers of one make of car that are serviced each week-day during a 4-week period.

8 6 5 2 8 10 8 9 6 5
5 9 10 8 6 5 8 9 10 15

a) Write down the modal number of cars serviced.

b) Find the median number of cars serviced.

c) Find the mean number of cars serviced per day.

4 This table shows the number of minutes late that the students in a class arrived for registration one Monday morning.

Minutes late	Frequency
1	0
2	1
3	5
4	7
5	11
6	2
7	1
8	0
9	1
10	0

a) What was the modal number of minutes late?
b) What is the mean number of minutes late?

5 Bagthorpe United football team recorded the number of goals it scored each week one football season.

Number of goals	Frequency
0	7
1	10
2	9
3	4

a) How many matches were played?
b) What was the total number of goals scored?
c) What was the mean number of goals scored?
d) What was the median number of goals scored?

6 Christine decides to count the number of items of 'junk' emails she receives each day during June.

Number of junk emails	Number of days
0	5
1	4
2	6
3	7
4	2
5	0
6	1

a) What was the mode of the number of junk emails?
b) What was the mean of the number of junk emails?

7 In Barnsfield, bus tickets cost 50p, £1.00, £1.50 or £2.00 depending on the length of the journey.
The frequency table shows the numbers of tickets sold on one Friday.
Calculate the mean fare paid on that Friday.

Price of ticket (£)	Number of tickets
0.50	140
1.00	207
1.50	96
2.00	57

8 For each of these sets of data
(i) find the mode.
(ii) find the median.
(iii) find the range.
(iv) calculate the mean.

a)

Score on dice	Number of times thrown
1	89
2	77
3	91
4	85
5	76
6	82
Total	500

b)

Number of matches	Number of boxes
47	78
48	82
49	62
50	97
51	86
52	95
Total	500

c)

Number of accidents	Number of drivers
0	65
1	103
2	86
3	29
4	14
5	3
Total	300

d)

Number of cars per house	Number of students
0	15
1	87
2	105
3	37
4	6
Total	250

9 800 people were asked how many cups of coffee they had bought one week.
The table shows the data.
Calculate the mean number of cups of coffee bought.

Number of coffees	Frequency
0	20
1	24
2	35
3	26
4	28
5	49
6	97
7	126
8	106
9	54
10	83
11	38
12	67
13	21
14	26

Representing continuous data

When data involve measurements, they are always grouped, even if they do not look like it. For instance, a length L given as 18 cm to the nearest centimetre means $17.5 \leqslant L < 18.5$. Any length between these values will count as 18 cm. Often, however, the groups are larger, to make handling the data easier. For example, when recording the heights, h cm, of 100 students in Year 11, groups such as $180 \leqslant h < 185$ may be used.

Bar graphs

Continuous data may be represented on a bar graph, using proper scales on both axes. When the data are continuous, there is no gap between the bars. In this chapter only bars of equal width are considered, and the height of each bar represents the frequency, as with the bar charts used for discrete data that you have studied before.

Example 2

Draw a bar graph to represent this information about the heights of students in Year 11 at Sandish School.

Height (h cm)	Frequency
$155 \leqslant h < 160$	2
$160 \leqslant h < 165$	6
$165 \leqslant h < 170$	18
$170 \leqslant h < 175$	25
$175 \leqslant h < 180$	9
$180 \leqslant h < 185$	4
$185 \leqslant h < 190$	1

Solution

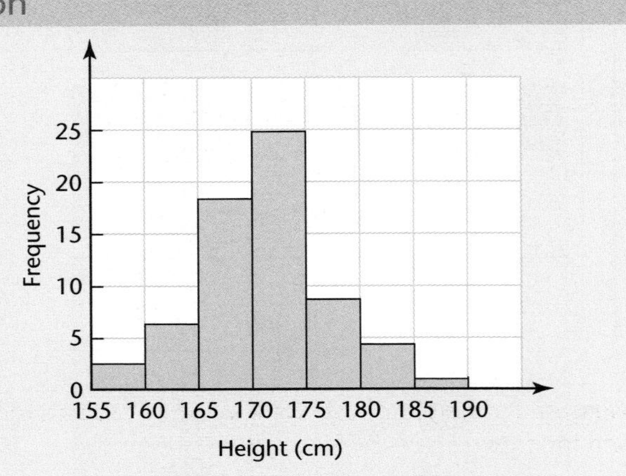

Exam tip

Check that you have labelled both scales carefully. The horizontal scale must be labelled continuously and not with the groups. Make sure that the boundaries of your bars match the boundaries of the groups.

Where the groups are not of the same width, the graph is called a **histogram**, and the **area** of each bar represents the frequency.

Exam tip

To work out the midpoint of a group, add together its boundary values and divide by 2.

Frequency polygons

A frequency polygon may also be used to represent the data. However, in this case, only one point is used to represent each group. The midpoint value of each group is chosen, as it is an average value for the group.

Example 3

Show the heights of the Year 11 students at Sandish School in a frequency polygon. The data is given in Example 2.

Solution

The midpoint of the $155 \leqslant h < 160$ group is $\frac{155 + 160}{2} = 157.5$.

So the points are plotted at h-values of 157.5, 162.5, 167.5 and so on.

Exam tip

It can be helpful to add a column for the midpoint to the frequency table.

Height (h cm)	Frequency	Midpoint
$155 \leqslant h < 160$	2	157.5
$160 \leqslant h < 165$	6	162.5
$165 \leqslant h < 170$	18	167.5
$170 \leqslant h < 175$	25	172.5
$175 \leqslant h < 180$	9	177.5
$180 \leqslant h < 185$	4	182.5
$185 \leqslant h < 190$	1	187.5

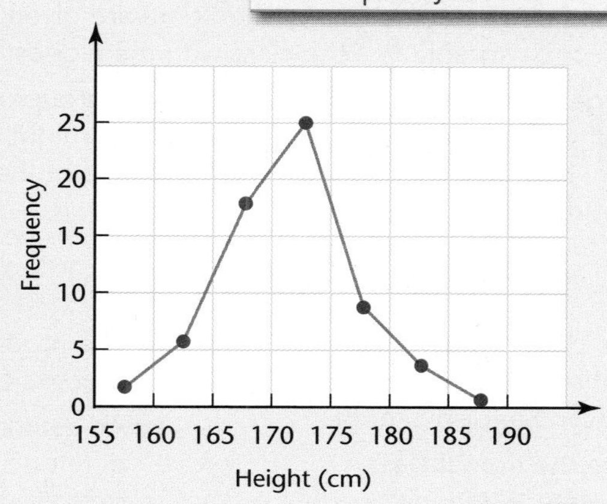

Activity 1

a) Measure the height of everyone in your class and record the data in two lists, one for boys and one for girls.

b) Choose suitable intervals for the data and organise the data in a two-way table.

c) Draw two frequency diagrams, one for the boys' data and one for the girls'.
Use the same scales for each diagram so that they can be compared easily.

d) Compare the two diagrams.
What do the shapes of the graphs tell you, in general, about the heights of the boys and girls in your class?

e) Compare your frequency diagrams with others in your class.
Have they used the same intervals for the data as you?
If they haven't, has this made a difference to their answers to part **d)**?
Which of the diagrams looks the best? Why?

Comparing distributions

When you compare two distributions, remember to make two types of comparison.

- Compare an average of each of the distributions. Use the mean, median or mode.
- Compare the spread of each of the distributions. One measure you can use is the range.

Make sure that your comments relate back to the context. For example, do not say 'the mean for the boys is larger' but say 'the boys are taller on average than the girls since the mean for the boys is larger'.

The mode and range of grouped continuous data

The **modal class** may be found when data are given as a table, frequency polygon or bar graph. It is the class with the highest frequency.

The **range** cannot be stated accurately from grouped data. For instance, the height of the tallest student in Example 3 might be 189.7 cm or 185.0 cm. However, you can use the midpoints of the groups to estimate the range.

Exam tip

When stating the modal class, take care to give its boundaries accurately.

Example 4

Use the data given in Example 3 about the heights of the Year 11 students in Sandish School.

a) State the modal class.
b) Estimate the range.

Solution

a) The modal class is the one with the largest frequency, which is the $170 \leqslant h < 175$ group.

b) An estimate of the range is found by finding the difference between the midpoint values of the top and bottom groups in the table.

Range = 187.5 – 157.5 = 30 cm

Exercise 19.3

1 The manager of a leisure centre recorded the ages of the women who used the swimming pool one morning. Here are his results.
Draw a bar graph to show these data.

Age (a years)	Frequency
$15 \leqslant a < 20$	4
$20 \leqslant a < 25$	12
$25 \leqslant a < 30$	17
$30 \leqslant a < 35$	6
$35 \leqslant a < 40$	8
$40 \leqslant a < 45$	3
$45 \leqslant a < 50$	12

2 **a)** Draw a bar graph to show this information.
b) State the modal class.

Length (y cm)	Frequency
$10 \leqslant y < 20$	2
$20 \leqslant y < 30$	6
$30 \leqslant y < 40$	9
$40 \leqslant y < 50$	5
$50 \leqslant y < 60$	3

3 a) Draw a bar graph to show this information.
b) State the modal class.

Mass (w kg)	Frequency
$30 \leqslant w < 40$	5
$40 \leqslant w < 50$	8
$50 \leqslant w < 60$	2
$60 \leqslant w < 70$	4
$70 \leqslant w < 80$	1

4 In a survey, the annual rainfall was measured at 100 different towns. Here are the results of the survey.

Rainfall (r cm)	Frequency
$50 \leqslant r < 70$	14
$70 \leqslant r < 90$	33
$90 \leqslant r < 110$	27
$110 \leqslant r < 130$	8
$130 \leqslant r < 150$	16
$150 \leqslant r < 170$	2

a) Draw a bar graph to show these data.
b) Which of the intervals is the modal class?
c) Which of the intervals contains the median value?

5 As part of a fitness campaign, a business measured the weight of all of its workers. Here are the results.

Weight (w kg)	Frequency
$60 \leqslant w < 70$	3
$70 \leqslant w < 80$	18
$80 \leqslant w < 90$	23
$90 \leqslant w < 100$	7
$100 \leqslant w < 110$	2

a) Draw a bar graph to show these data.
b) Which of the intervals is the modal class?
c) Which of the intervals contains the median value?

6 Here is a bar graph showing the lifetimes of the lightbulbs tested by a factory. Use the bar graph to make a grouped frequency table.

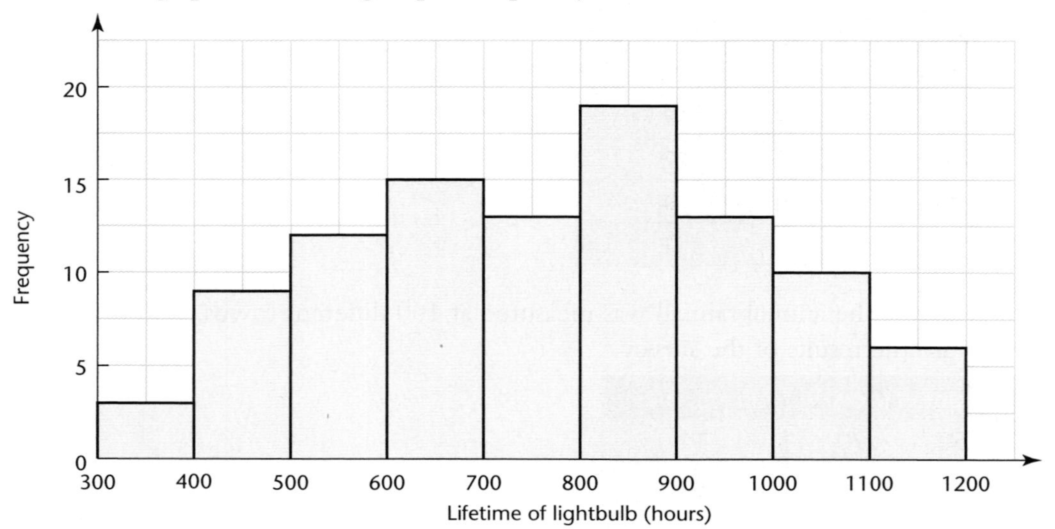

7 This bar graph shows the ages of a group of people travelling on the 08:00 train to London one Friday morning.

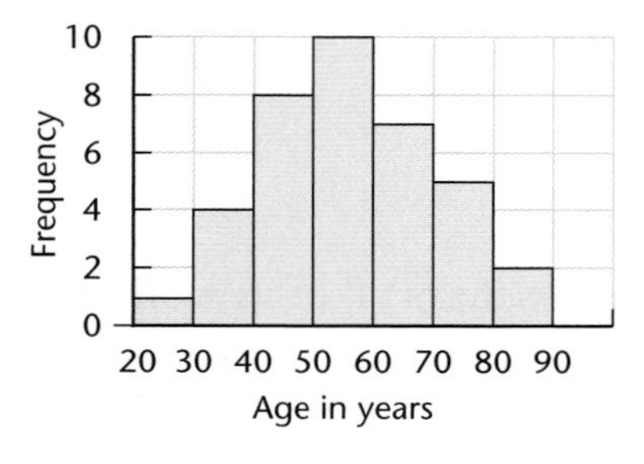

a) Make a frequency table for the data.
b) Estimate the range of the ages.

8 State the midpoint of each of these intervals.
a) $20\,\text{m} < \text{length} \leqslant 30\,\text{m}$
b) $3\,\text{litres} \leqslant \text{volume} \leqslant 3.5\,\text{litres}$
c) $45\,\text{kg} \leqslant \text{mass} < 50\,\text{kg}$
d) $3\,\text{mm} < \text{length} \leqslant 5\,\text{mm}$

9 State the midpoint of each of these intervals.
a) $25\,\text{cm} \leqslant \text{length} < 35\,\text{cm}$
b) $0.1\,\text{cm} \leqslant \text{length} < 0.2\,\text{cm}$
c) $500\,\text{kg} \leqslant \text{mass} < 520\,\text{kg}$
d) $10\,\text{ml} \leqslant \text{volume} < 14\,\text{ml}$
e) $0\,\text{seconds} \leqslant \text{time} < 5\,\text{seconds}$

10 Draw a frequency polygon to represent this distribution.

Height (cm)	Frequency
$50 \leq h < 60$	15
$60 \leq h < 70$	23
$70 \leq h < 80$	38
$80 \leq h < 90$	17
$90 \leq h < 100$	7

11 Draw a frequency polygon to represent this distribution.

Height (m)	Frequency
$0 \leq h < 2$	12
$2 \leq h < 4$	26
$4 \leq h < 6$	34
$6 \leq h < 8$	23
$8 \leq h < 10$	5

12 **a)** Draw a frequency polygon to show these data.
b) Estimate the range.

Length (x cm)	Frequency
$0 < x \leq 5$	8
$5 < x \leq 10$	6
$10 < x \leq 15$	2
$15 < x \leq 20$	5
$20 < x \leq 25$	1

13 The table shows the weight of packages sent out from a warehouse on one day.

Weight (w kg)	Frequency
$0 \leq w < 6$	8
$6 \leq w < 12$	14
$12 \leq w < 18$	19
$18 \leq w < 24$	15
$24 \leq w < 30$	10

Draw a frequency polygon to show these data.

14 The table shows the length of time that cars stayed in a car park one day.

Time (t minutes)	Frequency
$15 \leq t < 30$	56
$30 \leq t < 45$	63
$45 \leq t < 60$	87
$60 \leq t < 75$	123
$75 \leq t < 90$	67
$90 \leq t < 105$	22

Draw a frequency polygon to show these data.

15 The table shows the heights of 60 students.

Height (h cm)	Frequency
$168 \leq h < 172$	2
$172 \leq h < 176$	6
$176 \leq h < 180$	17
$180 \leq h < 184$	22
$184 \leq h < 188$	10
$188 \leq h < 192$	3

Draw a frequency polygon to show these data.

16 The table shows the number of words per sentence in the first 50 sentences of two books.

Number of words (w)	Frequency Book 1	Frequency Book 2
$0 < w \leq 10$	2	27
$10 < w \leq 20$	9	11
$20 < w \leq 30$	14	9
$30 < w \leq 40$	7	0
$40 < w \leq 50$	4	3
$50 < w \leq 60$	8	0
$60 < w \leq 70$	6	0

a) On the same grid, draw a frequency polygon for each book.

b) Use the frequency polygons to compare the number of words per sentence in each book.

17 The ages of all the people under 70 in a small village were recorded in 1985 and 2005.
The results are given in the table.

Age (a years)	Frequency in 1985	Frequency in 2005
$0 \leqslant a < 10$	85	50
$10 \leqslant a < 20$	78	51
$20 \leqslant a < 30$	70	78
$30 \leqslant a < 40$	53	76
$40 \leqslant a < 50$	40	62
$50 \leqslant a < 60$	28	64
$60 \leqslant a < 70$	18	56

a) On the same grid, draw a frequency polygon for each year.
b) Use the frequency polygons to compare the distribution of ages in the two years.

18 The results for a test for a group of boys and a group of girls in year 10 are shown below.

Girls									
30	32	68	47	52	63	27	58	94	39
65	81	67	58	43	58	74	42	76	90
25	65	49	80	75	73	32	57	66	48
Boys									
47	69	66	35	92	58	44	57	68	79
86	29	54	58	84	69	87	36	44	80
57	65	76	66	71					

The girls say they did better than the boys. Do you agree?
Use suitable calculations and diagrams to compare their results.
Give reasons for your choices.

Challenge 1

Work in groups.

Choose two different newspapers and compare the way they use words.

Question 16 in Exercise 19.3 gives one way of comparing, using the number of words per sentence. Discuss other suitable ways to compare them.

Carry out the investigation.

You could make a poster to show the results.

Key Ideas

- A stem-and-leaf diagram can be used to organise discrete data. The modal group and the median can be found easily from the diagram.
- Frequency tables are another way to organise discrete data. The mode can be found easily from the table.
- To calculate the mean from a frequency table, you multiply each of the numbers in the first column by its frequency and write the result in a third column. The mean is the total of the third column divided by the total of the frequencies in the second column.
- You can represent grouped continuous data as a bar graph. For equal width intervals, the height of each bar represents the frequency.
- You can also represent grouped continuous data as a frequency polygon. You plot the frequency of each group against the midpoint of the interval.
- The modal class is the class with the highest frequency.
- The range of grouped continuous data cannot be stated exactly. You can estimate the range by finding the difference between the midpoints of the first and last intervals.
- When comparing distributions, compare an average of each of the distributions, using the mean, median or mode, and compare the spread of each of the distributions, using the range.

Trial and improvement

YOU WILL LEARN ABOUT	YOU SHOULD ALREADY KNOW
○ Solving a problem by trial and improvement ○ Finding consecutive integers between which the solution of an equation lies ○ Using the method of trial and improvement to solve an equation to a given degree of accuracy	○ How to use your calculator to find powers and square roots ○ How to substitute numbers into algebraic expressions

Using trial and improvement to solve problems

Trial and improvement is one way of solving problems for which there is no straightforward method that you know. This is shown in the following examples.

Exam tip

If you know another method, e.g. using algebra, you should use it. You may lose marks otherwise.

Example 1

When a number is squared and 3 times the number is added the result is 154.

Find the number.

Solution

Try	Square	Add 3 times number	Result
20	400	+ 60 = 460	Too big, try smaller
10	100	+ 30 = 130	Too small but nearly correct, try a number bigger than 10
11	121	+ 33 = 154	Correct

The answer is 11.

Example 2

Joe is twice as old as Maria and their ages multiply to give 1922.

What are their ages?

Solution

You know that Joe is twice as old as Maria so start with a try at their ages. Remember that 'product' means the result of multiplying.

Joe's age	Maria's age	Product of ages	Result
40	20	800	Too small, try bigger
60	30	1800	Too small, try bigger
64	32	2048	Too big, try between 60 and 64
62	31	1922	Correct

Joe is 62 and Maria is 31.

Exam tip

The order in which you do the operations matters – at each step work out the result before doing the next step.

These examples had exact answers. Sometimes, you can only find the solution to a given number of decimal places, as in the following activity.

Activity 1

A cuboid has a square base. The height of the cuboid is 2 cm more than the sides of the base, and the volume of the cuboid is 500 cm^3.

Find the dimensions of the cuboid, correct to 2 decimal places. Use the table below to help you.

Length (cm)	Width (cm)	Height (cm)	Volume (cm^3)
5	5	7	$5 \times 5 \times 7 = 175$
10	10	12	
7	7	9	

Exercise 20.1

1 A number is squared and then subtracted. The result is 272.
Find the number.

2 A number is squared and then three times the number is added. The result is 340.
Find the number.

3 Two consecutive numbers have a product of 756.
Find the numbers.

4 Amy is six years older than Mark. The product of their age is 247.
What are their ages?

5 The length of a rectangle is 6 cm more than the width. The area is 247 cm^2.
Find the dimensions of the rectangle.

Using trial and improvement to solve equations

Often you are told that the solution of an equation lies between two given numbers. If you are not told this, your first step must be to find these two values. Then choose the number halfway between these values. Keep making improvements until the required accuracy is achieved.

Example 3

A solution of the equation $x^3 - 4x + 1 = 0$ lies between 1 and 2.

Use trial and improvement to find the solution correct to 1 decimal place.

Exam tip

Make sure you give the x-value (not the value on the right-hand side of the equation) to the required accuracy.

Solution

You have been given two values between which the solution lies, so for the first trial use the value halfway between these, that is 1.5. A table can be used to set out your trials.

x-value	Calculation	Comment
1.5	$1.5^3 - 4 \times 1.5 + 1 = ^-1.625$	Too small, so try a value between 1.5 and 2.0
1.8	$1.8^3 - 4 \times 1.8 + 1 = ^-0.368$	Too small, so try a value between 1.8 and 2.0
1.9	$1.9^3 - 4 \times 1.9 + 1 = 0.259$	Too big, so the solution lies between 1.8 and 1.9
1.85	$1.85^3 - 4 \times 1.85 + 1 = ^-0.0684$	Too small, so the solution lies between 1.85 and 1.9

The solution is greater than 1.85 so is nearer to 1.9 than to 1.8.

It is $x = 1.9$, correct to 1 decimal place.

Example 4

Show that $x^3 - 3x = 6$ has a solution between 2 and 3.

Find the solution correct to 1 decimal place.

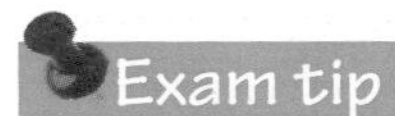

Exam tip

Always give the result of the calculation for the trial you have done as well as saying 'too big' or 'too small'.

Solution

$2^3 - 3 \times 2 = 2$

$3^3 - 3 \times 3 = 18$

Because 6 is between 2 and 18, there is a solution for x between 2 and 3.

For the first trial use the value halfway between these, that is 2.5

x-value	Calculation	Comment
2.5	$2.5^3 - 3 \times 2.5 = 8.125$	Too big, so try a value between 2 and 2.5
2.3	$2.3^3 - 3 \times 2.3 = 5.267$	Too small, so try a value between 2.3 and 2.5
2.4	$2.4^3 - 3 \times 2.4 = 6.624$	Too big, so the solution lies between 2.3 and 2.4
2.35	$2.35^3 - 3 \times 2.35 = 5.928$	Too small, so the solution lies between 2.35 and 2.4

The solution is nearer to 2.4 than to 2.3.

It is $x = 2.4$, correct to 1 decimal place.

Example 5

Show that the equation $\frac{6}{x} = x^2 + 3$ has a solution between $x = 1$ and $x = 2$.
Find this solution correct to 2 decimal places.

Solution

Since there are x's on both sides of this equation, it is often easiest to work with the left-hand side (LHS) and right-hand side (RHS) separately and then compare them. The solution is when they are equal, so the difference between them is zero.

x-value	LHS = $\frac{6}{x}$	RHS $= x^2 + 3$	LHS –RHS	Comments
1	$\frac{6}{1} = 6$	$1^2 + 3 = 4$	2	
2	$\frac{6}{2} = 3$	$2^2 + 3 = 7$	$^-4$	The sign of LHS – RHS has changed, so there is a solution between 1 and 2. Now try halfway between them.
1.5	$\frac{6}{1.5} = 4$	$1.5^2 + 3 = 5.25$	$^-1.25$	So solution is between 1 and 1.5. Try 1.3
1.3	$\frac{6}{1.3} = 4.615...$	$1.3^2 + 3 = 4.69$	$^-0.0746...$	So solution is between 1 and 1.3 but quite near 1.3. Try 1.2
1.2	$\frac{6}{1.2} = 5$	$1.2^2 + 3 = 4.44$	0.56	So solution is between 1.2 and 1.3 and near 1.3. Try 1.28
1.28	$\frac{6}{1.28} = 4.6875$	$1.28^2 + 3 = 4.6384$	0.0491	So solution is between 1.28 and 1.3. Try 1.29
1.29	$\frac{6}{1.29} = 4.651...$	$1.29^2 + 3 = 4.6641$	$^-0.012...$	So solution is between 1.28 and 1.29. Try 1.285
1.285	$\frac{6}{1.285} = 4.669...$	$1.285^2 + 3 = 4.651225$	0.018	So solution is between 1.285 and 1.29. So the solution is 1.29 correct to 2 decimal places

The solution is $x = 1.29$, correct to 2 d.p.

Exercise 20.2

1 a) $2^3 - 8 \times 2 = {}^-8$ and $3^3 - 8 \times 3 = 3$. Explain how this shows that there is a solution to the equation $x^3 - 8x = 0$ between 2 and 3.

b) Find the solution correct to 1 decimal place.

2 a) $1^3 - 2 \times 1 = {}^-1$ and $2^3 - 2 \times 2 = 4$. Explain how this shows that there is a solution to the equation $x^3 - 2x = 0$ between 1 and 2.

b) Find the solution correct to 1 decimal place.

3 Show that a solution of $x^3 - x = 90$ lies between 4 and 5.
Find it correct to 1 decimal place.

4 Show that a solution of $x^3 - x^2 = 30$ lies between 3 and 4.
Find it correct to 1 decimal place.

5 Show that a solution of $x^3 - 7x = 25$ lies between 3 and 4.
Find it correct to 1 decimal place.

6 Show that a solution of $x^3 + 2x = 2$ lies between 0 and 1.
Find it correct to 1 decimal place.

7 A solution of $x^3 - 2x + 6 = 0$ lies between $^-3$ and $^-2$.
Find it correct to 1 decimal place.

8 a) Find two consecutive integers between which the solution of $x^3 - 2x^2 = 4$ lies.
b) Find the solution correct to 2 decimal places.

9 A solution of $x^3 + 3x^2 + x = 0$ lies between $^-3$ and $^-2$.
Find it correct to 2 decimal places.

10 a) Find two consecutive integers between which a solution of $x^3 + 3x - 20 = 0$ lies.
b) Find the solution correct to 2 decimal places.

11 A solution of $x^3 - 5x^2 + 2x = 0$ lies between 0 and 1.
Find it correct to 2 decimal places.

12 The equation $x = \frac{2}{x^2} + 1$ has a solution between $x = 1$ and $x = 2$. Find this root correct to 1 decimal place.

13 Show that $x = {}^-2$ is a solution of the equation $x^2 - 5 = \frac{2}{x}$. Use trial and improvement to find, correct to 1 decimal place, the solution of this equation that is between 2 and 3.

14 Show that the equation $\frac{1}{x} = x^2 - 1$ has a solution between $x = 1$ and $x = 2$. Then find this solution, correct to 1 decimal place.

15 The equation $x^2 - 8 = \frac{3}{x}$ has a solution between $x = {}^-2$ and $x = {}^-3$. Find this solution, correct to 2 decimal places.

Key Ideas

- To solve a problem or an equation by trial and improvement try different values, deciding each time whether the test value should be bigger or smaller.
- If the answer is not exact, find two consecutive numbers that the answer lies between then choose the number halfway between these values and keep making improvements until the required accuracy is achieved.

CHAPTER 21

Transformations 1

YOU WILL LEARN ABOUT	YOU SHOULD ALREADY KNOW
○ Reflecting a shape in lines parallel to the axes and in the line $y = -x$ ○ Drawing, recognising and describing reflections, rotations and translations	○ The meaning of reflection and rotation symmetry ○ How to plot and read the coordinates of a point ○ How to reflect a simple shape in a vertical or horizontal line ○ How to draw the graph of a straight line given its equation

In this chapter, you will have more practice in using transformations which you have already learnt about. You may find tracing paper helpful with some of these transformations.

Reflections

You have already looked at reflections in vertical (up the page) and horizontal (across the page) lines, but you also need to be able to reflect a shape in a diagonal line.

When a shape is reflected it is 'turned over'. The reflection or image is exactly the same size and shape as the original, but in reverse. The shape and its image are **congruent**. For example the reflection of this shape could look like this.

Corresponding points are the same distance from the mirror line but on the opposite side.

Reflections can be drawn on plain paper, but you are often asked to draw them on squared paper.

In this diagram the shape has been reflected in the line PQ.

This means

- the image of point A is half a square diagonally on the other side of the line PQ
- the image of B is half a square diagonally on the other side
- the image of C is a full square diagonally on the other side
- the image of D is one and a half squares diagonally on the other side.

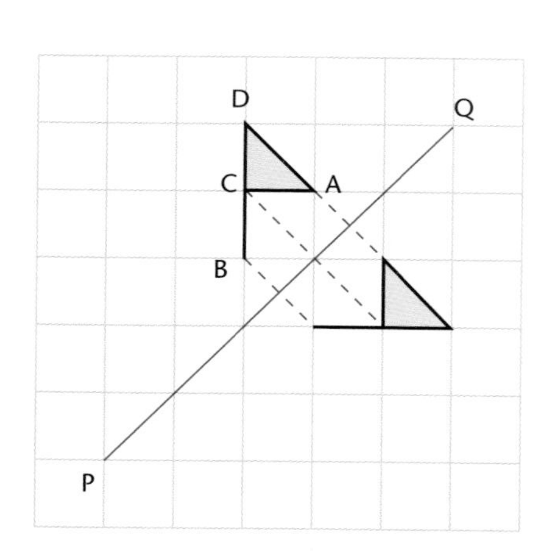

Instead of counting squares you could trace the shape and the mirror line, then turn the tracing paper over and line up the mirror line and your tracing of it.

Example 1

Reflect this shape in the line AB.

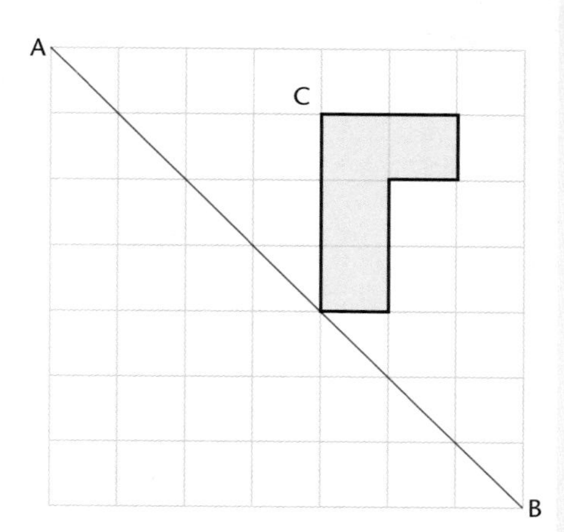

Solution

One point is on the line, so its image stays there. The images of the other points are each the same distance from the line as the original point, but on the opposite side. For example, point C and its image are both one and a half squares from the mirror line.

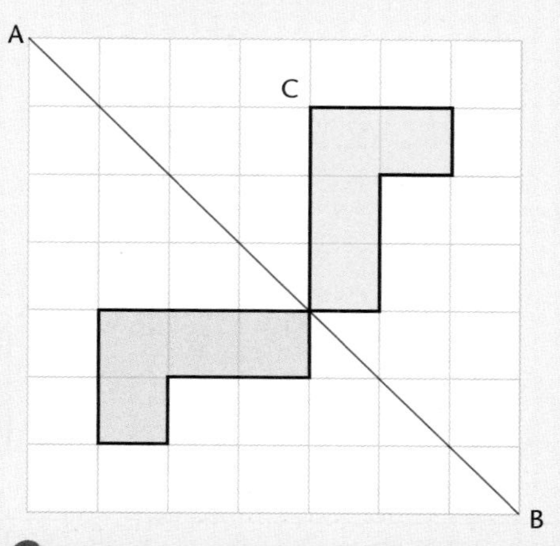

Exam tip

When you have drawn a reflection in a sloping line, check it by turning the page so the mirror line is vertical. Then you can easily see if it has been reflected correctly.

It is helpful to recall the equations of some simple lines which may be used for reflections.

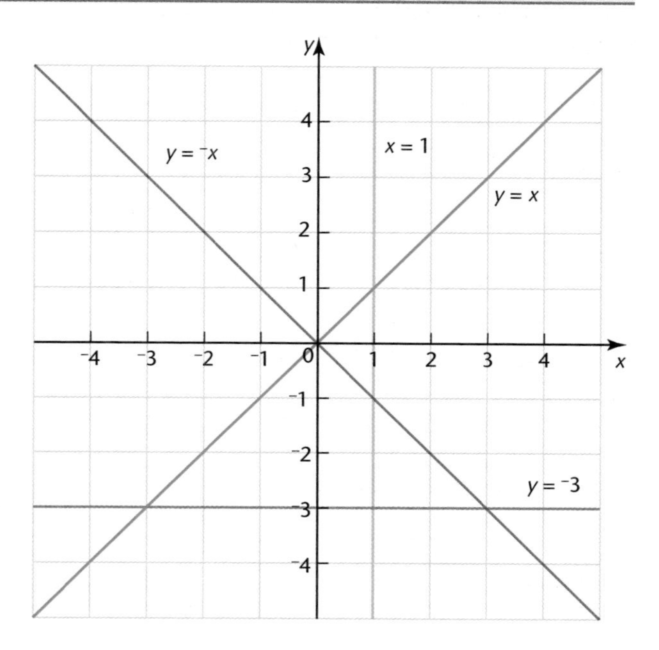

Example 2

Reflect triangle A in the line $x = 3$.

Label the image B.

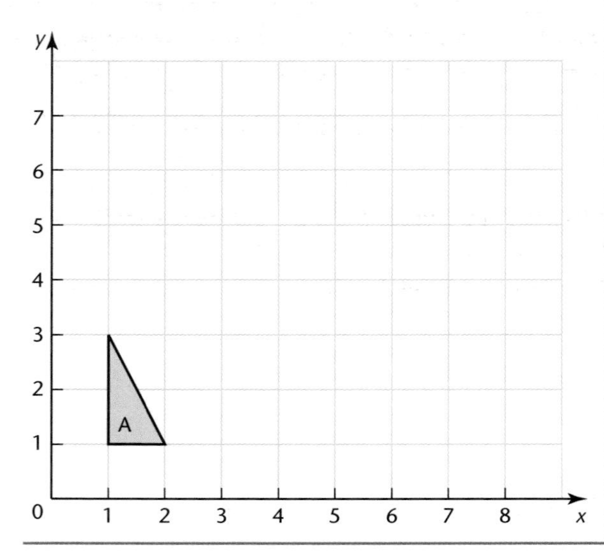

Solution

First draw the line $x = 3$.

Then reflect triangle A in the line.

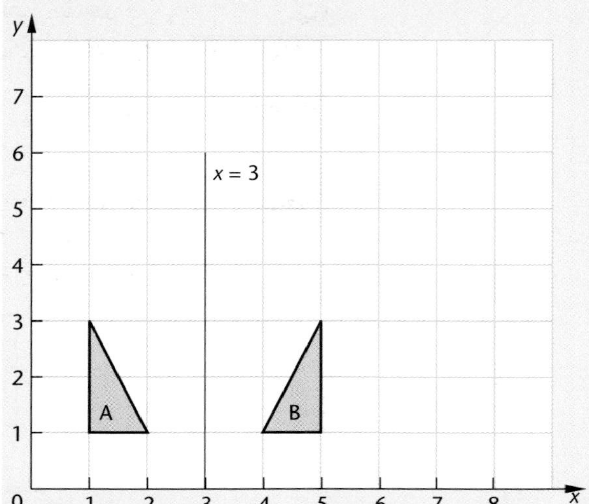

Exercise 21.1

1 For each part

- copy the diagram carefully, making it larger if you wish.
- reflect the shape in the mirror line.

a)

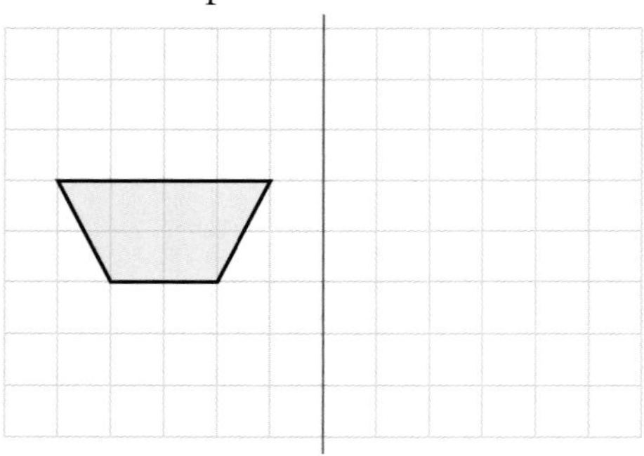

b)

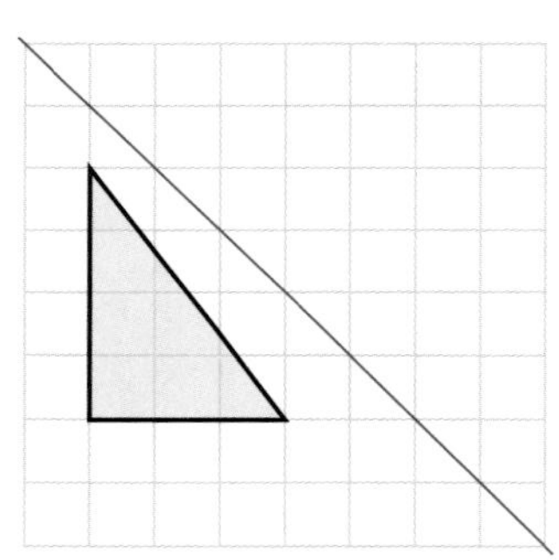

c)

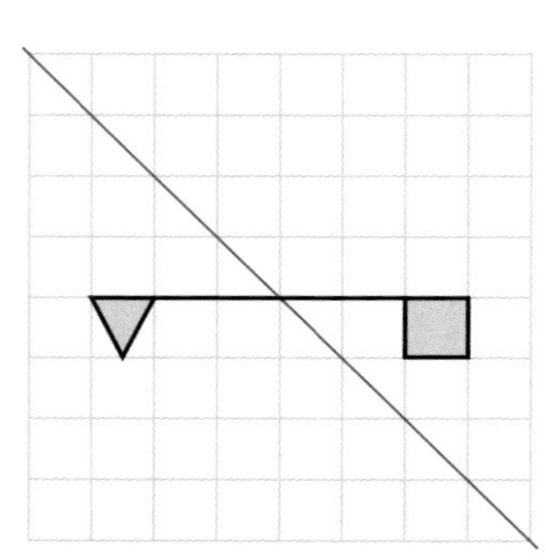

2 Copy the diagram.

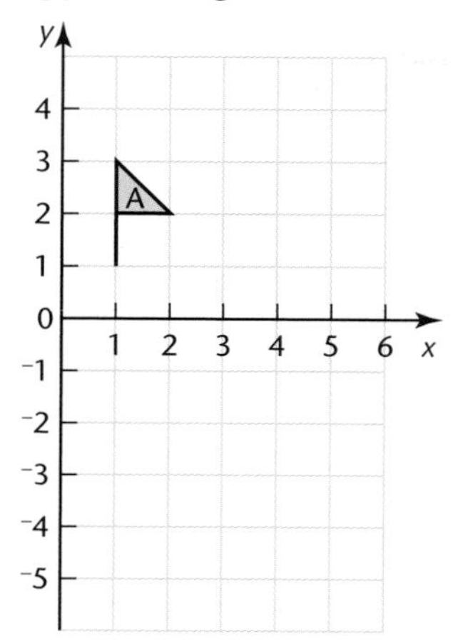

a) Reflect flag A in the line $x = 3$. Label the image B.
b) Reflect flag A in the line $y = ^{-}1$. Label the image C.

3 Copy the diagram.

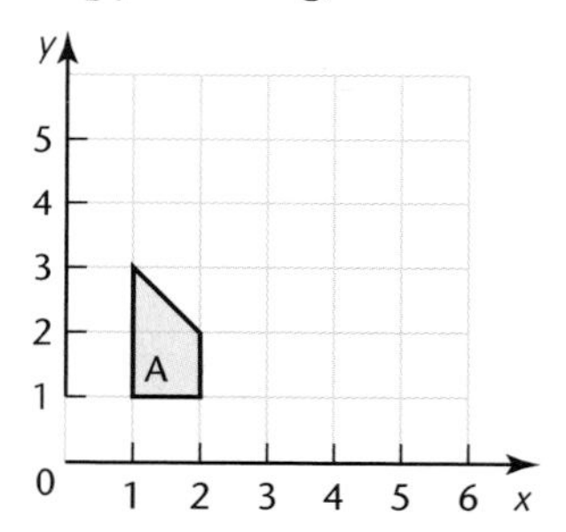

a) Reflect trapezium A in the line $y = 3$. Label the image B.
b) Reflect trapezium A in the line $x = 2$. Label the image C.

4 Draw a pair of axes and label them $^{-}4$ to 4 for x and y.
a) Draw a triangle with vertices at (1, 0), (1, $^{-}2$) and (2, $^{-}2$). Label it A.
b) Reflect triangle A in the line $y = 1$. Label it B.
c) Reflect triangle B in the line $y = x$. Label it C.

5 Draw a pair of axes and label them $^{-}4$ to 4 for x and y.
a) Draw a triangle with vertices at (1, 1), (2, 3) and (3, 3). Label it A.
b) Reflect triangle A in the line $y = 2$. Label it B.
c) Reflect triangle A in the line $y = ^{-}x$. Label it C.

Describing transformations

As well as carrying out transformations, you also need to be able to describe a transformation that has been done.

If it is a reflection, you need to describe the mirror line.

If it is a rotation, you need to give the centre of rotation and the angle, making it clear whether the rotation is clockwise or anticlockwise.

If it is a translation, you need to say how far the shape has moved and in what direction.

Example 3

For each of these diagrams, describe fully the transformation of T to T′.

a)

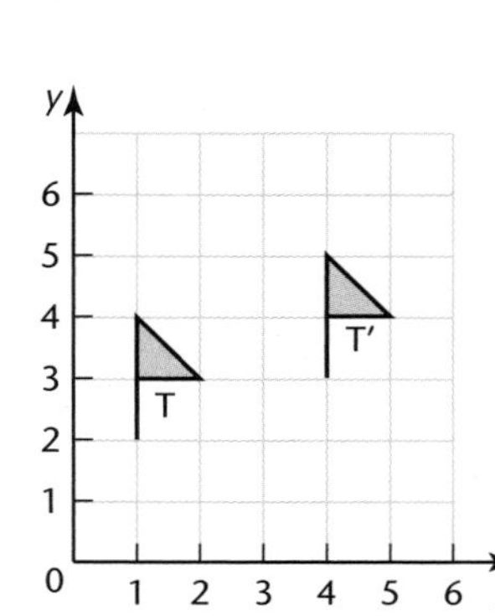

b)

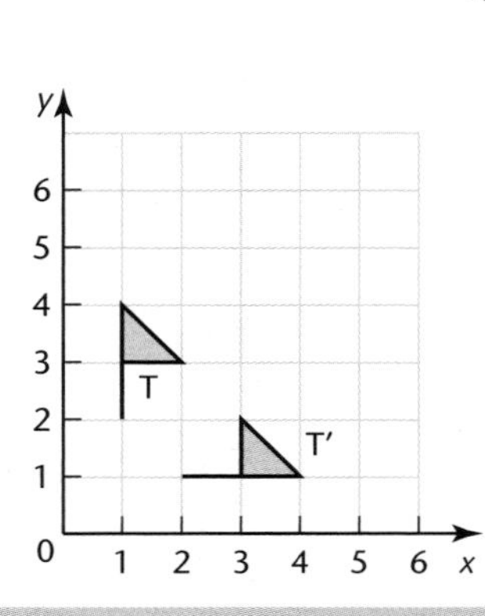

c)

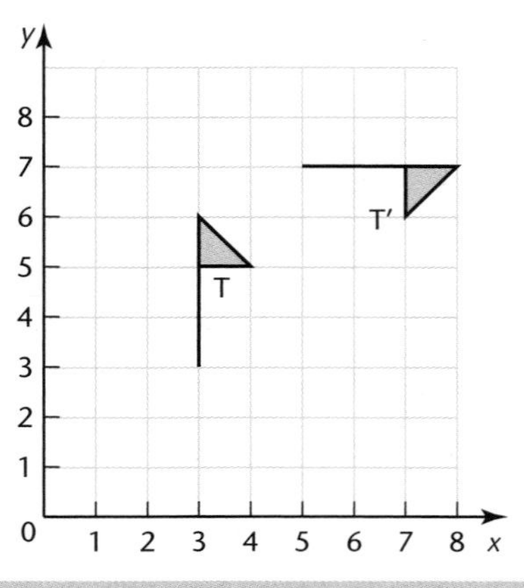

Solution

a) The image is the same way round and the same way up as the original shape, so it is a translation.

It has moved 3 squares to the right and 1 up. Check by counting squares.

This is a translation of three squares to the right and one up or $\begin{pmatrix} 3 \\ 1 \end{pmatrix}$.

b) The image is the opposite way round from the original shape, so it is a reflection.

To find the mirror line, draw lines between corresponding points in the original shape and the image.

Then draw a line through the midpoints of these joining lines.

Then label the line or give its equation.

This is a reflection in the line AB or y = x.

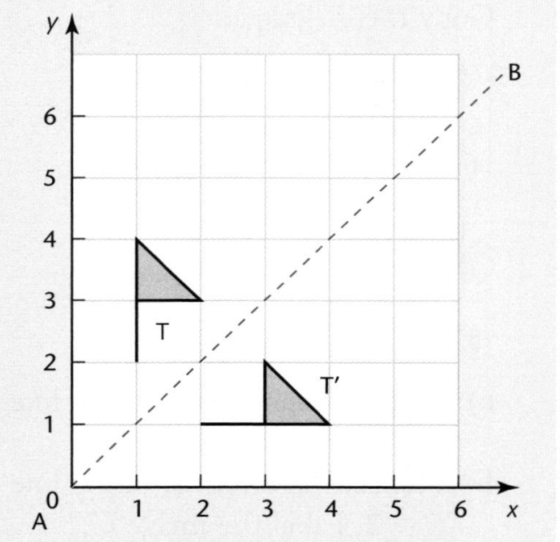

c) The image is the same way round but is not the same way up as the original shape, so it is a rotation.

The angle of rotation is 90° clockwise.

Clockwise rotations can be described as negative. You could say this is a rotation of –90°.

Use tracing paper and a pencil or compass point to find the centre of rotation. Trace flag T and use the pencil or compass point to hold the tracing to the diagram at a point.

Rotate the tracing paper and see if the tracing fits over flag T′.

Keep trying different points until you find the centre of rotation.

Here, the centre of rotation is (6, 4).

This is a rotation of 90° clockwise about (6, 4).

Foundation Silver/ Higher Initial

Creating patterns

Successive transformations can be made on a shape, either in a line or around a point, to create patterns.

Activity 1

a) Reflect the triangle in the line $y = 0$. Then reflect the image in $y = {}^-x$, its image in $x = 0$ and its image in $y = x$ and so on to make the full pattern.

T_1 is the reflection of T in $y = 0$, T_2 is the reflection of T_1 in $y = {}^-x$ and so on.

b) Use a similar method to create your own patterns or use the drawing menu of a word-processing program on a computer to draw shapes, then copy, paste and transform them.

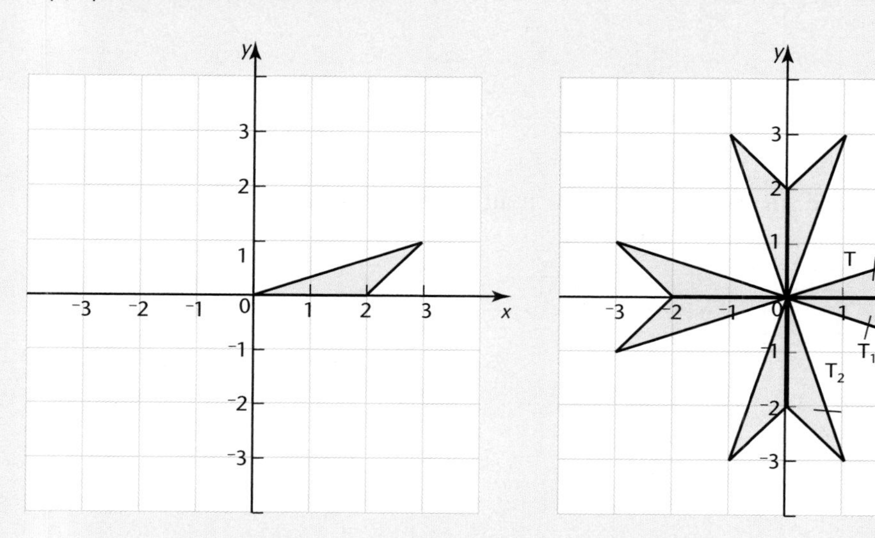

Activity 2

a) Reflect the shape in the line AB, then in a vertical line two squares to the right of AB, and so on to give a strip pattern.

Show at least eight shapes.

b) Draw a different shape and use reflection to create your own strip pattern.

Exercise 21.2

1 For each part of this question, first copy the diagram on to squared paper.

a) Rotate this shape through a half-turn about the origin.

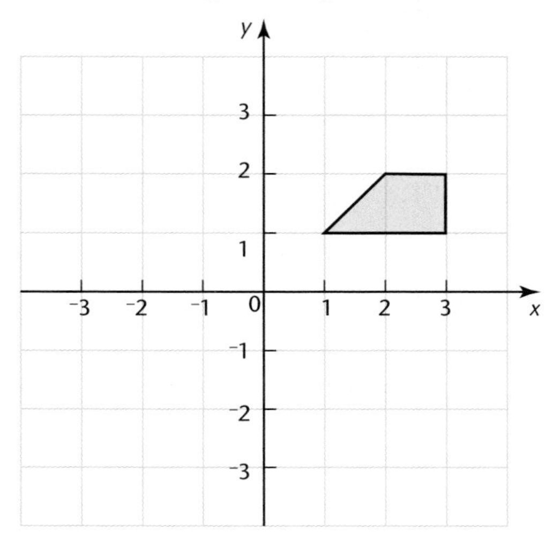

b) Rotate this shape through a quarter-turn anticlockwise about A.

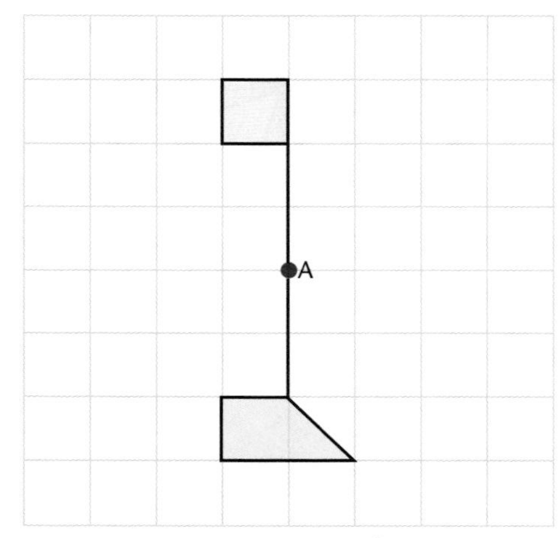

c) Translate this shape by $\begin{pmatrix} -2 \\ -1 \end{pmatrix}$.

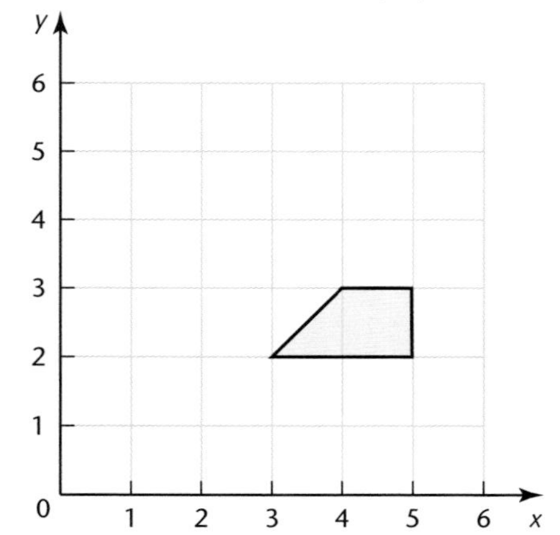

d) Translate this shape by $\begin{pmatrix} 5 \\ 0 \end{pmatrix}$.

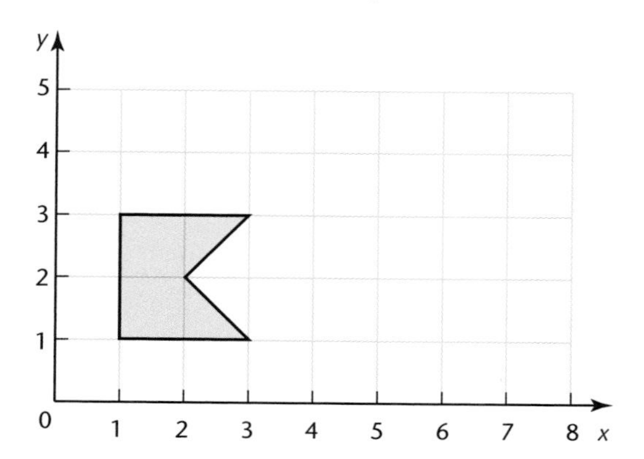

e) Rotate this shape through a three-quarter-turn anticlockwise about the origin.

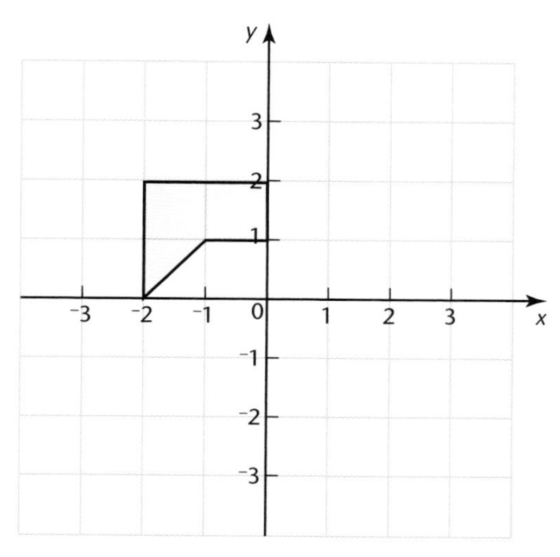

f) Translate this shape by $\begin{pmatrix} 3 \\ 2 \end{pmatrix}$.

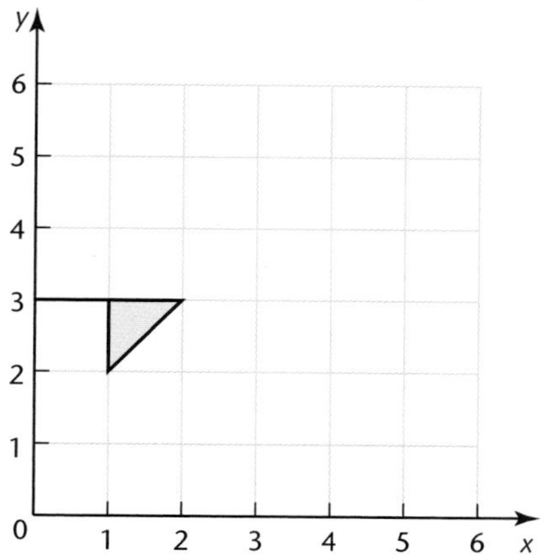

g) Rotate this shape through a quarter-turn anticlockwise about A.

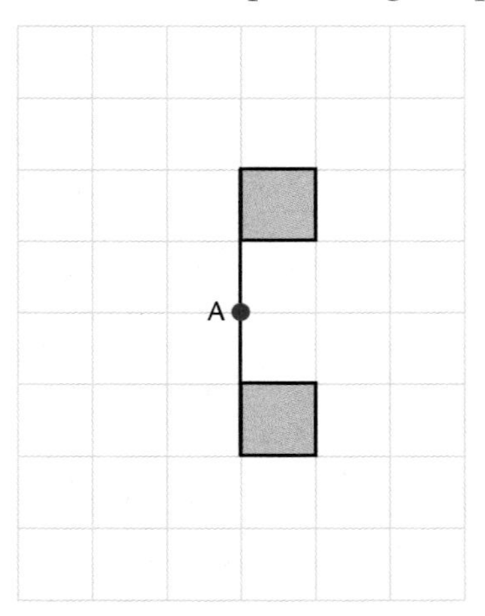

h) Translate this shape by $\begin{pmatrix} ^-2 \\ 4 \end{pmatrix}$.

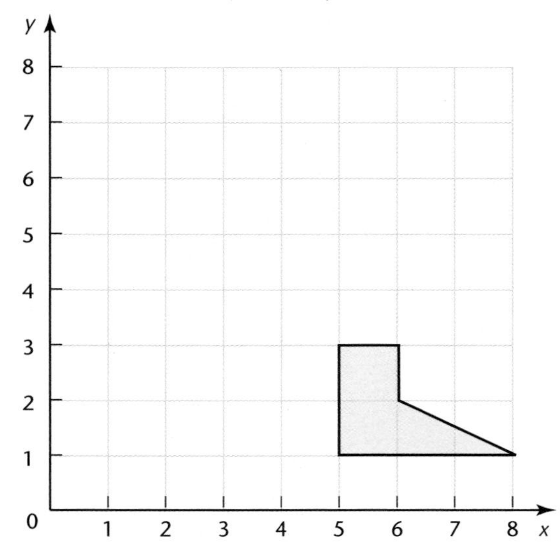

i) Rotate this shape through a quarter-turn anticlockwise about the origin.

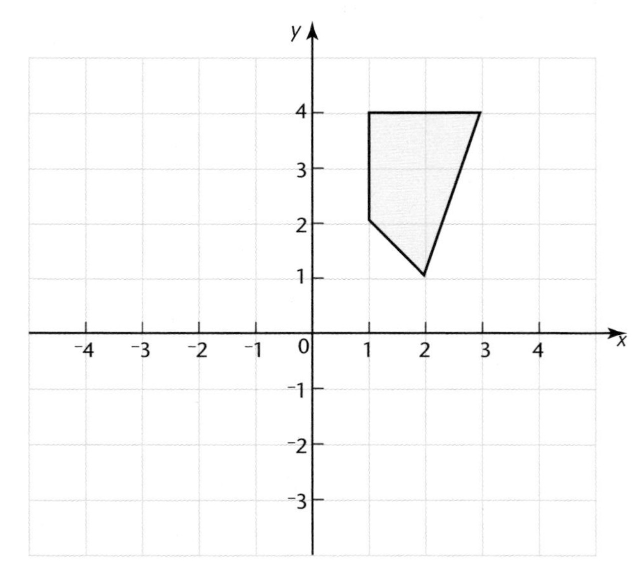

2 Describe fully the transformation that maps A on to B.

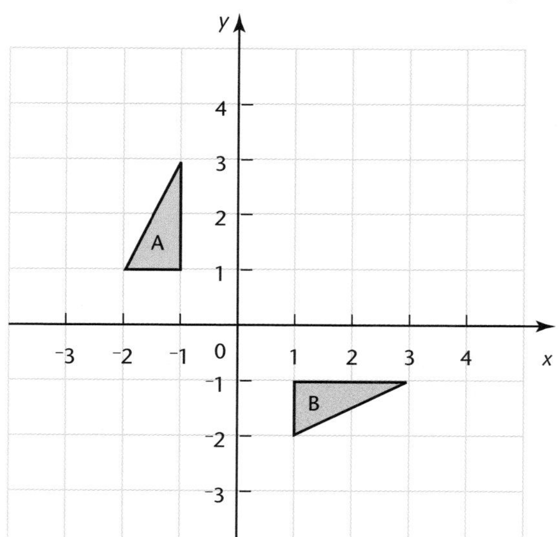

3 Describe fully the transformation that maps A on to B.

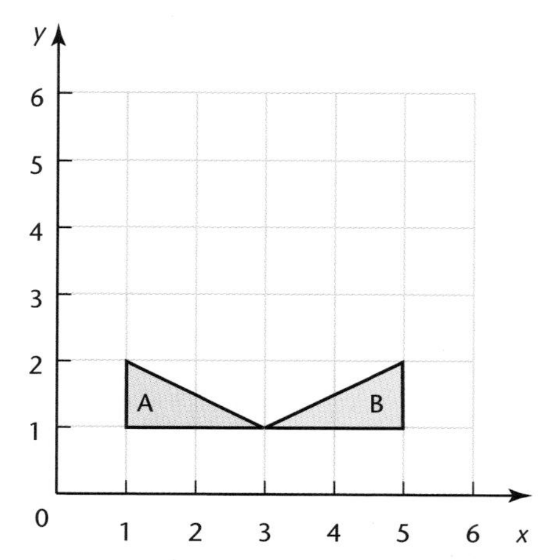

4 Describe fully the transformation that maps A on to B.

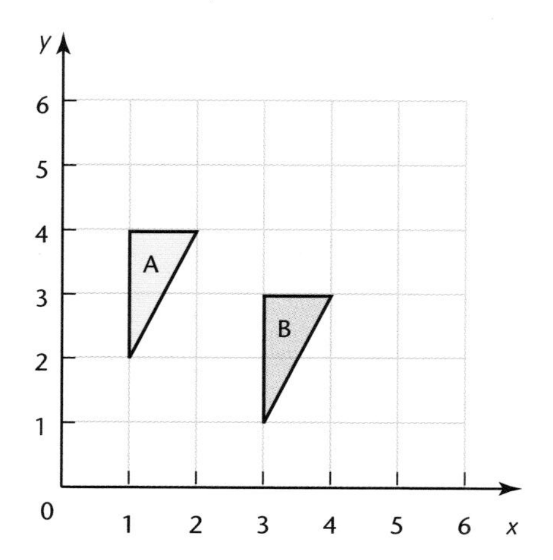

5 Draw a set of axes for x and y and label them both from $^-4$ to 4.

a) Plot the points ($^-2$, 1), ($^-1$, 1) and ($^-1$, 3), then join them to make a triangle. Label it A.

b) Rotate A through a half-turn about the origin and label the image B.

c) Reflect A in the y-axis and label the image C.

d) Describe fully the transformation that maps B on to C.

6 Draw a set of axes for x and y and label them both from $^-4$ to 4.

a) Plot the points (0, 0), (1, 3) and (1, 1), then join them to make a triangle.

b) Reflect the triangle in the line $y = x$.

c) Rotate the shape made by the two triangles by a quarter-turn anticlockwise about the origin.

d) Reflect all the shapes you now have in the x-axis to complete the pattern.

7 Describe fully the single transformation that maps

a) triangle A on to triangle B.

b) triangle A on to triangle C.

c) triangle A on to triangle D.

d) triangle B on to triangle D.

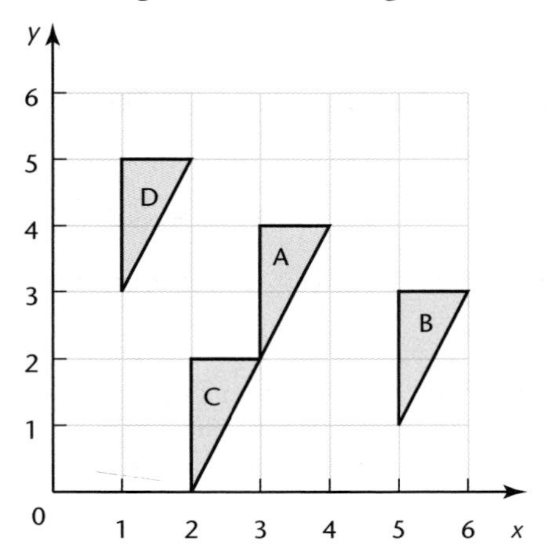

8 Describe fully the single transformation that maps

a) flag A on to flag B.

b) flag A on to flag C.

c) flag B on to flag D.

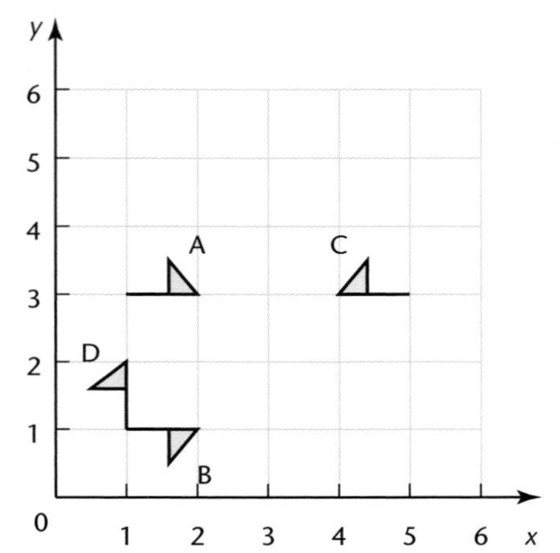

9 Describe fully the single transformation that maps
- **a)** flag A on to flag B.
- **b)** flag A on to flag C.
- **c)** flag A on to flag D.
- **d)** flag A on to flag E.
- **e)** flag A on to flag F.
- **f)** flag E on to flag G.
- **g)** flag B on to flag E.
- **h)** flag C on to flag D.

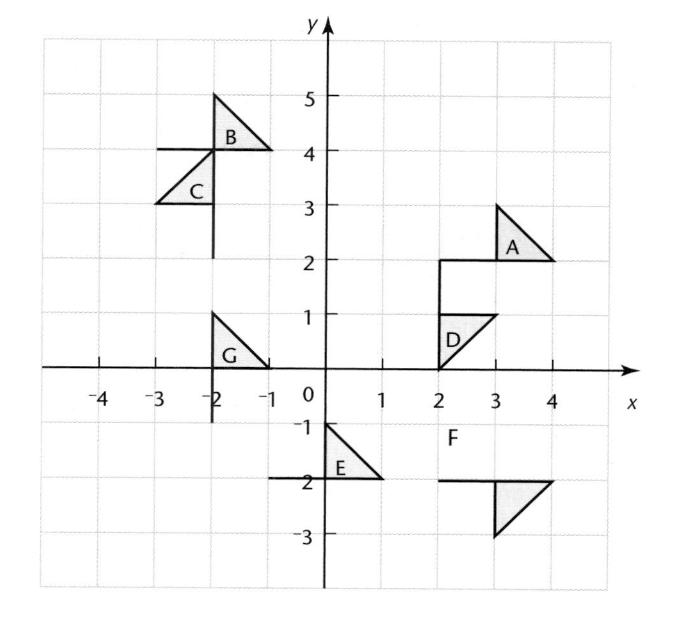

Key Ideas

- When a shape is reflected, rotated or translated, it stays exactly the same shape. The shape and its image are congruent.
- In a translation, every point on the shape moves the same distance, in the same direction.
- When describing a translation, you need to define the movement to the left or right and up or down. A column vector can be used.
- In a rotation, all the points move through the same angle about the same centre.
- When describing a rotation, you need to identify the centre, the angle and the direction (anticlockwise is positive).
- In a reflection, each point on the shape and the corresponding point on its image are the same distance from the mirror line, but on opposite sides.
- When describing a reflection, you need to define the mirror line.

CHAPTER 11

Enlargement

YOU WILL LEARN ABOUT	YOU SHOULD ALREADY KNOW
o Drawing and describing enlargements o Finding the centre and scale factor of enlargements	o How to plot points and read the coordinates of a point o How to use an angle measurer or protractor

Enlargements

If you draw an enlargement of a shape, this **image** is the same shape as the original **object**, but it is larger or smaller. You may just be asked to draw a '3 times enlargement' of a simple shape, without being given any other information. In this case copy the shape, making each side three times as long as the original one.

Example 1

Make a '2 times enlargement' of this shape.

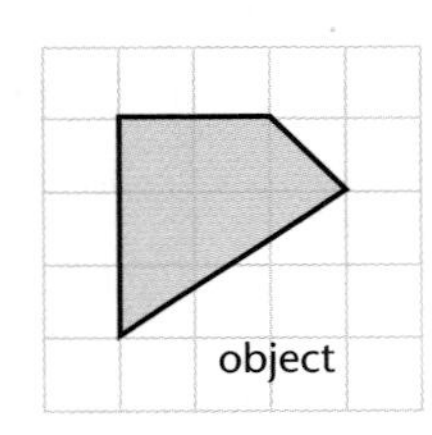

Solution

The enlargement can be drawn anywhere, but make sure it will fit on the grid. Make each of the sides twice as long as in the original shape.

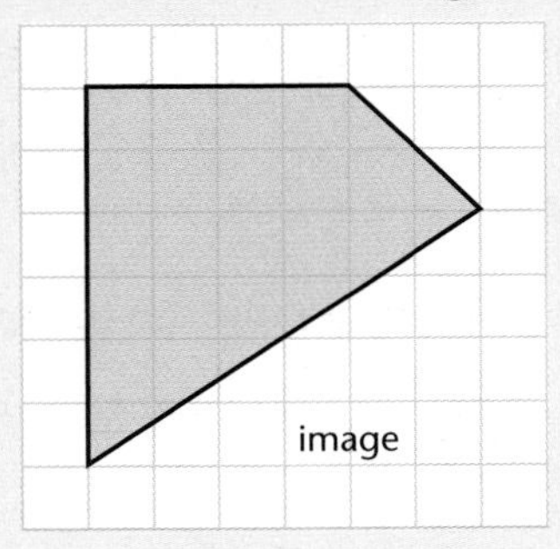

In Example 1, each side in the image is twice the length of the corresponding side in the original.

The enlargement and the original are not congruent, because they are different sizes, but they are **similar**.

More usually, you will be asked to draw an enlargement from a given point. In this case the enlargement has to be the correct size and also be in the correct position. The given point is the **centre of enlargement**.

For a '2 times enlargement', each point in the image must be twice as far from the centre as the corresponding point in the original is.

If the centre is on the original shape, that point will not move and the enlargement will overlap the original shape.

The number used to multiply the lengths for the enlargement is the **scale factor**.

You may be asked to describe an enlargement. To do this you must give the scale factor and the centre of enlargement.

Example 2

Enlarge this shape by scale factor 3 with the origin as the centre of enlargement.

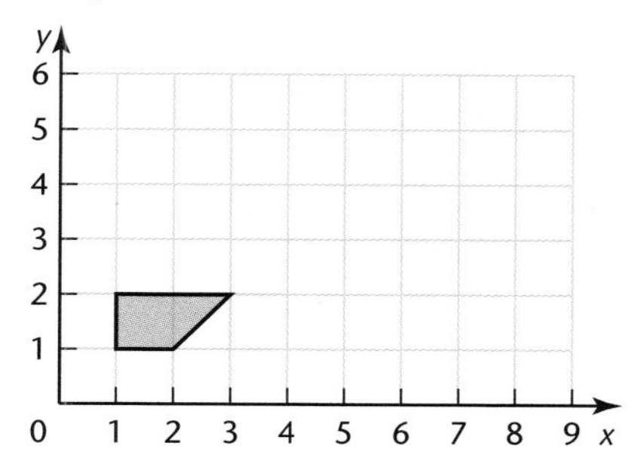

Solution

Scale factor 3 means this is a '3 times enlargement', so each point in the image is three times as far from the origin as the corresponding point in the original shape.

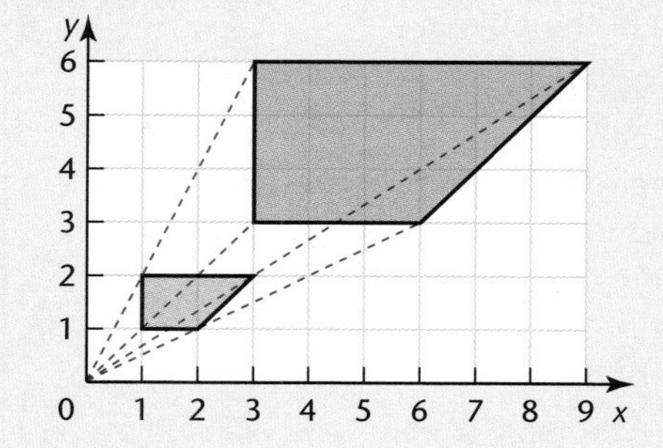

Activity 1

- On plain paper, draw a simple shape such as this.
- Choose a point on the paper to be the centre of enlargement.
- Draw lines from the centre to enlarge your shape by a scale factor of 2.
- Measure the distances from the centre accurately to achieve a good result, since any errors are enlarged!
- Compare your diagram with others in the class and look at how the position of the image changes depending on the position of the centre.

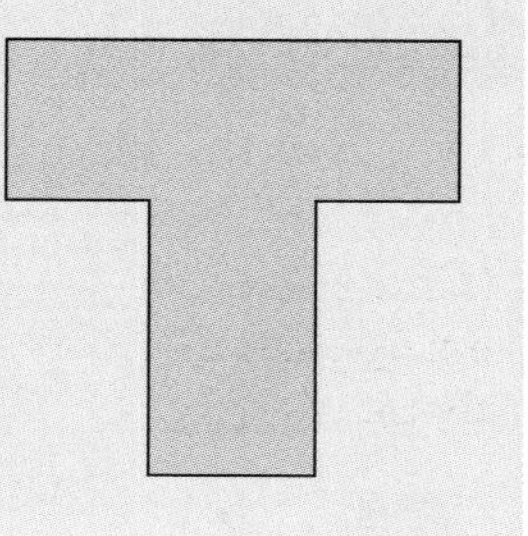

Example 3

Find the scale factor and the centre of enlargement of these shapes.

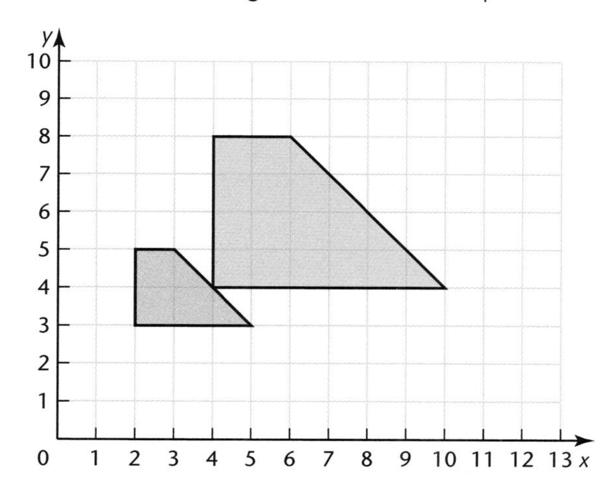

Solution

You find the scale factor by measuring corresponding sides in the object and image.

The lengths of the sides of the image are twice those of the object.

The scale factor is 2.

To find the centre of enlargement, you join the corresponding corners of the two shapes and extend the lines until they cross.
The point where they cross is the centre of the enlargement.

The centre of enlargement is the point (0, 2).

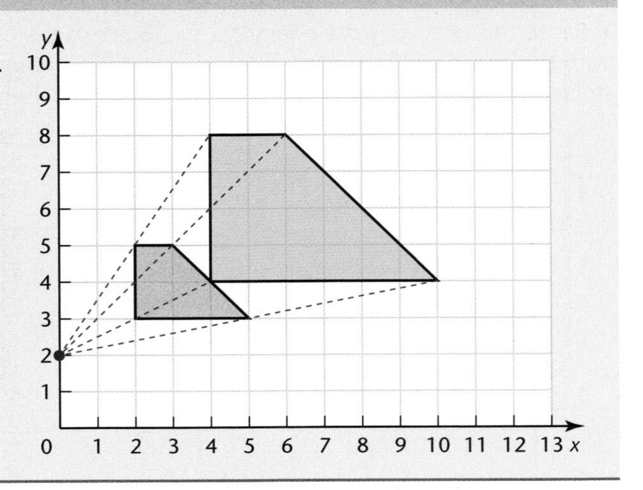

Exercise 1.1

1 Copy the diagram. Draw a '4 times enlargement' of the shape.

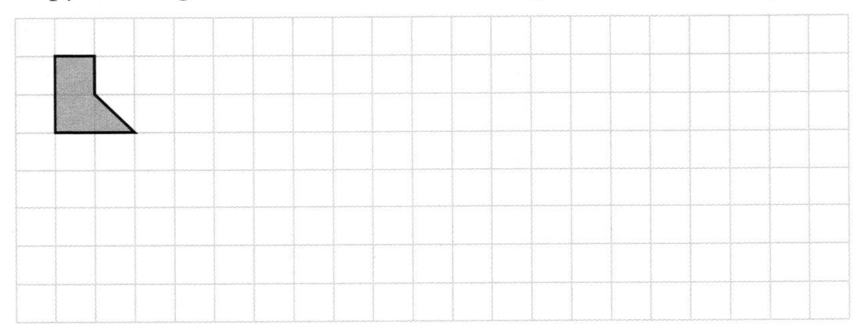

2 Copy the diagram. Draw a '2 times enlargement' of the shape.

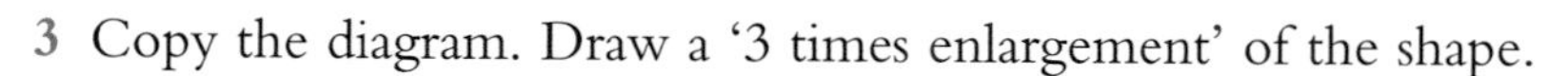

3 Copy the diagram. Draw a '3 times enlargement' of the shape.

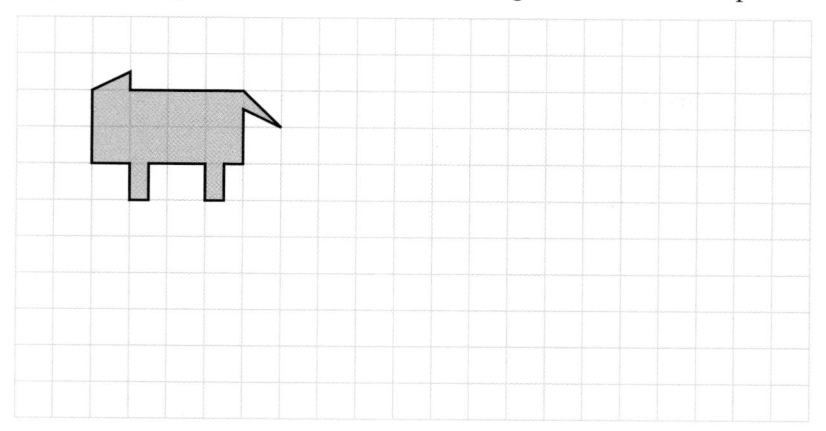

4 Copy the diagram. Draw a '3 times enlargement' of the shape.

5 Copy the diagram. Enlarge the shape by scale factor 2, centre the origin.

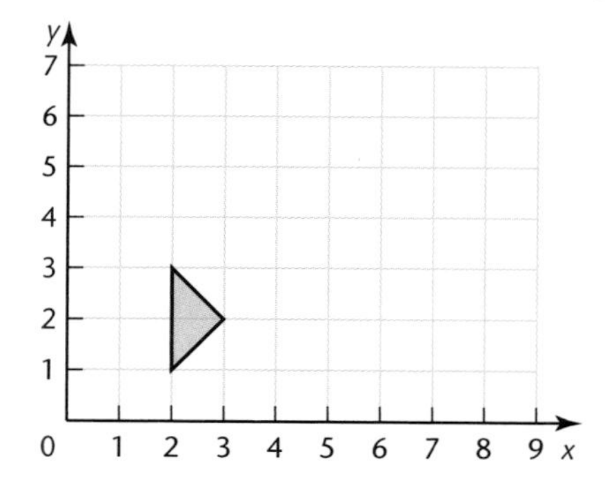

Foundation Gold/ Higher Bronze

6 Copy the diagram. Enlarge the shape by scale factor 3, centre the origin.

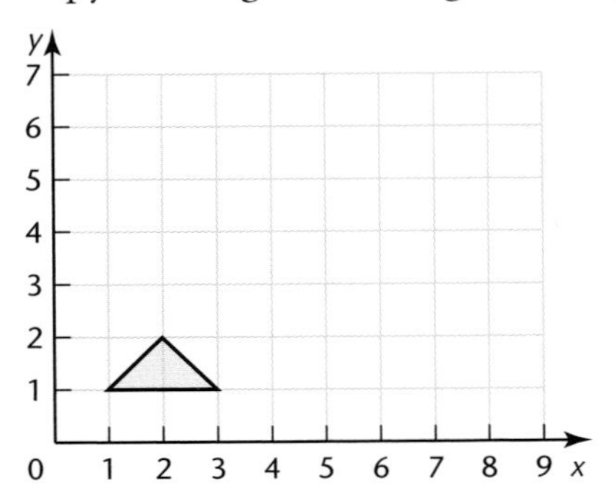

7 Copy the diagram. Enlarge the shape by scale factor 2, centre the point A.

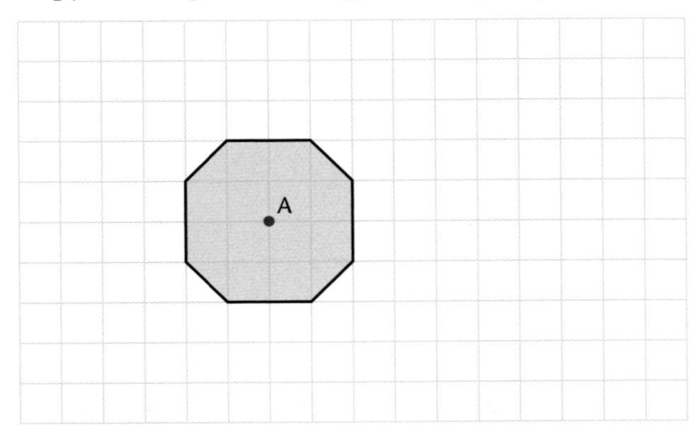

8 Copy the diagram. Enlarge the shape by scale factor 2, centre the point A.

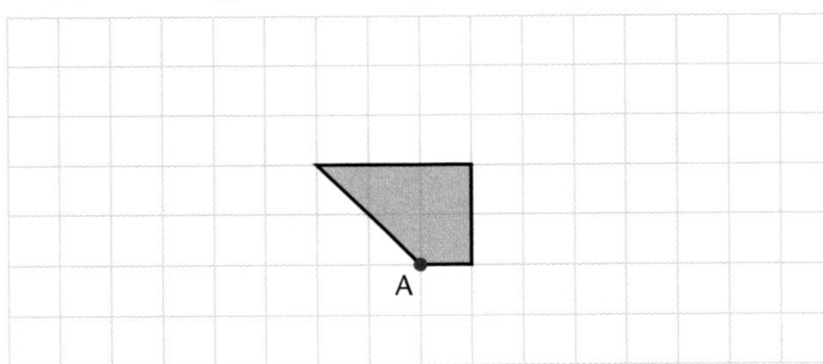

9 Copy the diagram. Enlarge the shape by scale factor 3, centre the point A.

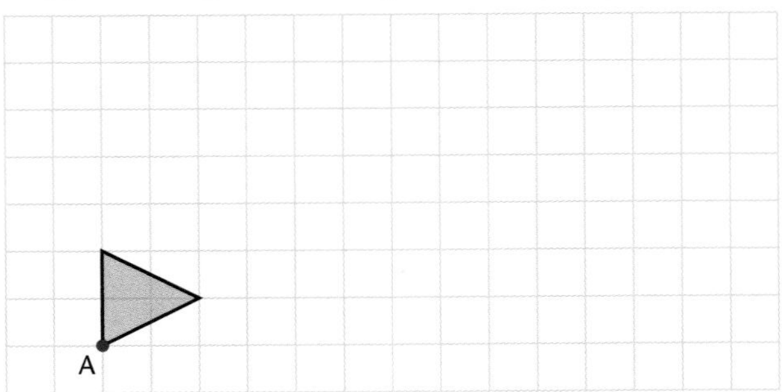

10 Copy the diagram. Enlarge the shape by scale factor 3, centre the point A.

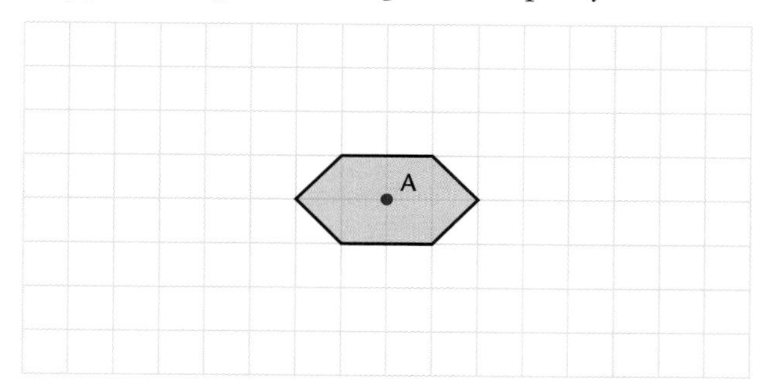

11 Copy the diagram. Enlarge the shape by scale factor 2, centre the point (2, 1).

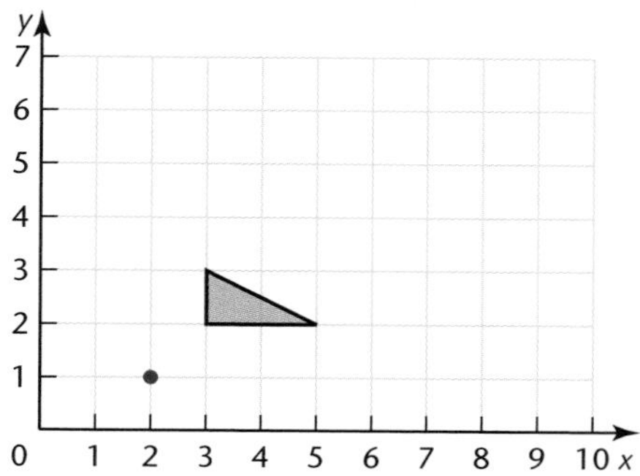

12 Copy the diagram. Enlarge the shape by scale factor 2, centre the point (2, 2).

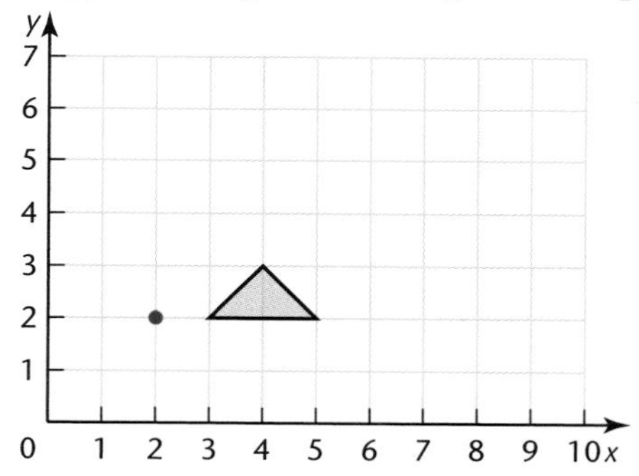

13 Describe the transformation that maps A on to B.

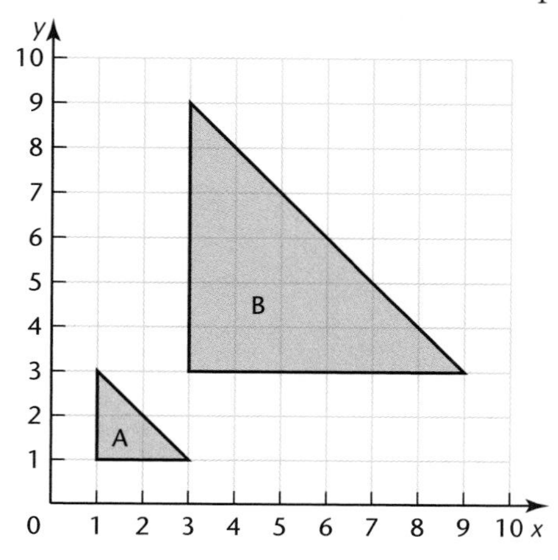

Foundation Gold/
Higher Bronze

14 Describe the transformation that maps A on to B.

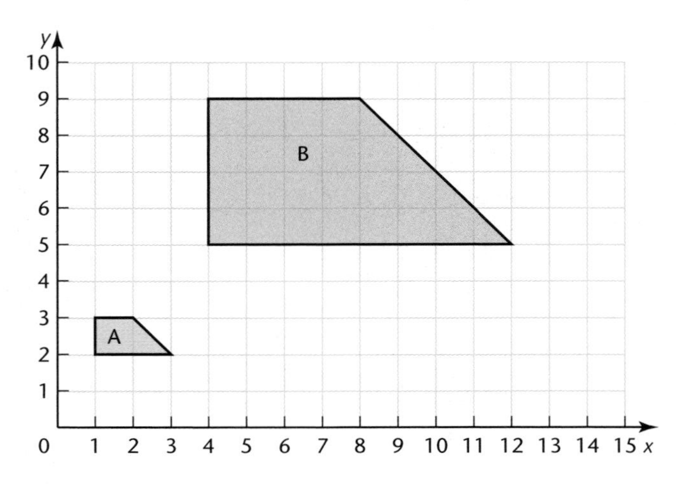

15 Describe the transformation that maps P on to Q.

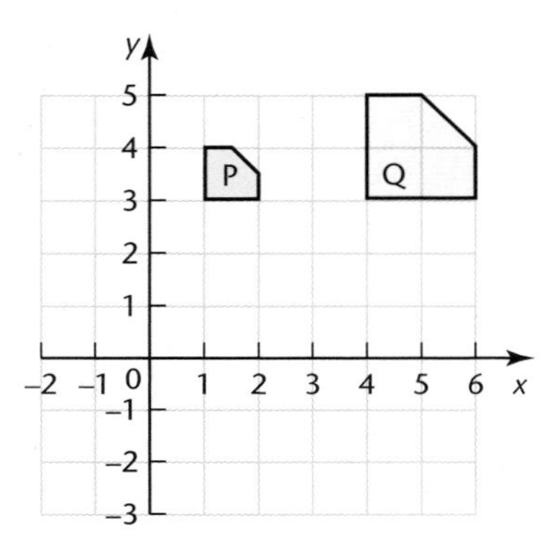

16 Describe the transformation that maps T on to W.

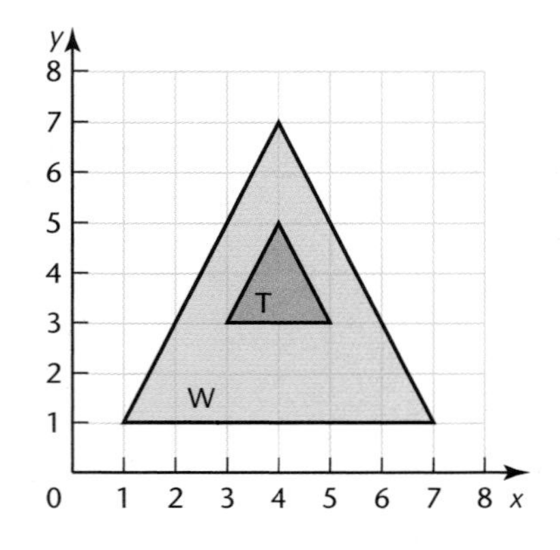

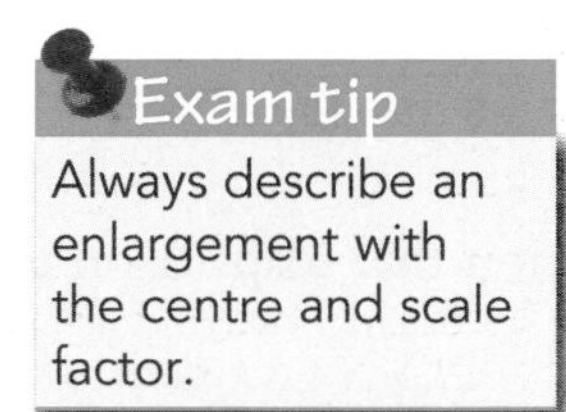

Exam tip

Always describe an enlargement with the centre and scale factor.

Activity 2

a) Draw a rectangle with side lengths of 2 cm and 3 cm.

Draw an enlargement of the rectangle with scale factor 2.

b) Find the perimeters of the original rectangle and its image. What do you notice?

c) Find the areas of the original rectangle and its image. What do you notice?

d) What would happen to the perimeter if the rectangle was enlarged with scale factor 3?

e) What would happen to the area if the rectangle was enlarged with scale factor 3?

Challenge 1

Two lemonade bottles are similar. One has height 30 cm and the other 15 cm.

a) The circumference of the larger bottle is 25 cm. What is the circumference of the smaller bottle?

b) The area of the label on the smaller bottle is 20 cm^2.

Explain why the area of the larger label is not 40 cm^2?

Activity 3

a) Draw a pair of axes and label them 0 to 6 for x and y.

Draw a triangle with vertices at (1, 2), (3, 2) and (3, 3). Label it A.

Enlarge the triangle by scale factor 2, with the origin as the centre of enlargement. Label it B.

b) **(i)** Think about what happens to the length of the sides of an object when it is enlarged by scale factor 2.

What do you think will happen to the length of the sides of an object if it is enlarged by scale factor $\frac{1}{2}$?

(ii) Think about the position of the image when an object is enlarged by scale factor 2. What happens to the distance between the centre of enlargement and the object?

What do you think will be the position of the image if an object is enlarged by scale factor $\frac{1}{2}$?

c) Draw a pair of axes and label them 0 to 6 for x and y.

Draw a triangle with vertices at (2, 4), (6, 4) and (6, 6). Label it A.

Enlarge the triangle by scale factor $\frac{1}{2}$, with the origin as the centre of enlargement. Label it B.

d) Compare your diagram with the diagram you drew in part **a)**.

What do you notice?

When the scale factor is a fraction, the image is smaller than the object.
This is still called an enlargement.

Example 4

Draw a pair of axes and label them 0 to 8 for both x and y.
Draw a triangle with vertices at P(5, 1), Q(5, 7) and R(8, 7).
Enlarge triangle PQR by scale factor $\frac{1}{3}$, centre C(2, 1).

Solution

The sides of the enlargement are $\frac{1}{3}$ the lengths of the original.
The distance from the centre of enlargement, C, to P is 3 across.
So the distance from C to P′ is $3 \times \frac{1}{3} = 1$ across.

The distance from C to Q is 3 across and 6 up.
So the distance from C to Q′ is $3 \times \frac{1}{3} = 1$ across and $6 \times \frac{1}{3} = 2$ up.

The distance from C to R is 6 across and 6 up.
So the distance from C to R′ is $6 \times \frac{1}{3} = 2$ across and $6 \times \frac{1}{3} = 2$ up.

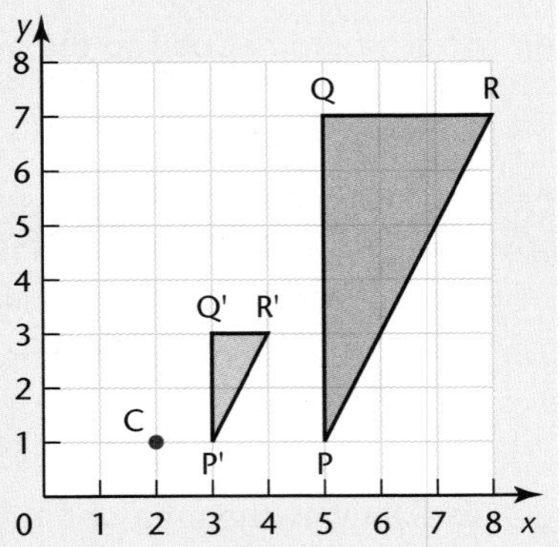

Exercise 1.2

1 Draw a pair of axes and label them 0 to 6 for both x and y.
   a) Draw a triangle with vertices at (4, 2), (6, 2) and (6, 6). Label it A.
   b) Enlarge triangle A by scale factor $\frac{1}{2}$, with the origin as the centre of enlargement. Label it B.
   c) Describe fully the single transformation that maps triangle B on to triangle A.

2 Draw a pair of axes and label them 0 to 8 for both x and y.
   a) Draw a triangle with vertices at (4, 5), (4, 8) and (7, 8). Label it A.
   b) Enlarge triangle A by scale factor $\frac{1}{3}$, with centre of enlargement (1, 2). Label it B.
   c) Describe fully the single transformation that maps triangle B on to triangle A.

3 Draw a pair of axes and label them 0 to 8 for both x and y.
   a) Draw a triangle with vertices at (0, 2), (1, 2) and (2, 1). Label it A.
   b) Enlarge triangle A by scale factor 4, with the origin as the centre of enlargement. Label it B.
   c) Describe fully the single transformation that maps triangle B on to triangle A.

4 Draw a pair of axes and label them 0 to 8 for both x and y.
   a) Draw a triangle with vertices at (4, 3), (4, 5) and (6, 2). Label it A.
   b) Enlarge triangle A by scale factor $1\frac{1}{2}$, with centre of enlargement (2, 1). Label it B.
   c) Describe fully the single transformation that maps triangle B on to triangle A.

5 At 'Photos-R-Us' photographs can be printed in three sizes.

Photo type	Print size (centimetres)
Mini	10 by 15
Mega	20 by 30
Jumbo	15 by 20

a) What is the scale factor of the enlargement from 'Mini' to 'Mega' print sizes?

b) Is 'Jumbo' print size an enlargement of 'Mini' print size? Show clearly how you decide.

c) 'Ultra' sized prints are 3 times enlargements of 'Mini' sized prints. What is the print size of 'Ultra' prints?

6 A shape is enlarged by a scale factor of 2. The image is then enlarged by a scale factor of 3.
What single enlargement will enlarge the original shape to the final image?

Key Ideas

- When a shape is enlarged by, for example, a scale factor of 2, from a given centre, each point on the enlargement will be 2 times as far away from the centre as the corresponding point on the original shape.
- A shape and its enlargement are **similar**.
- To find the scale factor of an enlargement, divide the length of a side of the image by the length of the corresponding side of the object.
- To find the centre of enlargement, join the corresponding corners of the two shapes and extend the lines until they cross. The point where they cross is the centre of the enlargement.
- The scale factor may be a fraction. In this case, the image may be smaller than the object.

CHAPTER 2

Percentages 2

YOU WILL LEARN ABOUT	YOU SHOULD ALREADY KNOW
o Changing between fractions, decimals and percentages o Using percentages to compare proportions o Finding the percentage of an increase or decrease o Increasing and decreasing a quantity by a percentage	o How to write and read fractions, decimals and percentages o How to multiply and divide numbers and decimals by 100

Percentages

You should also know that percentage means 'out of 100'. For example

'75% of cat owners surveyed said their cats preferred fish-flavoured treats.'

75% means that 75 out of every 100 people surveyed said their cats preferred fish-flavoured treats.

'40% of people asked said they drank coffee at breakfast.'

40% means that 40 people out of every 100 people drank coffee at breakfast.

Find as many examples of percentages as you can from newspapers, magazines and advertising material.

The equivalence of fractions, decimals and percentages

Seven out of 10 households buy their milk at supermarkets.

What is this as a percentage?

You can work this out either by using equivalent fractions

$\frac{7}{10} = \frac{70}{100} = 70\%$ (multiplying the numerator and the denominator by 10)

or by changing the fraction to a decimal

$\frac{7}{10} = 0.7 = 70\%$.

To change a fraction into a decimal, remember to divide the numerator by the denominator, as shown in Example 1.

Example 1	Solution
Convert $\frac{3}{8}$ to **a)** a decimal. **b)** a percentage.	**a)** $\frac{3}{8} = 3 \div 8 = 0.375$ **b)** $0.375 \times 100 = 37.5\%$ It is possible to go straight from the fraction to the percentage using multiplication of fractions. $\frac{3}{8} \times 100 = \frac{3}{8} \times \frac{100}{1} = \frac{300}{8} = 37.5\%$

Challenge 1

Use percentages to put each set of numbers in order, smallest first.

a) 0.3, $\frac{1}{3}$, 0.33, $\frac{4}{9}$, $\frac{5}{11}$, $\frac{2}{5}$, $\frac{2}{7}$

b) $\frac{7}{9}$, $\frac{2}{3}$, 0.7, $\frac{3}{5}$, $\frac{8}{11}$, 0.666

Exercise 2.1

1 Change each of these percentages to a fraction. Write your answers in their lowest terms.

a) 35% b) 65%
c) 8% d) 120%

2 Change each of these percentages to a decimal.

a) 16% b) 27%
c) 83% d) 7%
e) 31% f) 4%
g) 17% h) 2%
i) 150% j) 250%
k) 9% l) 12.5%

3 Change each of these decimals to a percentage.

a) 0.62 b) 0.56
c) 0.04 d) 0.165
e) 1.32 f) 0.37
g) 0.83 h) 0.08
i) 0.345 j) 1.25

4 Copy this table and complete it.

Fraction $\frac{a}{b}$	Decimal $a \div b$	Percentage = decimal × 100
$\frac{7}{10}$		
$\frac{2}{5}$		
$\frac{3}{4}$		
$\frac{1}{3}$		
$\frac{2}{3}$		

5 Change each of these fractions to a decimal.

a) $\frac{1}{100}$ b) $\frac{17}{100}$
c) $\frac{2}{50}$ d) $\frac{8}{5}$
e) $\frac{1}{8}$ f) $\frac{3}{8}$
g) $\frac{3}{20}$ h) $\frac{17}{40}$
i) $\frac{5}{16}$

6 Change each decimal you found in question **5** to a percentage.

7 Change each of these fractions into a percentage. Give your answers correct to 1 decimal place.

a) $\frac{1}{6}$ b) $\frac{5}{6}$
c) $\frac{5}{12}$ d) $\frac{1}{12}$
e) $\frac{3}{70}$

8 On Wednesday, 0.23 of the population watched EastEnders.
What percentage is this?

9 At a matinée, $\frac{4}{5}$ of the audience at the pantomime were children.
What percentage is this?

10 In a survey, $\frac{3}{10}$ of students liked cheese and onion crisps.
What percentage is this?

11 A kilometre is $\frac{5}{8}$ of a mile.
What percentage is this?

12 The winning candidate in an election gained $\frac{7}{12}$ of the votes.
What percentage is this? Give your answer correct to the nearest 1%.

13 Nicola spends $\frac{3}{7}$ of her allowance on sweets and drinks.
What percentage is this? Give your answer correct to the nearest 1%.

14 Imran has a part-time job.
He saves $\frac{2}{9}$ of his wages.
What percentage is this? Give your answer correct to the nearest 1%.

For the non-calculator section of the examination papers it is worth learning some basic equivalents.

$\frac{1}{2} = 0.5 = 50\%$	$\frac{1}{3} = 0.333\ldots = 33.3\ldots\%$	$\frac{1}{5} = 0.2 = 20\%$
$\frac{1}{4} = 0.25 = 25\%$	(33% to the nearest 1%)	$\frac{2}{5} = 0.4 = 40\%$
$\frac{3}{4} = 0.75 = 75\%$	$\frac{2}{3} = 0.666\ldots = 66.6\ldots\%$	$\frac{3}{5} = 0.6 = 60\%$
	(67% to the nearest 1%)	$\frac{4}{5} = 0.8 = 80\%$

Notice the rounding of $\frac{1}{3}$ and $\frac{2}{3}$. It is a common error to think that $\frac{1}{3} = 0.3 = 30\%$ and $\frac{2}{3} = 0.66 = 66\%$ or even $\frac{2}{3} = 0.6 = 60\%$.

Percentage increases and decreases

In this section you will use methods you have learned previously to solve problems.

Finding the percentage

Percentage increases and decreases are worked out as percentages of the original amount, not the new amount. Percentage profit or loss is worked out as a percentage of the cost price, not the selling price.

$$\text{Percentage change} = \frac{\text{change}}{\text{original amount}} \times 100$$

Example 2	Solution
An art dealer buys a painting for £45 and sells it for £72. What percentage profit is this?	Profit = £72 – £45 = £27 Percentage profit = $\frac{27}{45} \times 100 = 60\%$

Example 3	Solution
The value of a computer drops from £1200 to £700 in a year. What percentage decrease is this?	Decrease in value = £1200 – £700 = £500 Percentage decrease = $\frac{500}{1200} \times 100 = 41.7\%$ to 1 decimal place

Increase by a percentage

To increase £240 by 23% you first work out 23% of £240.

£240 × 0.23 = £55.20

Then you add £55.20 to £240.

£240 + £55.20 = £295.20

There is a quicker way to do the same calculation.

To increase a quantity by 23% you need to find the original quantity plus 23%.

You know that 23% of a number means $\frac{23}{100}$ × the number.

Similarly, 100% of a number means $\frac{100}{100}$ × the number, which is the same as the number itself.

So to increase £240 by 23% you need to find 100% of £240 plus 23% of £240.

This is the same as 123% of £240.

So the calculation can be done in one stage.

£240 × 1.23 = £295.20

The number that you multiply the original quantity by (here 1.23) is called the **multiplier**.

Example 4	Solution
Amir's salary is £17 000 per year. He receives a 3% increase. Find his new salary.	Amir's new salary is 103% of his original salary. So the multiplier is 1.03. £17 000 × 1.03 = £17 510

Decrease by a percentage

You can calculate a percentage decrease in a similar way.

Example 5	Solution
Kieran buys a DVD recorder in the sale. The original price was £225. Calculate the sale price.	A percentage decrease of 15% is the same as 100% – 15% = 85%. So the multiplier is 0.85. £225 × 0.85 = £191.25

Exercise 2.2

1 A shopkeeper buys an article for £10 and sells it for £12.
What percentage profit does she make?

2 Adam earned £5 per hour. His pay increased to £5.60 per hour.
What was the percentage increase in Adam's pay?

3 A season ticket for the car park normally costs £500 but if it is bought before 1 June it costs £360.
What percentage reduction is this?

4 Karl bought a CD for £12.50. A year later he sold it for £6.
What percentage of the value did he lose?

5 Will buys a signed photo for £22 and sells it for £28.
What is his percentage profit? Give your answer to the nearest 1%.

6 Hannah reduces the time it takes her to swim 30 lengths from 55 minutes to 47 minutes.
What percentage reduction is this? Give your answer to the nearest 1%.

7 Write down the multiplier that will increase an amount by each of these percentages.

a) 13% b) 20%
c) 68% d) 8%
e) 2% f) 17.5%
g) 150%

8 Write down the multiplier that will decrease an amount by each of these percentages.

a) 14% b) 20%
c) 45% d) 7%
e) 3% f) 23%
g) 16.5%

9 Sanjay used to earn £4.60 per hour from his Saturday job. He received a 4% increase.
How much does he earn now? Give your answer to the nearest penny.

10 In a sale, everything is reduced by 30%. Abi buys a pair of shoes in the sale. The original price was £42.
What is the sale price?

11 Ghalib's restaurant bill is £340 before he adds a tip.
What is the bill after he adds a 5% tip?

12 A shop increases its prices by 8%.
What is the new price of a skirt which previously cost £30?

13 In a sale, all the prices are reduced by 20%.
What is the sale price of an electric shaver which previously cost £27?

Activity I

VAT (Value Added Tax) is a tax added on to the price of goods and services.

a) Find out the current rate of VAT.

b) Write the current rate of VAT as a fraction and as a decimal.

Challenge 2

House prices rose by 12% in 2003, 11% in 2004 and 7% in 2005.
At the start of 2003 the price of a house was £120 000.

What was the price at the end of 2005? Give your answer to the nearest pound.

Express the price at the end of 2005 as a percentage of the price at the start of 2003.

Challenge 3

The value of an investment rose by 8% in 2004 and fell by 8% in 2005.

If the value of the investment was £3000 at the start of 2004, what was the value at the end of 2005? Express the final increase or decrease as a percentage of the original investment.

Challenge 4

The price of a camera is increased by 30%. Later, in a sale, the new price is reduced by 20%.
What percentage change is the sale price compared to the original price?

Key Ideas

- To change a fraction to a decimal, divide the numerator by the denominator.
  $\frac{a}{b} = a \div b$
- To change a decimal to a percentage, multiply by 100.
- To increase a quantity by 5%, for example, a quick way is to multiply the quantity by 1.05 (since 100 + 5 = 105).
- To reduce a quantity by 12%, for example, a quick way is to multiply the quantity by 0.88 (since 100 – 12 = 88).

CHAPTER 3 Probability 2

YOU WILL LEARN ABOUT	YOU SHOULD ALREADY KNOW
○ The link between relative frequency and probability	○ That probabilities can be expressed as fractions, decimals or percentages ○ That the total of the probabilities of all the mutually exclusive outcomes of an event is one ○ How to simplify fractions

Relative frequency and probability

It is not always possible to find probabilities from looking at equally likely outcomes. For example, you may have to work out the probability of throwing a 6 with a dice that may be biased, the probability of a young driver having an accident, or the probability that a person will visit a certain supermarket.

For this type of event you need to set up some sort of experiment, carry out a survey or look at past results.

Take the example of throwing a 6 with a dice that may be biased. For a fair (unbiased) dice, the probability of getting a 6 is $\frac{1}{6} = 0.166... = 0.17$ approximately.

If you were to throw it ten times and get four 6s, would this be evidence of bias?

The proportion of 6s is $\frac{4}{10} = 0.4$ which is very different from 0.17, but in a small sample of trials there may be runs of non-typical results. So you would not conclude that the dice was biased.

What about 10 times in 50 throws?

Here the proportion is $\frac{10}{50} = 0.2$ which is still quite a bit different from 0.17, but again you have not thrown it enough times to be sure. So you still would not conclude that the dice was biased.

What about 108 times in 600 throws?

You have thrown the dice a large number of times and the proportion of 6s is $\frac{108}{600} = 0.18$. This is too close to 0.17 to conclude that the dice was biased, now that you have thrown it so many times.

What about 100 times in 500 throws?

Now you have thrown the dice a large number of times and the proportion is $\frac{100}{500} = 0.2$. This is significantly different from 0.17, but not so much that you should conclude that the dice is biased.

The important question is, how many trials are necessary to ensure a representative result?

There is no fixed answer to this, other than 'the more the better'.

As a general rule, any event being examined should occur at least 100 times, but even this is probably a bare minimum.

In the case of a 6 occurring 100 times in 500 throws, the proportion of 6s is $\frac{100}{500}$. This fraction is called the **relative frequency.**

$$\text{Relative frequency} = \frac{\text{number of times an outcome occurs}}{\text{total number of trials}}$$

This is a measure of the proportion of the trials in which the outcome occurs. It is not itself a measure of probability. If, however, the number of trials is large enough, relative frequency can be used as an estimate of probability.

Remember that, no matter how many trials have taken place, relative frequency is still only an estimate, but in many cases it is the only method of estimating probability.

Activity 1

Copy this table and complete it by following the instructions below.

Number of trials	20	40	60	80	100
Number of heads					
Relative frequency $= \frac{\text{number of heads}}{\text{number of trials}}$					

- Toss a coin 20 times and use tally marks to record the number of times it lands heads.
- Now toss the coin another 20 times and enter the number of heads for all 40 tosses.
- Continue in groups of 20 and record the number of heads for 60, 80 and 100 tosses.
- Calculate the relative frequency of heads for 20, 40, 60, 80 and 100 tosses.

  Give your answers to 2 decimal places.

  a) What do you notice about the values of the relative frequencies?

  b) The probability of getting a head with one toss of a coin is $\frac{1}{2}$ or 0.5. Why is this?

  c) How does your final relative frequency value compare with 0.5?

Example 1

Ian carries out a survey on the colours of the cars passing his school.
His results are shown in this table.

Colour	Black	Red	Blue	White	Green	Other	Total
Number of cars	51	85	64	55	71	90	416

Use these figures to estimate the probability that the next car that passes will be

a) red. **b)** not red.

Solution

Since the number of trials is large, use relative frequency as an estimate of probability.

a) Relative frequency of a red car = $\frac{85}{416}$

Estimate of probability = $\frac{85}{416}$ or 0.204 (to 3 decimal places)

b) 416 – 85 = 331

Relative frequency = $\frac{331}{416}$

Estimate of probability = $\frac{331}{416}$ or 0.796 (to 3 decimal places)

Activity 2

Perform an experiment or conduct a survey to test one of these hypotheses.

- Every dice is unbiased so that the probability of getting any number is $\frac{1}{6}$.
- The probability of a boy or girl being born is the same, $\frac{1}{2}$.

Gather your data from your school by asking each student to state the genders of the members of their family (or your local hospital might give you information from the recent birth records). Estimate the probability of a boy or a girl being born.

Select a dice that you think has been very unfair to you, throw it as many times as you can and estimate the probability of getting a 1, a 2, a 3 …

Exercise 3.1

1 Use the figures from Ian's survey in Example 1 to estimate the probability that the next car will be

a) blue.

b) black or white.

Give your answers correct to 3 decimal places.

2 Kim Lee tossed a coin ten times and it came down heads eight times. Kim Lee said that the coin was biased towards heads.

Explain why she may not be right.

3 Solomon has a spinner in the shape of a pentagon with sides labelled 1, 2, 3, 4, 5.
Solomon spun the spinner 500 times.
The results are shown in the table.

Number on spinner	Number of times
1	102
2	103
3	98
4	96
5	101

a) What is the relative frequency of scoring
 (i) 2?
 (ii) 4?

b) Do the results suggest that Solomon's spinner is fair?
 Explain your answer.

4 In an experiment, a drawing pin is thrown. It can land either point up or point down. It lands point up 87 times in 210 throws.
Use these figures to estimate the probability that, the next time it is thrown, it will land

a) point up.

b) point down.

Give your answers correct to 2 decimal places.

5 Denise carried out a survey about crisps. She asked 400 people in the town where she lived which was their favourite flavour of crisps. The results are shown in this table.

Flavour	Number of people
Ready salted	150
Salt and vinegar	75
Cheese and onion	55
Prawn cocktail	50
Other	70

a) Explain why it is reasonable to use these figures to estimate the probability that the next person Denise asks will choose salt and vinegar.

b) Use the figures to estimate the probability that the next person Denise asks will choose
 (i) salt and vinegar.
 (ii) ready salted.
 Give your answers as fractions in their simplest form.

6 An insurance company finds that 203 drivers out of 572 in the age range 17–20 have an accident in the first year after passing their driving test.
Use these figures to estimate the probability that a driver aged 17–20 will have an accident in the first year after passing their test.

7 While Tom is standing at the bus stop he notices that five out of the 20 cars he sees passing are Fords. He says that therefore the probability that the next car will be a Ford is $\frac{1}{4}$.
Explain why he is wrong.

Foundation Gold/
Higher Bronze

8 Freya carries out a survey to find out how students in her school travel to school. She asks a random selection of 200 students. The results are shown in this table.

Method of travel	Number of students
Bus	34
Car	33
Train	23
Cycle	45
Walk	65

a) Explain why it is reasonable to estimate the probabilities of students travelling by the various methods from this survey.

b) Use these figures to estimate the probability that a student selected at random from the school
 (i) travels by bus.
 (ii) cycles.

9 Noel has two coins which he suspects may be biased.

a) He tosses the first coin 600 times and throws 312 heads.
Is there evidence to suggest that this coin is biased? Give your reasons.
If there is, estimate the probability that the next throw is a head.

b) He tosses the second coin 600 times and throws 420 heads.
Is there evidence to suggest that this coin is biased? Give your reasons.
If there is, estimate the probability that the next throw is a head.

10 The table shows the results of a survey on the type of detergent households use to do their washing.

Type of detergent	Number of households
Liquid	120
Powder	233
Tablets	85

Use these figures to estimate, correct to 2 decimal places, the probability that the next household surveyed will use

a) liquid.

b) tablets.

11 Stewart made a five-sided spinner. Unfortunately he did not make the pentagon regular. In order to find the probability of getting each of the numbers, he spun the spinner 400 times. His results are shown in this table.

Number	Frequency
1	63
2	84
3	101
4	57
5	95

Use the figures in the table to estimate the probability of the spinner landing on

a) 1.

b) 3.

c) an even number.

12 A shopkeeper noticed from his till roll that, out of 430 customers that day, 82 had spent over £10.

Use these figures to estimate the probability that his next customer will spend £10 or less.

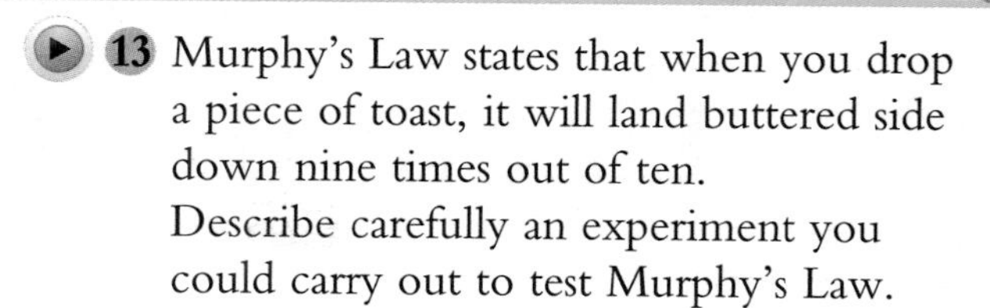

13 Murphy's Law states that when you drop a piece of toast, it will land buttered side down nine times out of ten.
Describe carefully an experiment you could carry out to test Murphy's Law.

Challenge 1

Work in pairs.
Put ten counters, some red and the rest white, into a bag.
Challenge your partner to work out how many counters there are of each colour.

You need to devise an experiment with 100 trials. At the start of each trial, all ten counters must be in the bag.

Key Ideas

- Relative frequency = $\dfrac{\text{number of times an event occurs}}{\text{total number of trials}}$
- If the number of trials is large enough, relative frequency can be used as an estimate of probability.

Linear graphs

YOU WILL LEARN ABOUT	YOU SHOULD ALREADY KNOW
○ Drawing straight-line graphs when you are given the equation of the line ○ Finding the gradient of a straight line	○ How to plot and read points in all four quadrants ○ How to substitute into equations ○ How to draw a straight-line graph using a table of values

Lines of the type $y = mx + c$

The most common straight-line graphs you need to draw have equations of the form $y = mx + c$, for example $y = 3x + 4$, $y = 2x - 5$ and $y = {}^{-}3x + 2$.

You need only two points to draw a straight line but you should work out a third point as a check.

It is best to use two values that are as far apart as possible together with $x = 0$, if it is in the range.

Example 1

Draw the graph of $y = 2x + 3$ for values of x from ⁻3 to 3.

Solution

You can use any three values of x from ⁻3 to 3. In this example the best values to use are ⁻3, 0 and 3.

When $x = {}^{-}3$ $\quad y = 2 \times {}^{-}3 + 3$
$\quad = {}^{-}6 + 3 = {}^{-}3.$

When $x = 0$, $\quad y = 2 \times 0 + 3$
$\quad = 0 + 3 = 3.$

When $x = 3$, $\quad y = 2 \times 3 + 3$
$\quad = 6 + 3 = 9.$

The graph needs to include x-values from ⁻3 to 3 and y-values from ⁻3 to 9.

Plot the points (⁻3, ⁻3), (0, 3) and (3, 9) and join them with a straight line.

Label the line.

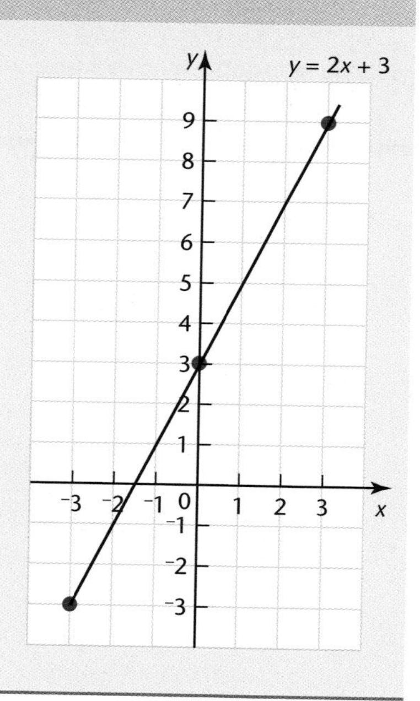

Exam tip

A common error is to say 2 × 0 = 2. Remember any number multiplied by zero is zero.

Exam tip

Write the values of x and y down clearly, as marks are often given for the correct calculation of the points even if they are plotted incorrectly.

Example 2

Draw the graph of $y = {}^-2x + 1$, for values of x from $^-4$ to 2.

Solution

When $x = {}^-4$, $y = {}^-2 \times {}^-4 + 1 = 8 + 1 = 9$.

When $x = 0$, $y = {}^-2 \times 0 + 1 = 0 + 1 = 1$.

When $x = 2$, $y = {}^-2 \times 2 + 1 = {}^-4 + 1 = {}^-3$.

In this case the y-values are from $^-3$ to 9.

Plot the points ($^-4$, 9), (0, 1) and (2, $^-3$) and join them with a straight line.

Notice that this line slopes the opposite way.

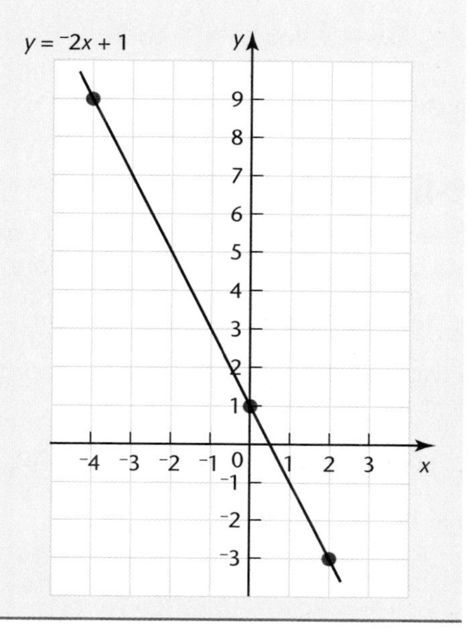

Exam tip

Any line with equation $y = ax + b$ will cross the y-axis at $y = b$.

Exam tip

If axes are already drawn for you in a question, check the scale carefully before plotting points or reading values.

Exercise 4.1

1 Draw the graph of $y = 2x$, for $x = {}^-3$ to 3.

2 Draw the graph of $y = 3x$, for $x = {}^-3$ to 3.

3 Draw the graph of $y = x + 3$, for $x = {}^-3$ to 3.

4 Draw the graph of $y = x + 6$, for $x = {}^-4$ to 2.

5 Draw the graph of $y = 3x + 2$, for $x = {}^-4$ to 2.

6 Draw the graph of $y = 4x + 2$, for $x = {}^-3$ to 3.

7 Draw the graph of $y = 2x - 3$, for $x = {}^-2$ to 4.

8 Draw the graph of $y = 3x + 5$, for $x = {}^-4$ to 2.

9 Draw the graph of $y = 4x - 4$, for $x = {}^-2$ to 4.

10 Draw the graph of $y = x - 5$, for $x = {}^-1$ to 6.

11 Draw the graph of $y = x - 2$, for $x = {}^-3$ to 3.

12 Draw the graph of $y = 2x - 5$, for $x = {}^{-}1$ to 5.

13 Draw the graph of $y = {}^{-}2x + 5$, for $x = {}^{-}2$ to 4.

14 Draw the graph of $y = {}^{-}x + 1$, for $x = {}^{-}3$ to 3.

15 Draw the graph of $y = {}^{-}3x - 4$, for $x = {}^{-}4$ to 2.

16 Draw the graph of $y = {}^{-}3x - 2$, for $x = {}^{-}4$ to 2.

17 Draw the graph of $y = {}^{-}2x - 4$, for $x = {}^{-}4$ to 2.

Harder straight-line equations

Sometimes you could be asked to draw a straight line with an equation of the form $3x + 2y = 12$, where both the x-term and the y-term are on the same side of the equation. In this case, it is easier to work out the value of x when y is 0, and the value of y when x is 0.

Draw the line joining these two points. Then check that the coordinates of another point on the graph fit the equation.

Example 3

Draw the graph of $4x + 3y = 12$.

Solution

When $x = 0$, $3y = 12$ so $y = 4$.
When $y = 0$, $4x = 12$ so $x = 3$.

The axes on the graph need to be labelled from 0 to 3 for x, and from 0 to 4 for y.

Plot the points (0, 4) and (3, 0) and join them with a straight line.

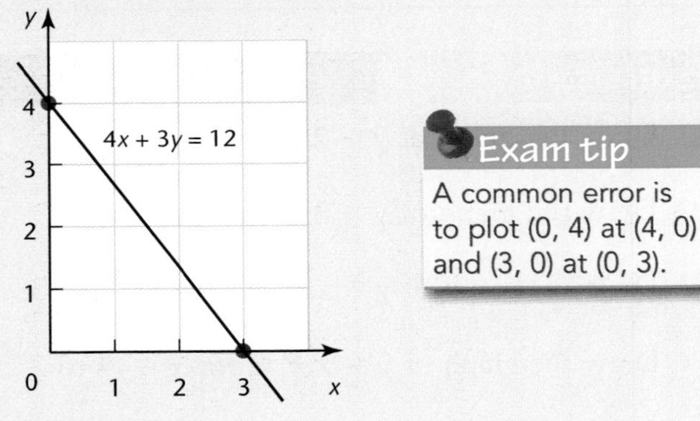

Check with a point on the line, such as $x = 1.5$, $y = 2$.
$4x + 3y = 4 \times 1.5 + 3 \times 2 = 6 + 6 = 12$, which is correct.

Exam tip

A common error is to plot (0, 4) at (4, 0) and (3, 0) at (0, 3).

Other types of equation include $2y = 3x + 5$, where you need to find $2y$, then divide by 2 to find y.

Exercise 4.2

1 Draw the graph of each of these lines.

a) $3x + 5y = 15$
b) $2x + 5y = 10$
c) $7x + 2y = 14$
d) $3x + 2y = 15$
e) $2y = 5x + 3$, for $x = {}^{-}3$ to 3
f) $3y = 2x + 6$, for $x = {}^{-}3$ to 3
g) $2y = 3x - 5$, for $x = {}^{-}2$ to 4
h) $2y = 5x - 8$, for $x = {}^{-}2$ to 4

2 a) On the same grid, draw the graphs of $y = 8$ and $y = 4x + 2$, for $x = {}^-3$ to 3.

b) Write down the coordinates of the point where the two lines cross.

3 a) On the same grid, draw the graphs of $y = x + 3$ and $y = 4x - 3$, for $x = {}^-2$ to 3.

b) Write down the coordinates of the point where the two lines cross.

4 Find the coordinates of the point where these two lines cross.

$y = 2x + 3$

$y = 7 - 2x$

5 Find the coordinates of the point where these two lines cross.

$y = 3x - 2$

$y = 12 - 4x$

Gradient of straight-line graphs

The gradient of a graph is the mathematical way of measuring its steepness or rate of change.

$$\text{Gradient} = \frac{\text{increase in } y}{\text{increase in } x}$$

To find the gradient of a line, mark two points on the graph, then draw in the horizontal and the vertical to form a triangle as shown.

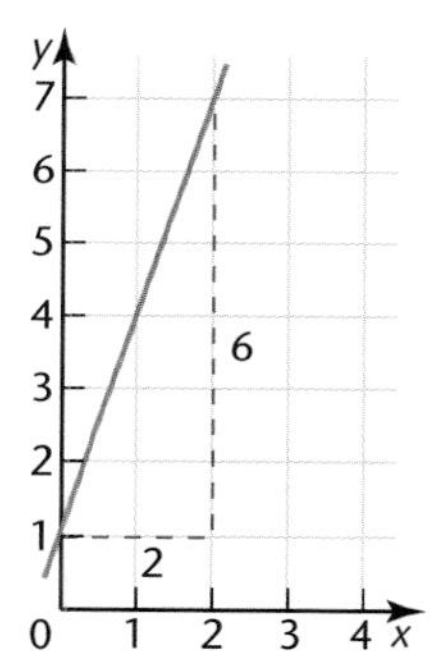

Gradient $= \frac{6}{2} = 3$

Exam tip

Choose two points far apart on the graph, so that the x-distance between them is an integer. If possible, choose points where the graph crosses gridlines. This makes reading values and dividing easier.

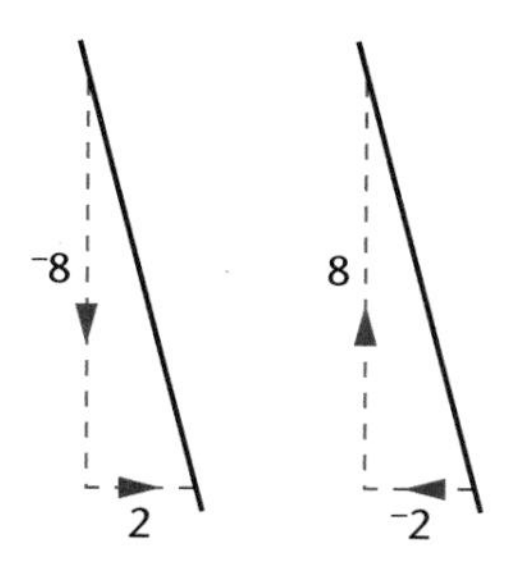

Here the gradient $= \frac{^-8}{2}$ or $\frac{8}{^-2}$.
Both give the answer ⁻4.

Exam tip

Check you have the correct sign, positive or negative, for the slope of the line.

Lines with a positive gradient slope forwards.

Lines with a negative gradient slope backwards.

Horizontal lines have a gradient of zero.

You can do this type of example without drawing a diagram, but draw one if you prefer, so that you can see the triangle.

Example 4

Find the gradient of the line joining the points (3, 5) and (8, 7).

Solution

Increase in $x = 5$ Subtract $8 - 3 = 5$

Increase in $y = 2$ Subtract $7 - 5 = 2$. Remember to subtract in the same order.

Gradient $= \frac{2}{5} = 0.4$

When interpreting graphs about physical situations, the gradient tells you the rate of change.

> **Exam tip**
> When calculating gradients from a graph, use the scale to work out the increase in x and y, rather than just counting the number of squares on the grid.

Example 5

For a distance–time graph, the gradient gives the velocity. Find the velocity in this graph.

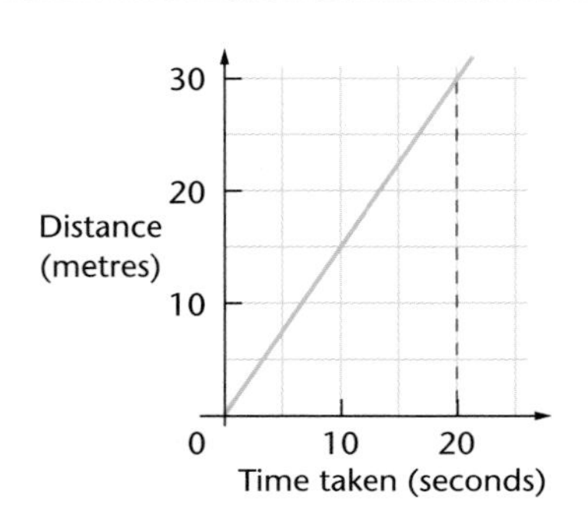

Solution

Gradient $= \frac{30}{20}$ m/s $= 1.5$ m/s

Velocity = 1.5 m/s

> **Exam tip**
> Use the units on the axes to help you to recognise what the rate of change represents. 'Velocity' is sometimes called 'speed'.

Exercise 4.3

1 Find the gradient of each of these lines.

a)

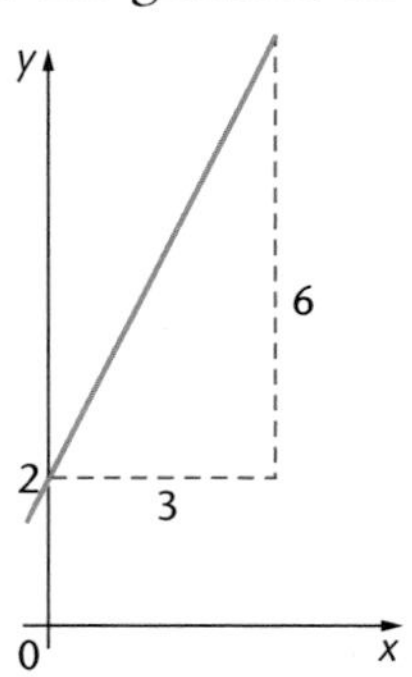

b)

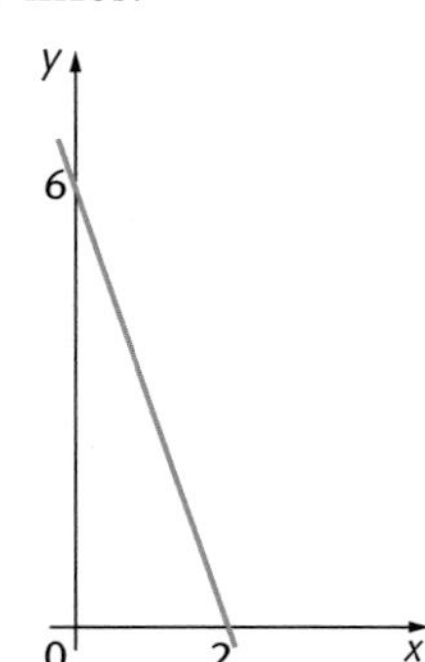

c)

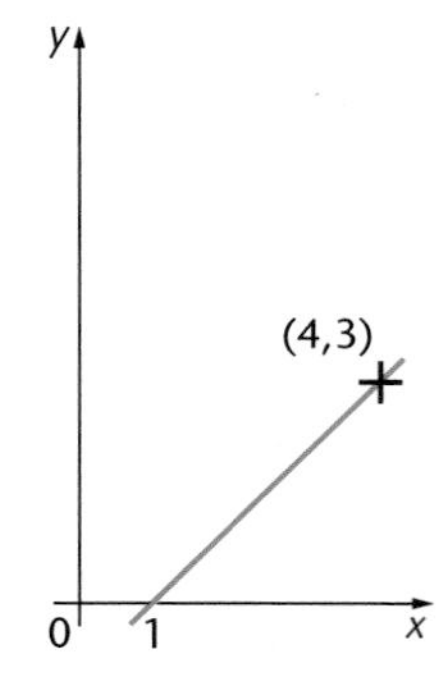

d)

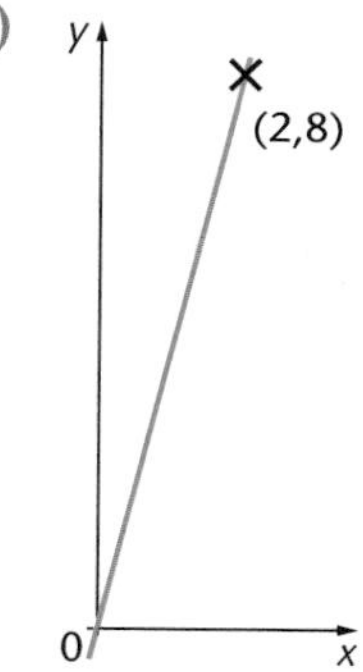

e)

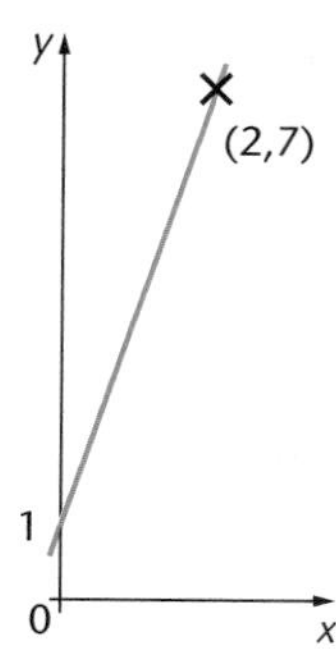

f)

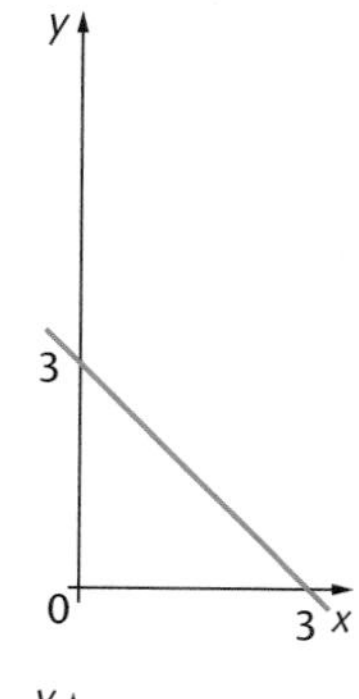

g)

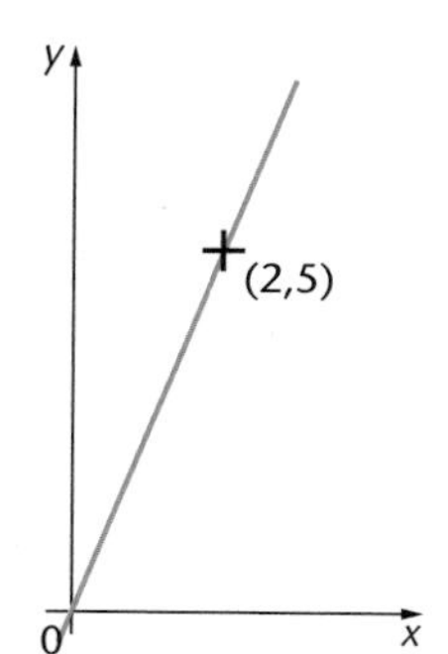

h)

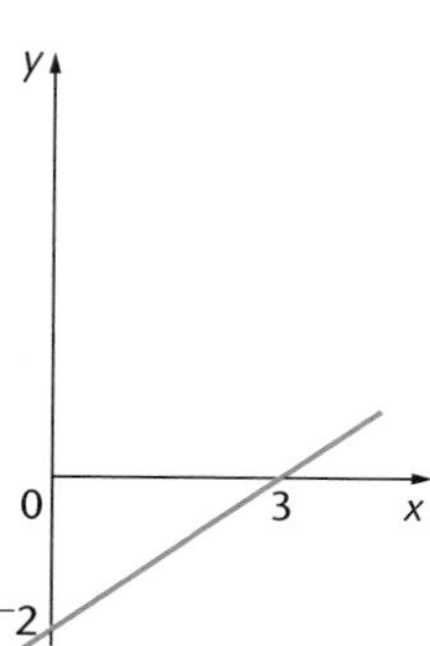

i)

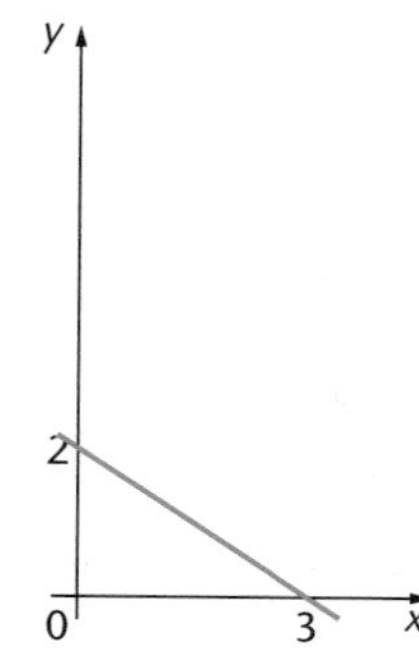

j)

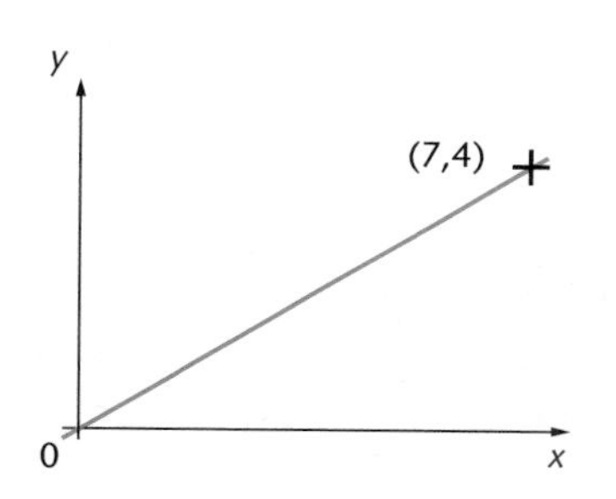

k)

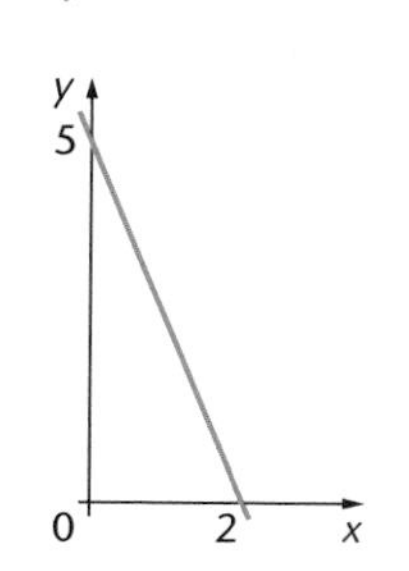

l)

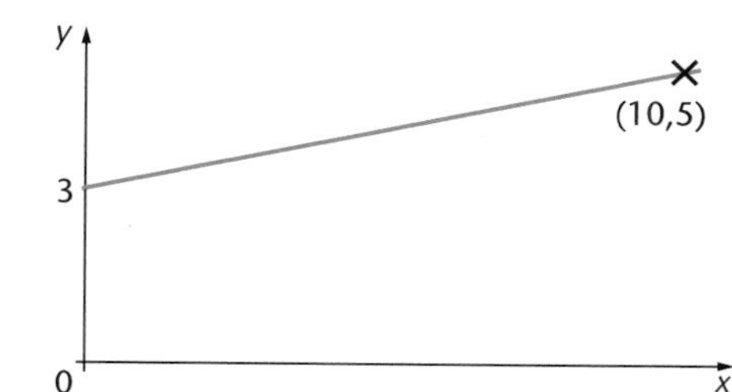

2 Calculate the gradient of the line joining each of these pairs of points.

a) (3, 2) and (4, 8) **b)** (4, 0) and (6, 8)
c) (5, 3) and (7, 3) **d)** ($^-1$, 4) and (7, 2)
e) (0, 4) and (2, $^-6$) **f)** (1, 5) and (3, 5)
g) ($^-1$, 1) and (3, 2) **h)** ($^-2$, 6) and (0, 4)

3 Calculate the gradient of the line joining each of these pairs of points.

a) (1, 8) and (5, 6) **b)** (2, 10) and (10, 30)
c) ($^-3$, 0) and ($^-1$, 5) **d)** ($^-3$, 6) and ($^-1$, $^-2$)
e) (3, $^-1$) and ($^-1$, $^-5$) **f)** (0.6, 3) and (3.6, $^-9$)
g) (2.5, 4) and (3.7, 4.9) **h)** (2.5, 7) and (4, 2.2)

4 A ball bearing rolls in a straight groove. The graph shows its distance from a point P in the groove.
Find the gradient of the line in this graph. What information does it give?

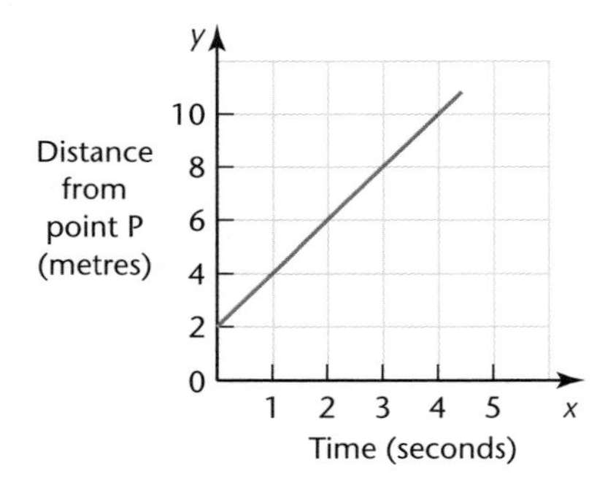

5 Find the gradient of the line in this graph. What information does it give?

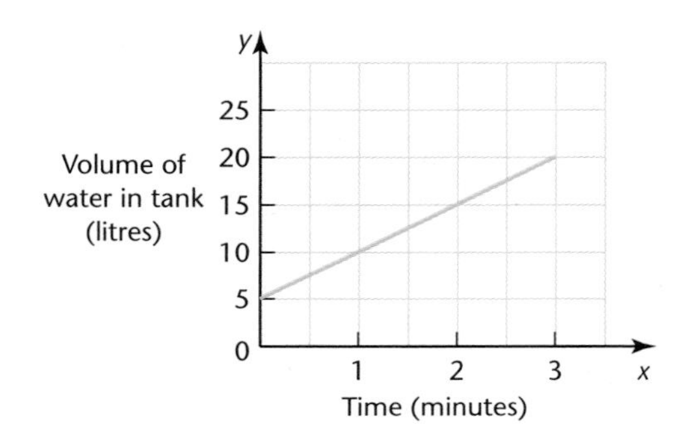

6 Find the gradient of each of the sides of triangle ABC.

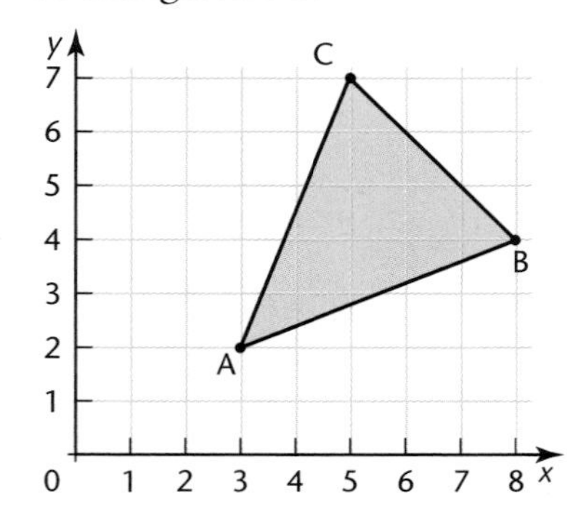

7 The table shows the cost of x minutes of calls on a mobile phone.

Number of minutes (x)	5	12	20	23
Cost (£C)	1.30	3.12	5.20	5.98

Find the gradient of the graph of C against x, and say what this gradient represents.

8 The table gives the cost when x metres of string are sold.

Number of metres (x)	0.25	0.5	1.75	3.00
Cost (C pence)	20	42	147	252

Find the gradient of the graph of C against x and say what this gradient represents.

9 A lorry is refuelling with diesel.
a) Draw a graph of these data.

Time (t min)	0	1	2	3
Diesel in tank (d litres)	20	30	40	50

b) Find the gradient of the graph and state what it represents.

10 On the same diagram, draw these straight lines and find their gradients.
a) $y = 2x$ **b)** $y = 2x + 1$
c) $y = 3x$ **d)** $y = 3x + 2$

11 On the same diagram, draw these straight lines and find their gradients.
a) $y = x + 1$ **b)** $y = 2x + 1$
c) $y = 2x - 3$ **d)** $y = 5x - 2$
e) $y = 4x$ **f)** $y = 4x + 3$

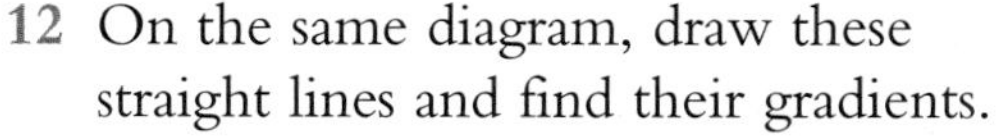

12 On the same diagram, draw these straight lines and find their gradients.

a) $y = {}^{-}x + 3$ b) $y = {}^{-}2x + 1$

c) $y = {}^{-}3x$ d) $y = {}^{-}3x + 2$

e) $y = {}^{-}x$ f) $y = {}^{-}2x - 5$

13 Draw these lines on the same diagram.

$y = 2x + 1$

$2y = 3 - x$

What do you notice about the two lines?

Find the gradients of the two lines.

How are the gradients linked?

Exam tip

Label axes clearly. Check you understand the scales when axes are drawn for you.

Challenge 1

When a road goes up a steep slope, a road sign gives the gradient of the slope.

Investigate the different ways that gradients are given on road signs.

Is this the same on road signs in the rest of Europe?

How do these gradients relate to the mathematical gradient used in this chapter?

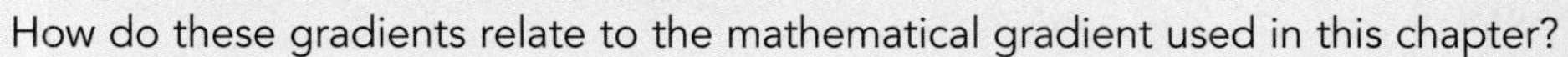

Key Ideas

- You only need two points to draw a straight line but plot a third as a check.
- Gradient = $\frac{\text{increase in y}}{\text{increase in x}}$.
- Lines with a positive gradient slope forwards.
- Lines with a negative gradient slope backwards.
- Horizontal lines have a gradient of zero.
- For graphs about physical situations, the gradient gives the rate of change.

Chapter 5 Angles in polygons

YOU WILL LEARN ABOUT	YOU SHOULD ALREADY KNOW
○ Angle facts relating to polygons ○ Solving problems using angle facts	○ Angle facts relating to points, lines, triangles and quadrilaterals ○ Angle facts relating to parallel lines

Angles in polygons

Activity 1

Draw a polygon with five sides.
This is called a pentagon.

Put a pencil along the base of the pentagon, as shown in the diagram, then slide it to the right until its end reaches the end of the base.

Now carefully rotate the pencil about its end until it is along the next side of the pentagon.

Slide it up until its end reaches the top of that side.

Continue in this way until you reach the base again.

When you get back to the beginning, the pencil will have turned through 360°.

It will have turned through the angles shown in the bottom diagram.

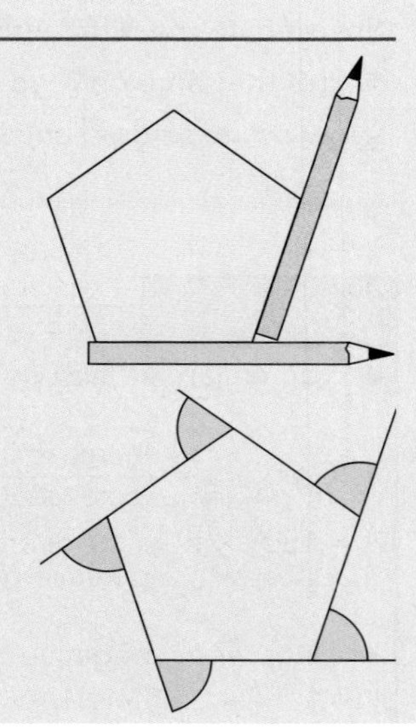

These angles are called **exterior angles**.
Activity 1 shows that the sum of the exterior angles of the pentagon is 360°.
This method could have been used for any convex polygon.

The sum of the exterior angles of any convex polygon is 360°.

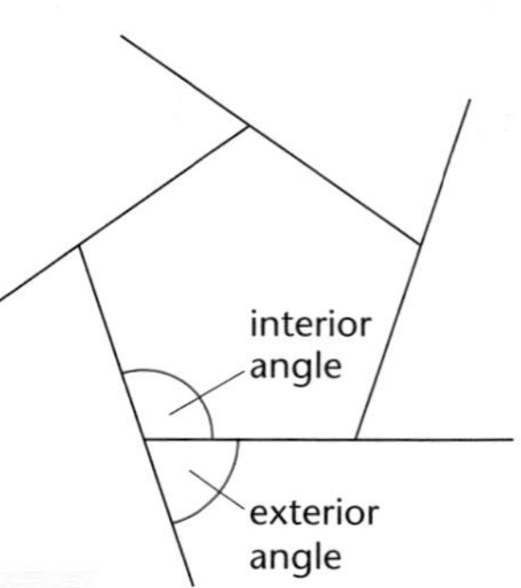

At each vertex, the interior and exterior angles make a straight line.

For any convex polygon, at any vertex: interior angle + exterior angle = 180°.

Example 1

The two unlabelled exterior angles of this pentagon are equal.
Find their size.

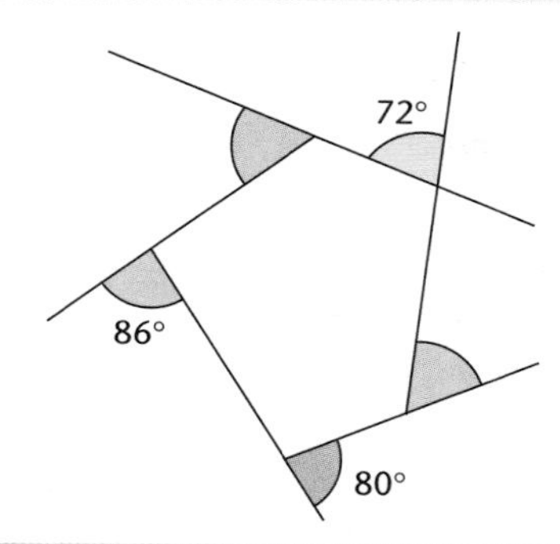

Solution

First, find the sum of the angles that are given.

$72° + 80° + 86° = 238°$

The sum of all the exterior angles is 360°.

The sum of the remaining two angles = $360° - 238° = 122°$.

So each angle is $122 \div 2 = 61°$.

In many problems about polygon angles, the easiest way to solve them is to use the fact that the sum of the exterior angles is 360°.

Sometimes, however, it is useful to find the sum of the interior angles.

At each vertex of a convex polygon: interior angle + exterior angle = 180°.

> Sum of (interior + exterior angles) for the polygon
> = 180° × number of angles
> = 180° × number of sides.

But the sum of the exterior angles is 360°.

So the sum of the interior angles of the polygon = 180° × number of sides – 360°.

Putting this algebraically, for an n-sided convex polygon

> the sum of the interior angles = $(180n - 360)°$.

Another way you can find the sum of the interior angles of any polygon is to divide it into triangles.

Challenge 1

A quadrilateral is divided into two triangles, as shown in the diagram.

Prove that the interior angles of a quadrilateral add up to 360°.

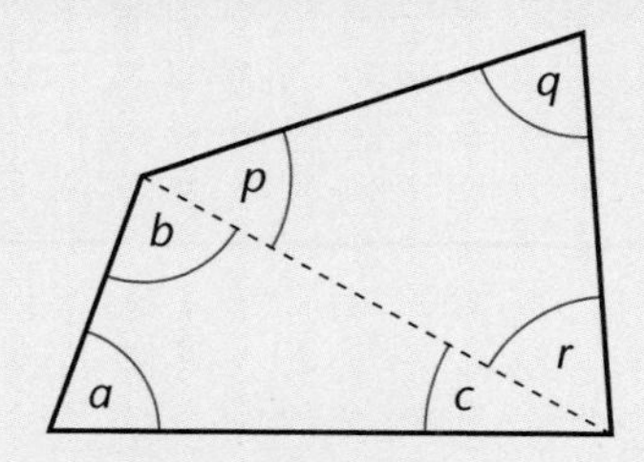

Foundation Gold/ Higher Bronze

For any convex polygon

- The sum of all the exterior angles is 360°.
- The interior angle + the exterior angle = 180°.

interior angle

exterior angle

For an *n*-sided polygon

- The sum of the interior angles = 180° × (n – 2)

This means that

- The sum of the interior angles of a triangle is 180°.
- The sum of the interior angles of a quadrilateral is 360°.
- The sum of the interior angles of a pentagon is 540°.
- The sum of the interior angles of a hexagon is 720°.

For any regular polygon

- All the interior angles are equal, all the exterior angles are equal, and all the sides are equal.
- Each exterior angle = $\dfrac{360°}{\text{number of sides}}$
- Each interior angle = 180° – exterior angle
- The angle at the centre = $\dfrac{360°}{\text{number of sides}}$

Example 2

Find the value of x in this hexagon.

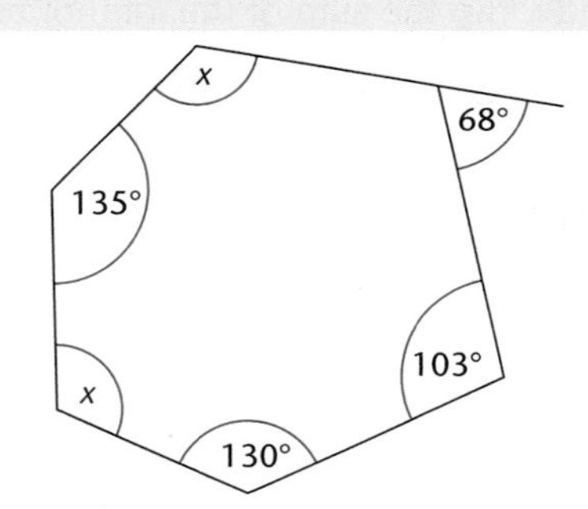

Solution

First, find the size of the missing interior angle.

Interior angle + exterior angle = 180°
Interior angle = 180° – 68°
= 112°

Next, find the sum of the interior angles.

Sum of interior angles = 112° + 103° + 130° + 135° + 2x
= 480° + 2x

The sum of the interior angles of a hexagon = 720°, therefore

480° + 2x = 720°
2x = 720° – 480°
= 240°
x = 120°

Example 3

Find the interior angle of a regular hexagon.

Solution

For a regular hexagon, the exterior angle $= 360° \div 6 = 60°$.
So the interior angle $= 180° - 60° = 120°$.

Example 4

Find the number of sides of a regular polygon with an interior angle of 144°.

Solution

If the interior angle $= 144°$,
the exterior angle $= 180° - 144°$
$= 36°$.
The sum of the exterior angles $= 360°$.
Therefore the number of sides $= 360° \div 36° = 10$.

Challenge 2

A shape will tessellate if a number of copies of the shape will fit together on a flat surface without any gaps. Regular octagons do not tessellate but, as this picture of a tiled floor shows, regular octagons and squares may be combined to cover a surface.

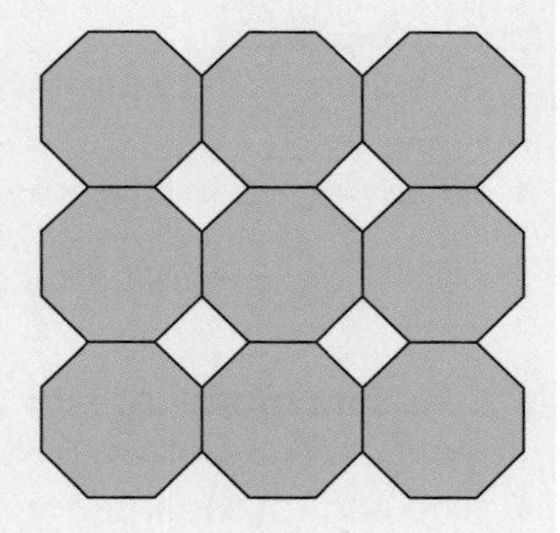

Can you find

a) which regular polygons tessellate?

b) which combinations of regular polygons will cover a surface?

Exercise 5.1

1 Three of the exterior angles of a quadrilateral are 90°, 52° and 87°. Find the size of the other exterior angle.

2 Four of the exterior angles of a pentagon are 70°, 59°, 83° and 90°.
 a) Find the size of the other exterior angle.
 b) Find the size of each interior angle.

3 Five of the exterior angles of a hexagon are 54°, 48°, 65°, 35° and 80°.
 a) Find the size of the other exterior angle.
 b) Find the size of each interior angle.

4 Three of the exterior angles of a quadrilateral are 110°, 61° and 74°. Find the size of the other exterior angle.

5 Four of the exterior angles of a pentagon are 68°, 49°, 82° and 77°.
 a) Find the size of the other exterior angle.
 b) Find the size of each interior angle.

6 Four of the exterior angles of a hexagon are 67°, 43°, 91° and 37°.
 a) Find the size of the other exterior angles, given that they are equal.
 b) Find the size of each interior angle.

7 A regular polygon has nine sides.
Find the size of each of its exterior and interior angles.

8 Find the interior angle of a regular dodecagon (12 sides).

9 A regular polygon has an exterior angle of 24°.
How many sides does it have?

10 What is the sum of the interior angles of
a) a hexagon?
b) a decagon?

11 Six of the angles of a heptagon are 122°, 141°, 137°, 103°, 164° and 126°.
Calculate the size of the remaining angle.

12 A regular polygon has 15 sides.
Find the size of each of its exterior and interior angles.

13 Find the size of the interior angle of a regular 20-sided polygon.

14 A regular polygon has an exterior angle of 30°.
How many sides does it have?

15 What is the sum of the interior angles of
a) an octagon?
b) a nonagon?

16 A polygon has 11 sides. Ten of its interior angles add up to 1490°.
Find the size of the remaining angle.

Key Ideas

- The sum of the exterior angles of any convex polygon is 360°.
- In any polygon, the interior angle and the exterior angle add up to 180°.
- The sum of the interior angles of a triangle is 180°.
- The sum of the interior angles of a quadrilateral is 360°.
- The sum of the interior angles of a pentagon is 540°.
- The sum of the interior angles of a hexagon is 720°.
- The exterior angle of a regular polygon = $\frac{360°}{\text{number of sides}}$
- The angle at the centre of a regular polygon = $\frac{360°}{\text{number of sides}}$

Real-life graphs

YOU WILL LEARN ABOUT	YOU SHOULD ALREADY KNOW
○ Drawing and interpreting graphs showing real-life situations	○ How to draw straight-line graphs

Some graphs tell a story – they show what happened in an event. To find out what is happening, first look at the labels on the axes. They tell you what the graph is about.

Look for important features on the graph. For instance, does it increase or decrease at a steady rate (a straight line) or is it curved?

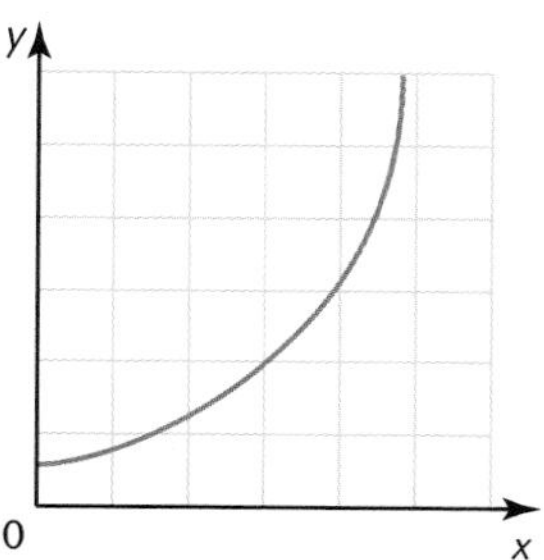

The rate of change is increasing.

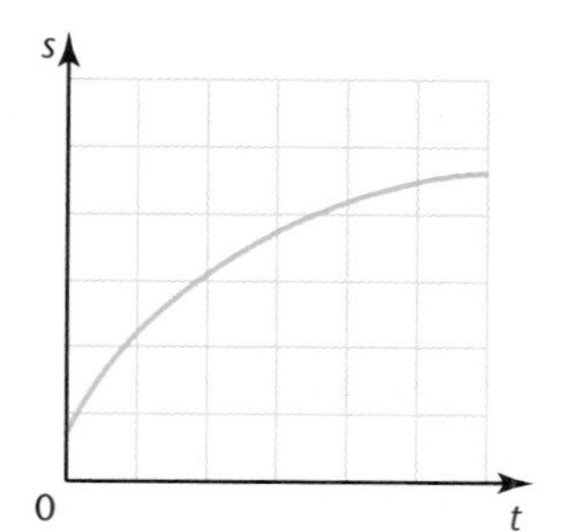

The rate of change is decreasing.

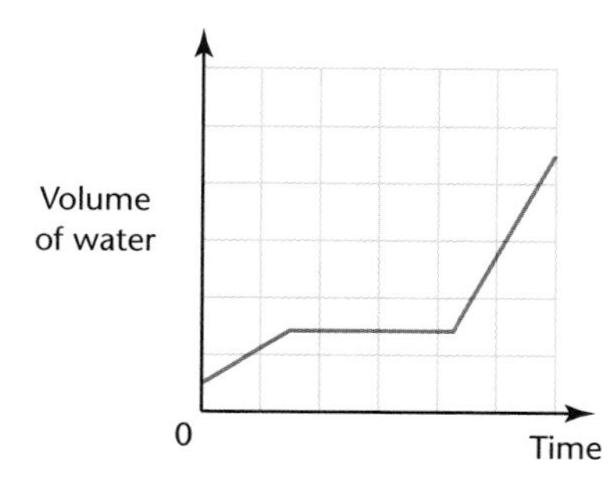

A flat part of the graph – no change for the variable on the vertical axis.

Foundation Gold/ Higher Bronze

Example 1

This graph shows the number of people at a football stadium one afternoon.

Describe what may have happened at the various stages.

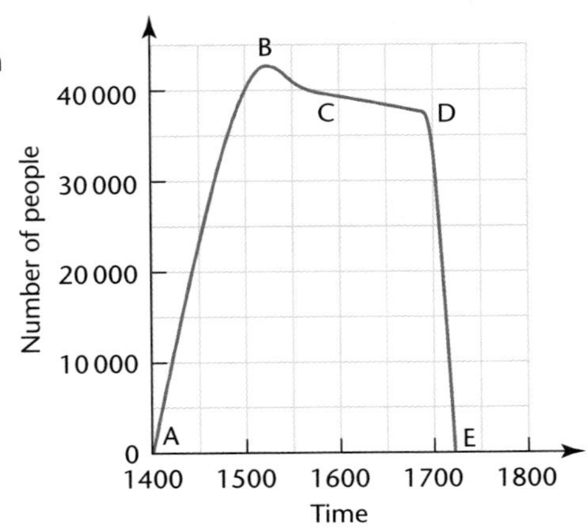

Solution

A: The gates open.

B: Kick off.

C: Some people leave during the match.

D: The match ends and people start to leave.

E: The stadium is empty.

Example 2

John ran the first two miles to school at a speed of 8 mph. He then waited 5 minutes for his friend. They walked the last mile to school together, taking 20 minutes.

Draw a graph to show his journey.

Solution

If John is travelling at 8 mph, it will take him $\frac{1}{4}$ hour (15 minutes) to run 2 miles.

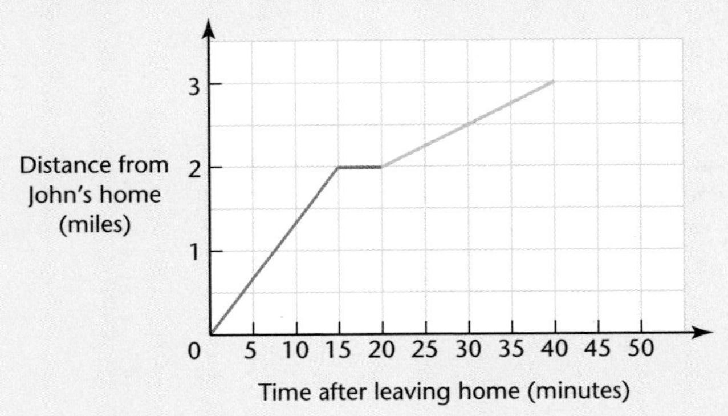

The first part of the graph in Example 2 is steeper than the last part. This shows that John went faster in the first 15 minutes than he did in the last 20 minutes. The flat part of the graph shows where John stayed in the same place for 5 minutes.

Exam tip

If you are asked to describe a story graph, try to include numerical information. For example, instead of 'stopped' write 'stopped at 10:14 p.m. for 6 minutes'.

Activity 1

Think of some labels for the axes and an appropriate 'story' description to fit this graph.

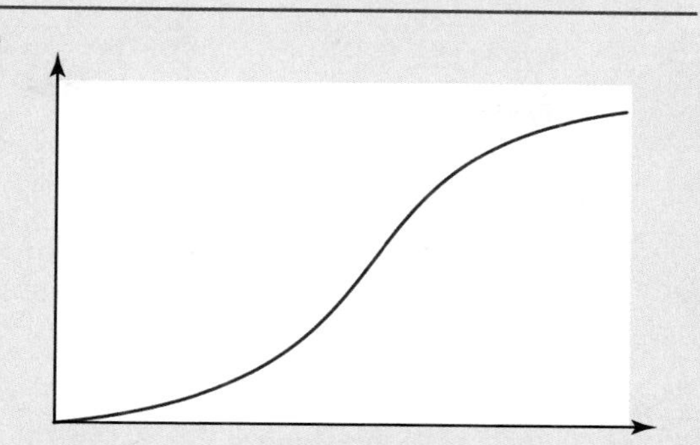

Exam tip

When drawing a graph, don't forget to label the axes.

Exercise 6.1

1 This line graph shows the average monthly day-time temperature in Leeds.

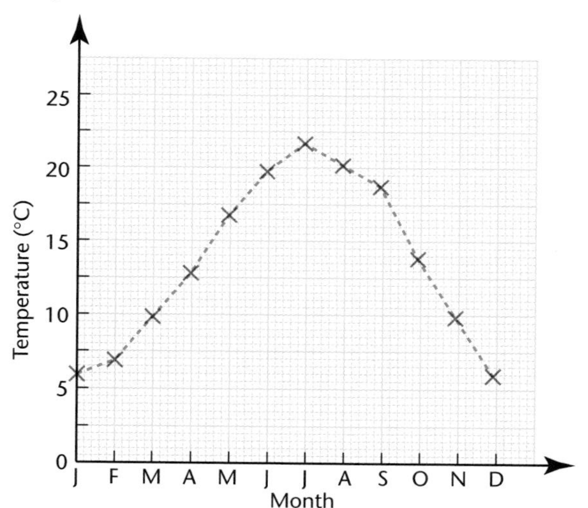

a) Which is the hottest month and what is its average temperature?

b) What is the average temperature in October?

c) After which month did the temperature start to decrease rapidly?

Exam tip

When there can be no values between the marked points, the graph is drawn with a dotted line to show the trend.

2 This graph shows the average monthly rainfall in Birmingham and Brussels.

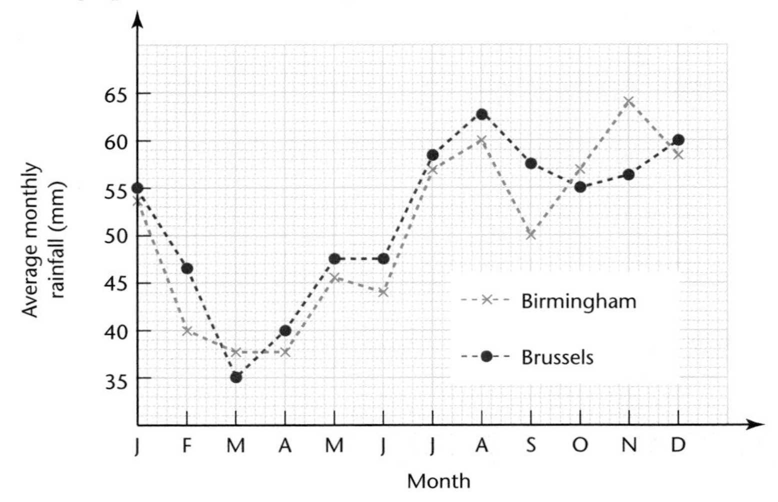

a) Which city is usually wetter in June?
b) What is the average rainfall in Birmingham in January?
c) For which months are there, on average, more rain in Birmingham than Brussels?

3 This graph shows the volume of water in a bath.

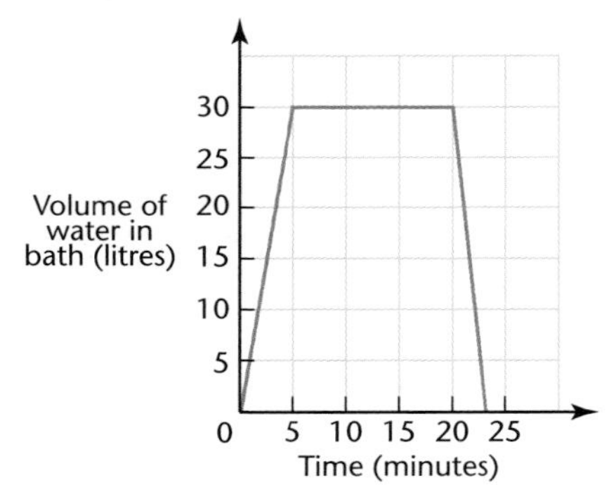

a) How long did the bath take to fill?
b) How much water was in the bath when the taps were turned off?
c) How many litres per minute went down the plughole when the bath was emptied?

4 Amy drew this sketch graph to show what happened to the volume of water in her bath.

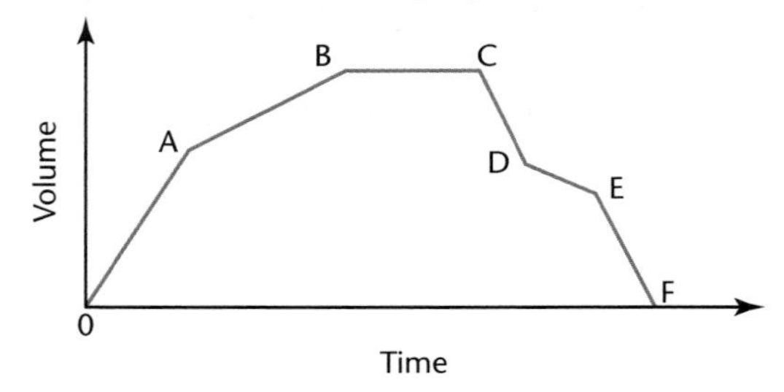

a) Both taps were 'on' at 0. What happened at A?
b) What happened at B?
c) What happened at C?
d) What happened between D and E?

5 This graph shows the temperature in a school during a 24-hour period.

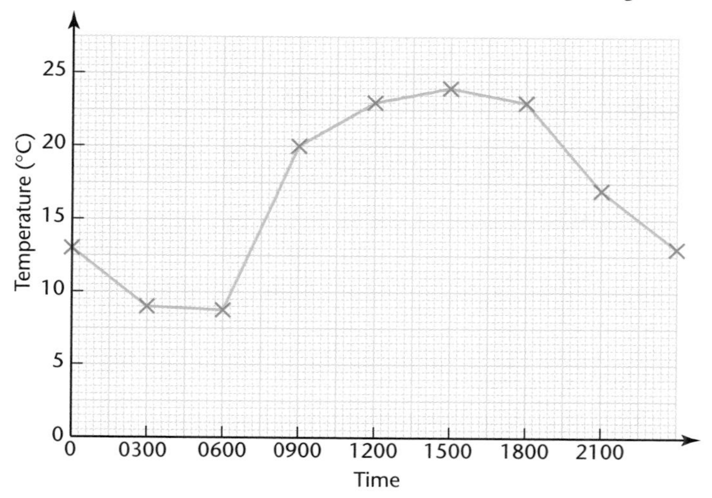

a) At what time did the heating switch on?
b) What was the temperature at 9 o'clock in the morning?
c) What is the difference between the lowest and highest temperature?

6 This graph shows the number of people at a theme park one bank holiday.

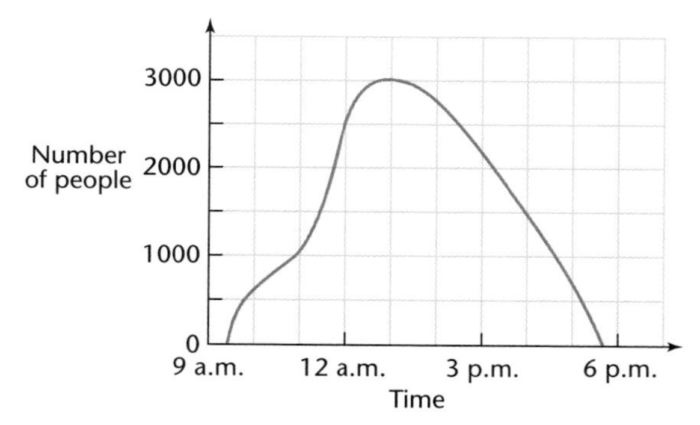

a) When did the park open?
b) During which hour did most people go into the park?

7 At a rock concert, the gates opened at 5 p.m. People came in fairly slowly at first, but then quite steadily from 5:45 p.m. until the start at 7 p.m. There were then 50000 people in the stadium. The concert lasted until 10 p.m. At the end people left quickly and the stadium was almost empty by 10:30 p.m.
Sketch a graph to show how the number of people in the stadium for this rock concert changed.

Foundation Gold/ Higher Bronze

8 This graph shows the amount of fuel in a car's petrol tank.

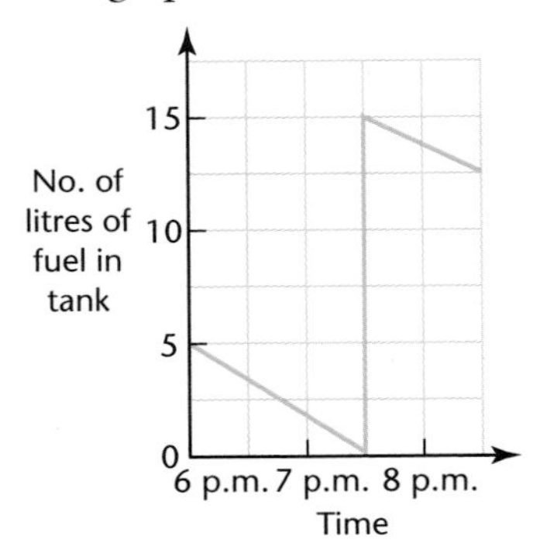

a) How many litres were used between 6 and 7 p.m.?
b) Describe what happened between 7:30 and 8 p.m.

9 The graph below shows how the depth of water in a water tank varies as the tank is filled.

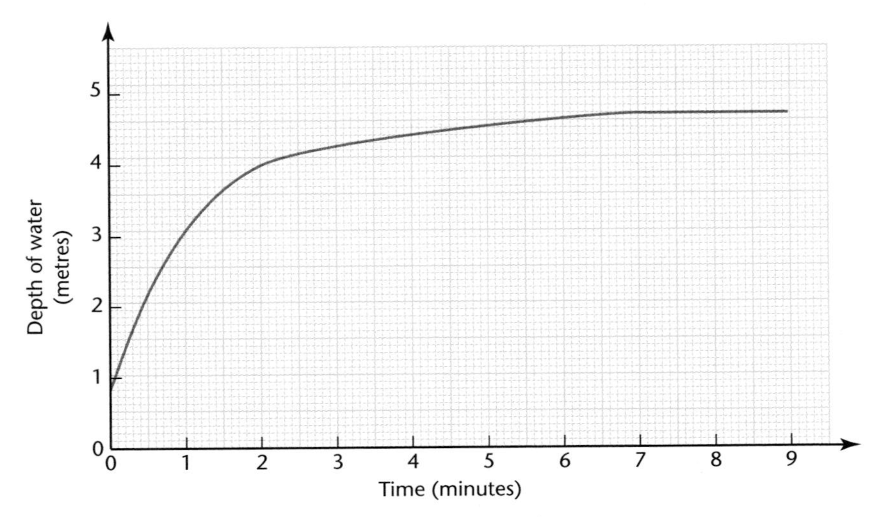

a) What is the depth of water to start with?
b) By how much does the water rise in the first minute?
c) At what time does the water reach the maximum depth? What is this depth?

10 Some water was heated over a burner. Its temperature rose quickly at first and then more and more slowly. The burner was turned off and the temperature fell rapidly at first and then more and more slowly.
Draw a sketch graph to show how the temperature changes.

11 The speed of a car at the start of a journey is shown on this graph.

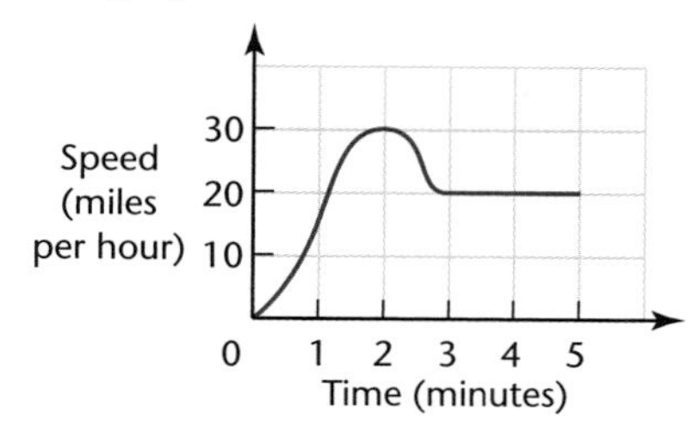

a) What is happening on the flat portion of the graph?
b) Between which times is the car slowing down?

12 Alix ran hot water into a bath for 4 minutes at a rate of 15 litres per minute. She then turned on the cold tap too so that the bath filled at 20 litres a minute for another 2 minutes.
a) Draw a graph to show how the volume of water in the bath changed.
b) How much water was there in the bath at the end of this time?

13 This graph shows the cost of hiring a car for a day.

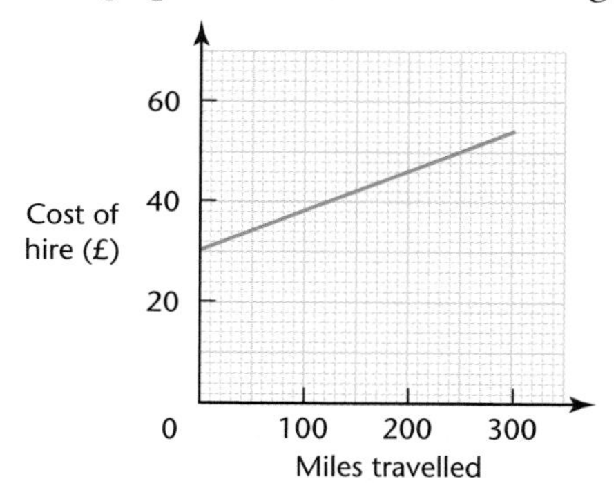

a) Pedro travelled 150 miles.
How much did he pay for this car hire?
b) Jane paid £48 for her car hire.
How many miles did she travel?
c) What was
(i) the basic hire charge?
(ii) the charge per mile?

Foundation Gold/ Higher Bronze

14 This graph shows the monthly bill for a mobile phone for different numbers of minutes used.

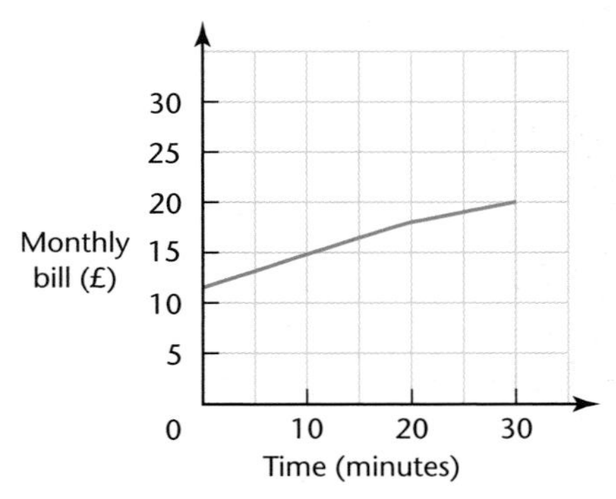

a) How many minutes have been used if the bill is £15?

b) There are two line segments on the graph.
What do they show?

15 A water company charges £8 each quarter for a water meter, then 50p per cubic metre for the first 100 cubic metres used and 70p per cubic metre for water used above this amount.
Draw a graph to show the total bill for different amounts of water used, up to 200 cubic metres.

Challenge 1

Water is poured at a steady rate into this conical glass until it is full.

Sketch a graph to show how the depth of water in the glass changes with time.

Key Ideas

- The labels on the axes tell you what a graph is about.
- Each feature on the graph is part of the story. For instance, does it increase or decrease at a steady rate (a straight line) or is it curved?
- Be specific and include numerical information when describing a graph that tells a story.

Chapter 7 Checking solutions and calculations

YOU WILL LEARN ABOUT	YOU SHOULD ALREADY KNOW
○ Using estimates to check the accuracy of your answer ○ Finding an estimate of the answer to a problem	○ Basic numerical methods such as finding ratios and percentages ○ How to round numbers to 1 significant figure

Foundation Gold/ Higher Bronze

Checking your work

When you have solved a problem, how do you know if your answer is right? Sometimes, accuracy is vital. There was a story in the news of a doctor treating a baby. The doctor put the decimal point in the wrong place in a calculation and prescribed 100 times as much of the drug as she intended.

Using common sense

Does your answer sound sensible in the context of the question? In practical problems, your own experience often gives you an idea of the size of the answer you expect. For example, in the case of a shopping bill, you would probably react and check if it came to more than you expected!

Using number facts

Activity 1

Multiplying

Starting with the number 400, multiply it by the numbers in each of these sets.

Set A:	5	1.1	3.2	1.003	1.4	1.2
Set B:	0.5	0.999	0.6	0.9	0.7	0.95

What happens to the number 400? Does it get bigger or smaller? What conclusions can you come to?

In particular, what happens when any number is multiplied by a number which is

a) greater than 1? **b)** less than 1? **c)** 1 itself?

Division

Repeat the activity, but divide instead of multiplying. Write down your conclusions.

In particular, what happens when any number is divided by a number which is

a) greater than 1? **b)** less than 1? **c)** 1 itself?

Look at the answer to this calculation.

$752 \div 24 = 18048$

When 752 is divided by a number that is greater than 1, the result should be less than 752.

Instead, it is more. It looks as if the × button was pressed by mistake, instead of the ÷ button. Using number facts can help to spot errors like this.

Starting with any positive number,

- multiplying by a number greater than 1 gives a result that is larger than the number.
- multiplying by a positive number smaller than 1 gives a smaller result.
- dividing by a number greater than 1 gives a smaller result.
- dividing by a positive number smaller than 1 gives a larger result.

Here are some other facts that you can use to check a calculation.

- odd × odd = even even × odd = odd even × even = even
- $+ \times + = +$ $+ \times - = -$ $- \times - = +$
  and similarly for division
- Multiplying any number by 5 will give a result that ends in 0 or 5.
- The last digit in a multiplication comes from multiplying the last digits of the numbers.

Using inverse operations

Without a calculator, it is difficult to work out the square root of most numbers. However, if you know the square numbers, you can tell whether an answer is sensible.

Example 1

Show how you can tell that the answer to this calculation is wrong.

$\sqrt{35} = 9.52$ to 2 decimal places

Solution

$6^2 = 36$ so $\sqrt{35}$ must be less than 6.

Using inverse operations is also useful when you are using your calculator.

For example, if you work out $6.9 \div 750 = 0.0092$ using a calculator, you can check the answer by using your calculator to work out $0.0092 \times 750 = 6.9$.

Using estimates

Many people make estimates when they are shopping. They round prices to the nearest pound to make calculations easier.

Example 2

Kate has £25 birthday money to spend. She sees CDs at £7.99.

How many of them can she buy?

Solution

You do not need to know the exact answer to $25 \div 7.99$ so do a quick estimate.

Use £8 instead of £7.99.

$3 \times 8 = 24$ so $25 \div 8 = 3$ 'and a bit'.

So Kate can buy three CDs.

You can extend this idea to your maths lessons, and any other subjects where you use calculations.

Exercise 7.1

1 Which of these answers might be correct and which are definitely wrong? Show how you decided.

a) $39.6 \times 18.1 = 716.76$
b) $175 \div 1.013 = 177.275$
c) $8400 \times 9 = 756\,000$
d) A lift takes 9 people, so a party of 110 people will need 12 trips.
e) Henry has £100 birthday money to spend and reckons that he can afford 5 DVDs costing £17.99 each.

2 Look at these calculations. The answers are all wrong.
For each calculation, show how you can tell this quickly, without using a calculator to work it out.

a) $^{-}6.2 \div {}^{-}2 = {}^{-}3.1$
b) $12.4 \times 0.7 = 86.8$
c) $31.2 \times 40 = 124.8$
d) $\sqrt{72} = 9.49$ to 2 d.p.
e) $0.3^2 = 0.9$
f) $16.2 \div 8.1 = 20$
g) $125 \div 0.5 = 25$
h) $6.4 \times {}^{-}4 = 25.6$
i) $24.7 + 6.2 = 30.8$
j) $76 \div 0.5 = 38$
k) $({}^{-}0.9)^2 = {}^{-}0.81$
l) $\sqrt{1000} = 10$
m) $1.56 \times 2.5 = 0.39$
n) $360 \div 15 = 2400$

3 Use estimates to calculate a rough total cost for each of these.

a) Seven packs of crisps at 22p each
b) Nine DVDs at £13.25 each
c) 39 theatre tickets at £7.20 each
d) Five CDs at £5.99 and two posters at £1.99 each.
e) Three meals at £5.70 and two drinks at 99p each.

Rounding numbers to 1 significant figure

We often use rounded numbers instead of exact ones.

Activity 2

Which of the numbers in these statements are likely to be exact, and which have been rounded?

a) Yesterday, I spent £14.62.

b) My height is 180 cm.

c) Her new dress cost £40.

d) The attendance at the Arsenal match was 32 000.

e) The cost of building the new school is £27 million.

f) The value of π is 3.142.

g) The Olympic Games were held in Athens in 2004.

h) There were 87 people at the meeting.

Activity 3

Look for some examples of rounded and unrounded numbers in newspapers, magazines, advertising material and on TV.

Which, if any, of the rounded numbers do you think have been rounded up, and which have been rounded down? Why?

Can you think of any cases when someone might want to round figures up rather than down?

You should already know how to round numbers to 1 significant figure (1 s.f.). This means giving just one non-zero figure, with zeros as placeholders to make the number the correct size.

Example 3

Round each of these numbers to 1 significant figure.

a) £29.95 **b)** 48 235 **c)** 0.072

Solution

a) £29.95 = £30 to 1 s.f.

The second non-zero digit is 9, so round the 2 up to 3.
Looking at place value, the 2 represents 20.
You use one zero to show the size.

b) 48 235 = 50 000 to 1 s.f.

The second non-zero digit is 8, so round the 4 up to 5.
Looking at place value, the 4 represents 40 000.
You use zeros to show the size.

c) 0.072 = 0.07 to 1 s.f.

The second non-zero digit is 2, so the 7 stays as it is.
Looking at place value, the 7 is 0.07, which stays as it is.

Exercise 7.2

Round each of these numbers to 1 significant figure.

1 8.2	9 0.68	17 6027
2 6.9	10 3812	18 0.013
3 17	11 4199	19 0.58
4 25.1	12 3.09	20 0.037
5 493	13 14.9	21 1.0042
6 7.0	14 167	22 20 053
7 967	15 21.2	23 0.069
8 0.43	16 794	24 1942

Foundation Gold/
Higher Bronze

Checking answers by rounding to 1 significant figure

It is important to be able to check calculations quickly, without using a calculator. One way to do this is to round the numbers to 1 significant figure and find an approximate answer to the calculation.

Example 4

Find an approximate answer to the calculation 5.13×4.83.

Solution

$5.13 \times 4.83 = 5 \times 5 = 25$ — Round each number to 1 s.f. to give a much simpler calculation.

Exam tip

In a multiplication it may be possible to round one number up and another number down. This might give an estimate that is closer to the exact answer.

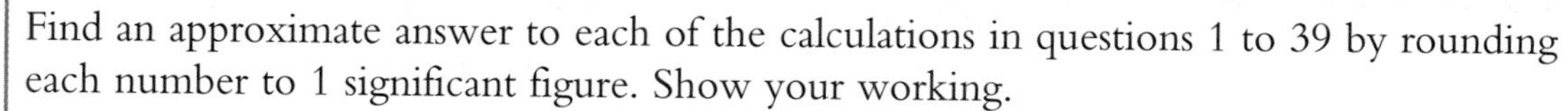

Exercise 7.3

Find an approximate answer to each of the calculations in questions 1 to 39 by rounding each number to 1 significant figure. Show your working.

1 $31.3 \div 4.85$

2 113.5×2.99

3 $44.669 \div 8.77$

4 $3.6 \times 14.9 \times 21.5$

5 48.67×12.69

6 0.89×5.2

7 61.33×11.79

8 $(1.8 \times 2.9) \div 3.2$

9 $\dfrac{14.56 \times 22.4}{59.78}$

10 $\dfrac{4.9^2 \times 49.3}{96.7}$

11 $\sqrt{4.9 \times 5.2}$

12 $\dfrac{3.99}{0.8 \times 1.64}$

13 $198.5 \times 63.1 \times 2.8$

14 $\dfrac{\sqrt{8.1 \times 1.9}}{1.9}$

15 $(0.35 \times 86.3) \div 7.9$

16 $\sqrt{103.5} \div \sqrt{37.2}$

17 9.87×0.0657

18 $0.95 \div 4.8$

19 $32 \times \sqrt{124}$

20 $\dfrac{62 \times 9.7}{10.12 \times 5.1}$

21 0.246×0.789

22 $44.555 \div 0.086$

23 46×82

24 $\sqrt{84}$

25 $\dfrac{1083}{8.2}$

26 7.05^2

27 $43.7 \times 18.9 \times 29.3$

28 $\dfrac{2.46}{18.5}$

29 $\dfrac{29}{41.6}$

30 917×38

31 $\dfrac{283 \times 97}{724}$

32 $\dfrac{614 \times 0.83}{3.7 \times 2.18}$

33 $\dfrac{6.72}{0.051 \times 39.7}$

34 $\sqrt{39 \times 80}$

35 $65.4 \div 3.9$

36 $\dfrac{194.4 \div 3.9}{27.3}$

37 $\dfrac{49.7}{4.1 \times 7.9}$

38 3.1×14.9

39 $47 \times (21.7 + 39.2)$

40 At the school fete, Jack sold 245 ice-creams at 85p each.
Estimate his takings.

41 A rectangle measures 5.8 cm by 9.4 cm.
Estimate its area.

42 A circle has diameter 6.7 cm.
Estimate its circumference. ($\pi = 3.142$)

43 A new car is priced at £14 995 excluding VAT. VAT at 17.5% must be paid on it.
Estimate the amount of VAT to be paid.

44 A cube has sides of 3.7 cm.
Estimate its volume.

45 Pedro drove 415 miles in 7 hours 51 minutes.
Estimate his average speed.

Activity 4

Use a calculator to see how close your approximations in Exercise 7.3 questions 1 to 14 are to the correct answers.

Challenge 1

Make up some multiplication and division calculations to test whether this statement is true.

'In multiplication and division calculations, rounding each number to 1 significant figure will always give an answer which is correct to 1 significant figure.'

Key Ideas

- Check your work by doing at least one of these checks.
- Does the answer sound sensible in the context of the question?
- Do a quick estimate, for example by rounding each number to 1 significant figure.
- Do the calculation another way to check.
- Work backwards from your answer.

CHAPTER

8 Scatter diagrams and correlation

YOU WILL LEARN ABOUT	YOU SHOULD ALREADY KNOW
○ Interpreting a scatter diagram ○ The meaning of correlation ○ Drawing a line of best fit ○ Using a line of best fit	○ How to plot points

Drawing scatter diagrams

Scatter diagrams (also called scatter graphs) are used to investigate any possible link or relationship between two features or **variables**. Values of the two features are plotted as points on a graph. If these points tend to lie in a straight line then there is a relationship or **correlation** between the two features.

Example 1

The table shows the mean annual temperature for 12 cities which lie north of the equator.

City	Latitude (degrees)	Mean temperature (°C)
Bombay	19	31
Casablanca	34	22
Dublin	53	13
Hong Kong	22	25
Istanbul	41	18
St Petersburg	60	8
Manila	15	32
Oslo	60	10
Paris	49	15
London	51	12
New Orleans	30	22
Calcutta	22	26

a) Draw a scatter diagram to show these data.

b) Comment on the graph.

Solution

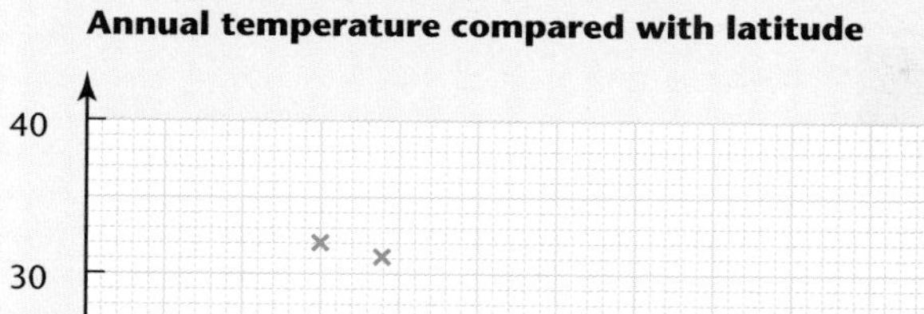

a) Plot the latitude as the x-coordinate and the mean temperature as the y-coordinate.

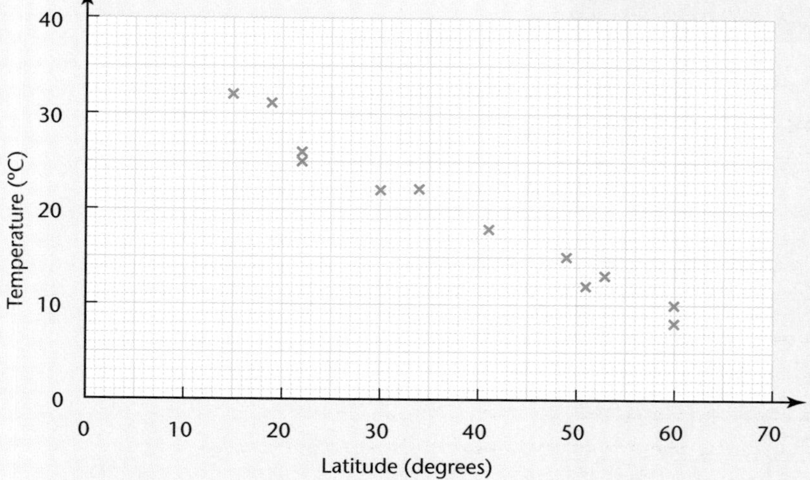

b) You can see that the points tend to lie in a straight line.

There appears to be a relationship between temperature and latitude: the farther north the city, the lower the temperature tends to be.

Exercise 8.1

1 The marks of ten students in the two papers of a French exam are shown in the table.

Paper 1	20	32	40	45	60	67	71	80	85	91
Paper 2	15	25	40	40	50	60	64	75	76	84

a) Plot these marks on a scatter diagram.
b) Comment on the graph.

2 Twelve people were chosen at random and asked to solve a problem. The time, in minutes, each person took and their age were recorded.

Age	14	45	18	74	60	23	21	56	20	39	30	40
Time	20	12	12	23	15	6	5	18	6	11	7	10

a) Draw a scatter diagram.
b) Comment on the graph.

3 In the Key Stage 2 Mathematics National Tests, students take two written tests, test A and test B, and a mental mathematics test. The marks for the 26 students in a class are shown in the table.

Test A	Test B	Mental mathematics
28	32	16
36	37	19
6	5	3
17	18	5
24	31	16
21	29	17
24	27	14
12	11	9
19	21	17
27	37	16
30	28	16
13	22	6
19	29	17
9	6	5
18	33	12
35	36	18
12	17	6
26	33	15
9	11	4
27	31	19
13	22	9
35	37	18
15	23	12
8	12	3
25	37	19
14	24	19

a) Draw a scatter diagram to compare the marks gained on test A and the mental test score.

b) Draw a scatter diagram to compare the marks gained on test A and those gained on test B.

4 The table gives the height and the resting pulse rate for 18 people.

Height (cm)	Pulse rate (beats per minute)
160	68
162	64
180	80
173	92
170	80
163	80
148	82
160	84
180	90
165	84
172	116
163	95
168	90
182	76
170	84
155	80
175	104
180	68

Draw a scatter diagram and investigate if there is a relationship between height and pulse rate.

5 Sam is convinced that the more chocolate bars he eats the heavier he will become.
Sketch a scatter diagram to show this.

Activity 1

Do tall people have large feet?

Survey people in your class or in your year and record their height and foot size (foot length might be better than shoe size but either could be done).

Plot height on the horizontal axis and foot size on the vertical axis.

What do you notice? Try other year groups.

Correlation

A scatter diagram is used to find out whether there is a **correlation**, or relationship, between two sets of data. The data are presented as pairs of values, each of which is plotted as a coordinate point on a graph.

Here are some examples of what a scatter diagram could look like and how we might interpret them.

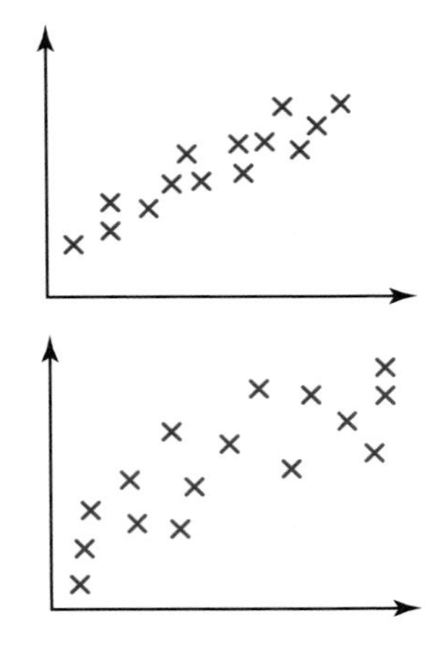

Strong positive correlation

Here, one quantity increases as the other increases.

This is called **positive correlation**.

The trend is from bottom left to top right.

When the points are closely in line, we say that the correlation is **strong**.

Weak positive correlation

These points also display positive correlation.

The points are more scattered, so we say that the correlation is **weak**.

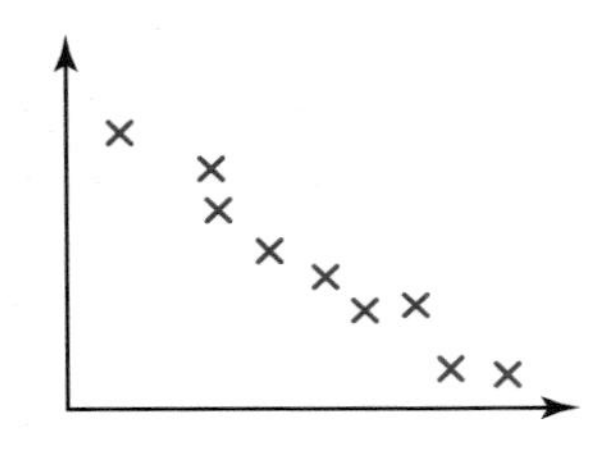

Strong negative correlation

Here, one quantity decreases as the other increases.

This is called **negative correlation**.

The trend is from top left to bottom right.

Again, the points are closely in line, so we say that the correlation is **strong**.

Weak negative correlation

These points also display negative correlation.

The points are more scattered, so the correlation is **weak**.

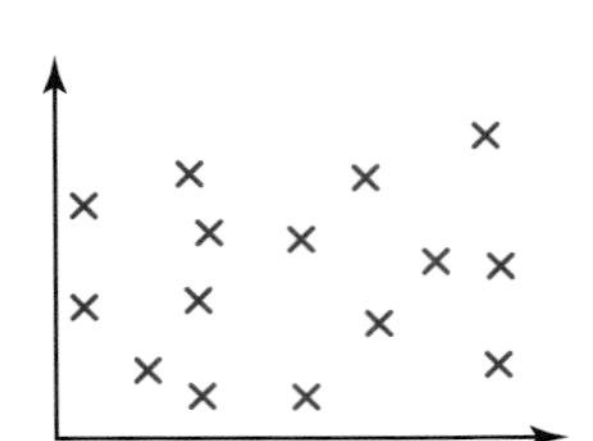

No correlation

When the points are totally scattered and there is no clear pattern, we say that there is **no correlation** between the two quantities.

Lines of best fit

If a scatter diagram shows correlation, you can draw a **line of best fit** on it.

Try putting your ruler in various positions on the scatter diagram until you have a slope which matches the general slope of the points. There should be roughly the same number of points on each side of the line.

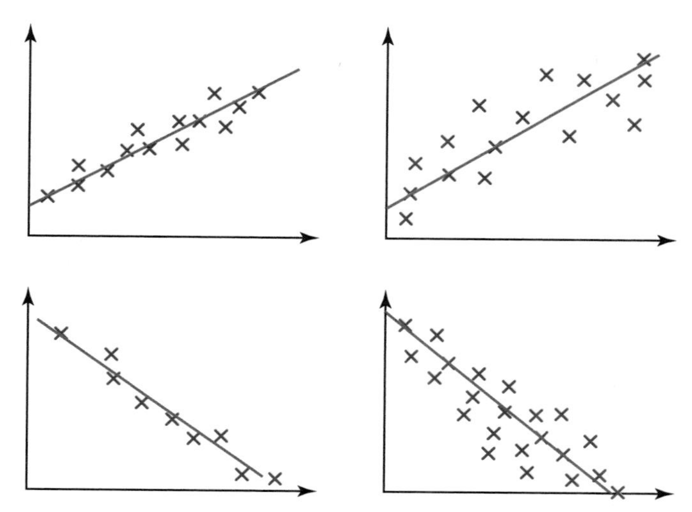

You cannot draw a line of best fit on a scatter diagram with no correlation.

You can use the line of best fit to predict a value when only one of the pair of quantities is known.

Example 2

The table shows the heights and weights of 12 people.

Height (cm)	150	152	155	158	158	160	163	165	170	175	178	180
Weight (kg)	56	62	63	64	57	62	65	66	65	70	66	67

a) Draw a scatter diagram to show these data.

b) Comment on the strength and type of correlation between these heights and weights.

c) Draw a line of best fit on your scatter diagram.

d) Tom is 162 cm tall. Use your line of best fit to estimate his weight.

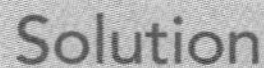

Solution

a), c)

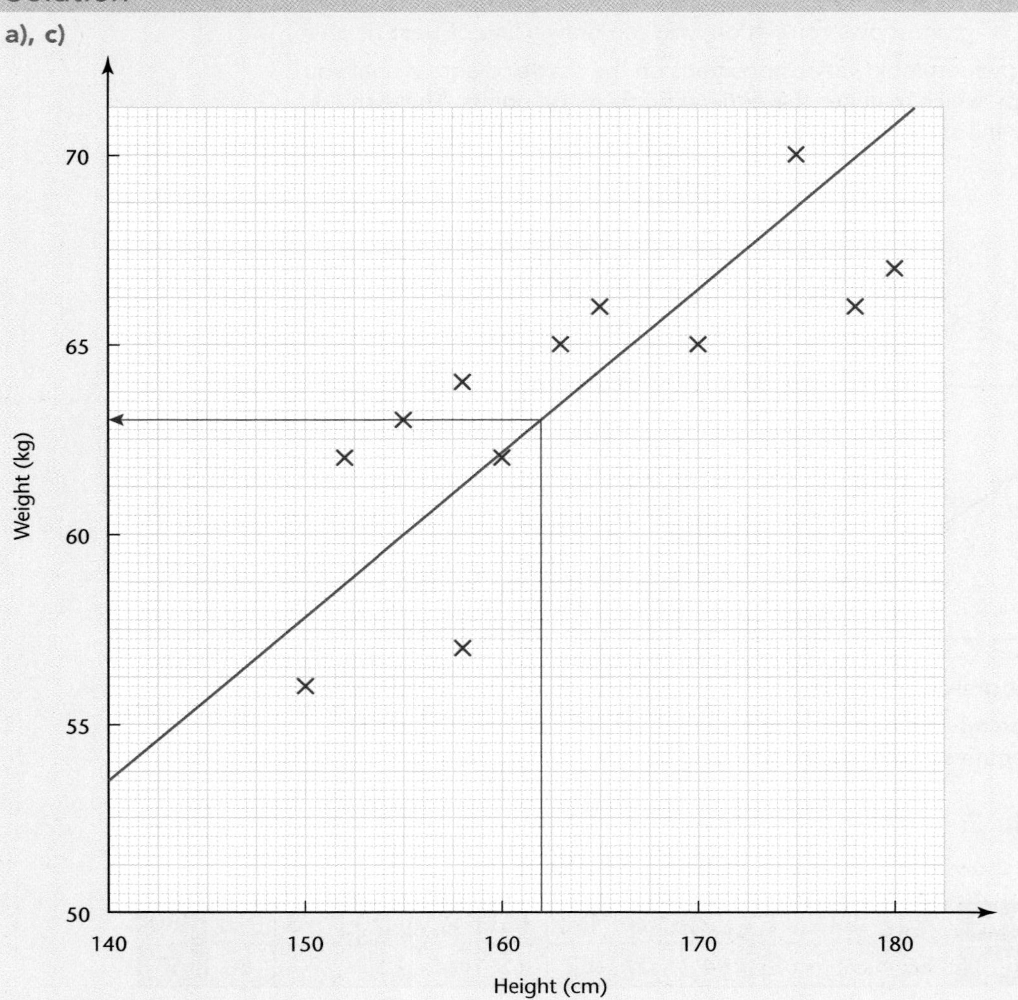

b) Weak positive correlation.

d) Draw a line up from 162 cm on the height axis to meet your line of best fit.
Now draw a horizontal line and read off the value where it meets the weight axis.
Tom's probable weight is about 63 kg.

Exercise 8.2

1 The scatter diagram shows the numbers of sun beds hired out and the hours of sunshine at Brightsea. Comment on the results shown by the scatter diagram.

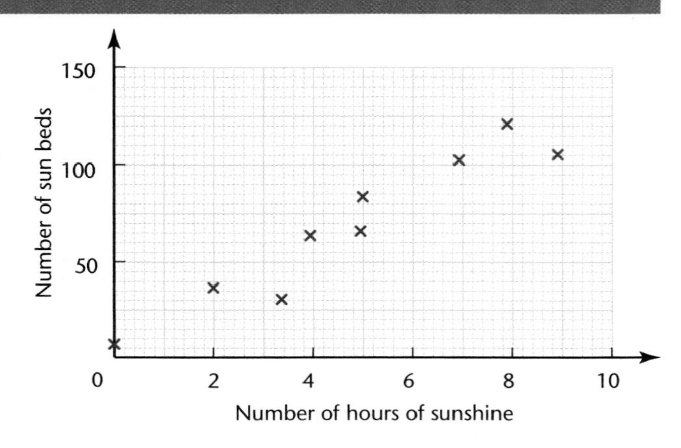

2 A firm noted the numbers of days of 'sick-leave' taken by its employees in a year, and their ages. The results are shown in the graph. Comment on the results shown by the scatter diagram.

3 A teacher thinks that there is a correlation between how far back in class a student sits and how well they do at maths. To test this, she plotted their last maths grade against the row they sit in. Here is the graph she drew.
Was the teacher right? Give your reasons.

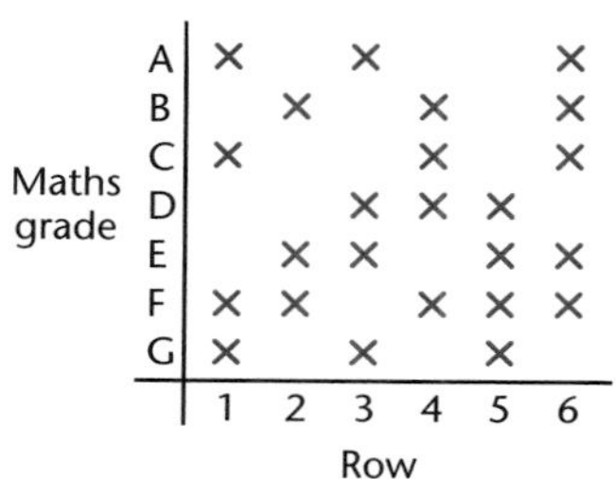

4 The scatter diagram shows the positions of football teams in the league and their mean crowd numbers, in thousands.
Comment on the scatter diagram.

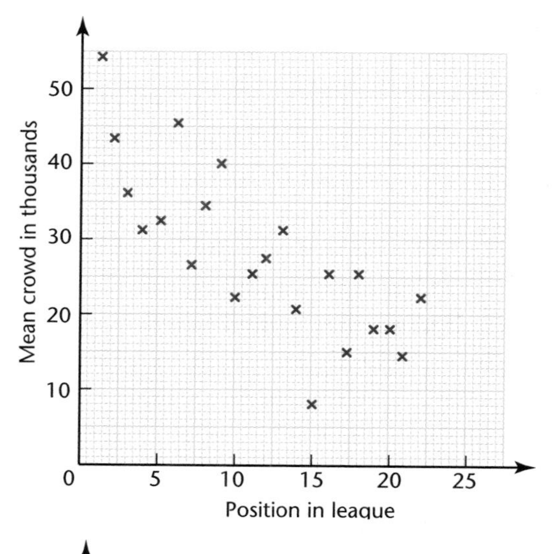

5 The scatter diagram shows the ages of people and the numbers of lessons they took before they passed their driving tests.
Comment on the scatter diagram.

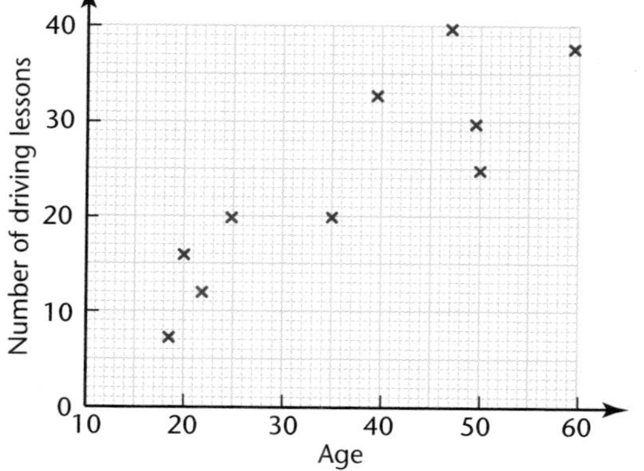

Foundation Gold/ Higher Bronze

6 The table shows the maths and science marks of eight students in their last examination.
  a) Draw a scatter diagram to show this information, with the maths mark on the horizontal axis.
  b) Comment on the diagram.
  c) Draw a line of best fit.
  d) Use your line of best fit to estimate
    (i) the science mark of a student who scored 40 in maths.
    (ii) the maths mark of a student who scored 75 in science.

Students	Maths mark	Science mark
A	10	30
B	20	28
C	96	80
D	50	55
E	80	62
F	70	70
G	26	38
H	58	48

7 The table shows the amount of petrol left in a fuel tank after different numbers of miles travelled.
  a) Draw a scatter diagram to show this information, with the number of miles on the horizontal axis.
  b) Comment on the diagram.
  c) Draw a line of best fit.
  d) Use your line of best fit to estimate the number of gallons left after 170 miles.

Number of miles	Number of gallons
50	7
100	5.2
150	4.2
200	2.6
250	1.2
300	0.4

8 In Kim's game, 20 objects are placed on a table and you are given a certain time to look at them. They are then removed or covered up and you have to recall as many as possible. The table shows the lengths of time given to nine people and the numbers of items they remembered.
  a) Draw a scatter diagram to show this information, with time in seconds on the horizontal axis.
  b) Comment on the diagram.
  c) Draw a line of best fit.
  d) Use your line of best fit to estimate the number of items remembered if 32 seconds are allowed.
  e) Why should the diagram not be used to estimate the number of items remembered in 3 seconds?

Time in seconds	Number of items
20	9
25	8
30	12
35	10
40	12
45	15
50	13
55	16
60	18

9 In Lily's class, a number of students have part-time jobs. Lily thinks that the more time they spend on their jobs, the worse they will do at school. She asked ten of them how many hours a week they spent on their jobs, and found their mean marks in the last examinations. Her results are shown in the table.

a) Plot a scatter diagram to show Lily's results, with time in hours on the horizontal axis.

b) Do the results confirm Lily's views? Are there any exceptions?

c) Draw a line of best fit for the relevant points.

d) Estimate the mean score of a student who spent 12 hours on their part-time job.

Student	Time on part-time job (hours)	Mean mark in examination
A	9	50
B	19	92
C	13	52
D	3	70
E	15	26
F	20	10
G	5	80
H	17	36
I	6	74
J	22	24

10 In an experiment, Tom's reaction time is tested after he has undergone vigorous exercise. The table shows Tom's reaction times and the lengths of time spent in exercise.

a) Draw a scatter diagram to show this information, with the number of minutes of exercise on the horizontal axis.

b) Comment on the diagram.

c) Draw a line of best fit.

d) Use your line of best fit to estimate Tom's reaction time after 35 minutes of exercise.

Amount of exercise (minutes)	Reaction time (seconds)
0	0.34
10	0.46
20	0.52
30	0.67
40	0.82
50	0.91

11 The table shows the number of bad peaches per box after different delivery times.

a) Draw a scatter diagram to show this information.

b) Describe the correlation shown in the scatter diagram.

c) Draw a line of best fit on your scatter diagram.

d) Use your line of best fit to estimate the number of bad peaches expected after a 12 hour delivery time.

Delivery time (hours)	Number of bad peaches
10	2
4	0
14	4
18	5
6	2

12 The table shows the marks of 15 students taking Paper 1 and Paper 2 of a maths examination. Both papers were marked out of 40.

a) Draw a scatter diagram to show this information.

b) Describe the correlation shown in the scatter diagram.

c) Draw a line of best fit on your scatter diagram.

d) Joe scored 32 on Paper 1 but was absent for Paper 2.
Use your line of best fit to estimate his score on Paper 2.

Paper 1	Paper 2
36	39
34	36
23	27
24	20
30	33
40	35
25	27
35	32
20	28
15	20
35	37
34	35
23	25
35	33
27	30

13 The table shows the engine size and petrol consumption of nine cars.

a) Draw a scatter diagram to show this information.

b) Describe the correlation shown in the scatter diagram.

c) Draw a line of best fit on your scatter diagram.

d) A tenth car has an engine size of 2.8 litres.
Use your line of best fit to estimate the petrol consumption of this car.

Engine size (litres)	Petrol consumption (mpg)
1.9	34
1.1	42
4.0	23
3.2	28
5.0	18
1.4	42
3.9	27
1.1	48
2.4	34

14 Tracy thinks that the larger your head, the cleverer you are. The table shows the number of marks scored in a test by each of ten students, and the circumference of their head.

a) Draw a scatter diagram to show this information.

b) Describe the correlation shown in the scatter diagram.

c) Is Tracy correct?

d) Can you think of any reasons why the data may not be valid?

Circumference of head (cm)	Mark
600	43
500	33
480	45
570	31
450	25
550	42
600	23
460	36
540	24
430	39

Challenge 1

a) Dan thinks that the more time he spends on his school work, the less money he will spend. Sketch a scatter diagram that shows this.

b) Fiona thinks that the more she practises, the more goals she will score at hockey. Sketch a scatter diagram to show this.

Challenge 2

Investigate one of these.

- Petrol consumption and the size of a car's engine
- A person's height and their head circumference
- House price and the distance from the town centre
- Age of car and second-hand price

Key Ideas

- Scatter diagrams show the correlation between two variables.

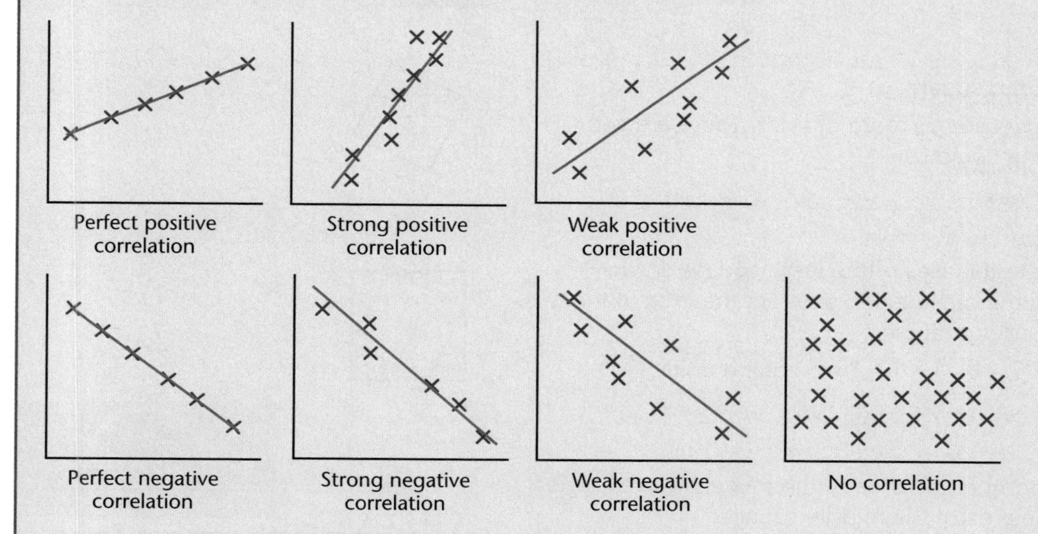

- If there is reasonable correlation, a line of best fit can be drawn.
- The line of best fit should reflect the slope of the points.
- There should be approximately the same number of points on each side of the line.
- The line of best fit can be used to estimate the value of one variable if the other is known.
- The line of best fit can only be used to estimate values within the range of the given data.

Pythagoras' theorem

YOU WILL LEARN ABOUT	YOU SHOULD ALREADY KNOW
o Using Pythagoras' theorem to find the length of one side of a right-angled triangle o Using Pythagoras' theorem to solve problems o Using Pythagoras' theorem to show that a triangle is right-angled	o How to find the area of a triangle o How to find squares and square roots on a calculator

Pythagoras' theorem

This is a square drawn on squared paper.

Its area is 4 square units.

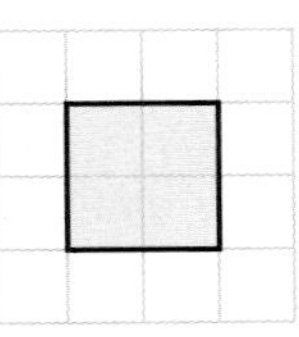

Here is a tilted square.

Calculating its area is more difficult. There are two methods you could use.

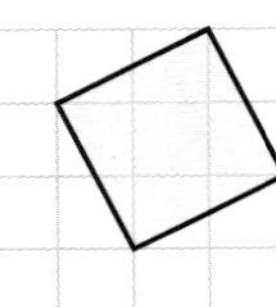

1 Calculate the area of the large square drawn around the outside and subtract the area of the four shaded triangles.

Area = 9 – 1 – 1 – 1 – 1 = 5 square units

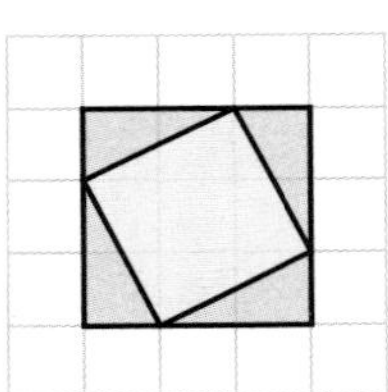

2 Add together the area of the four shaded triangles and the area of the middle square.

Area = 1 + 1 + 1 + 1 + 1 = 5 square units

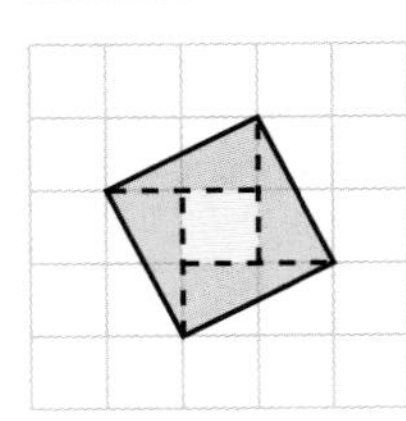

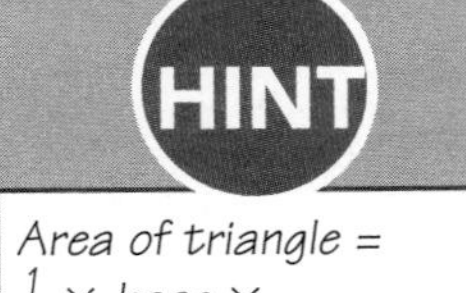

Area of triangle = $\frac{1}{2}$ × base × perpendicular height

Activity 1

a) Use either method 1 or method 2 to calculate the area of the squares in the diagram opposite.

b) Draw some more tilted squares of your own on squared paper and find their areas.

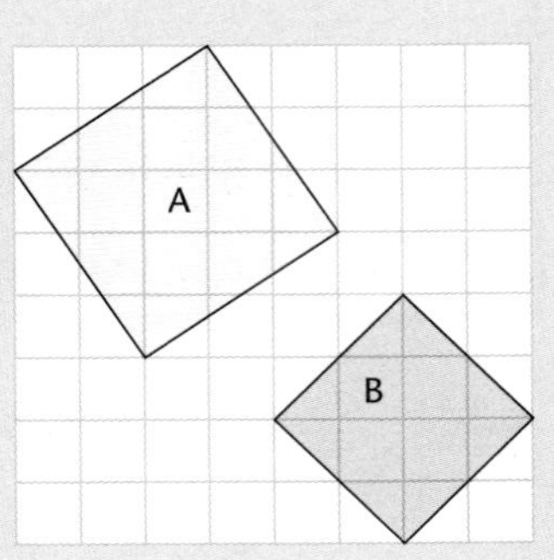

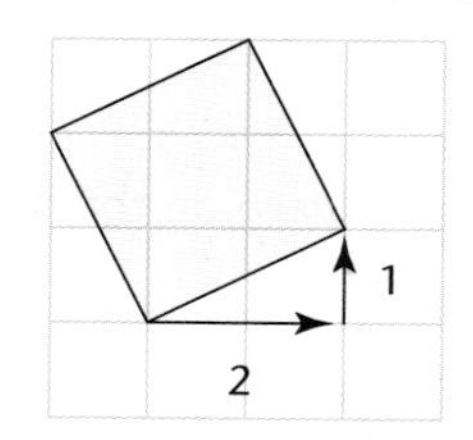

Look at all the tilted squares you drew in Activity 1.

Code the tilt by drawing a triangle at the base and writing down the length of its sides, like this.

In this diagram the code is (2, 1). In the diagram in Activity 1, square A has a code of (3, 2) and square B has a code of (2, 2).

Check that you agree.

Code all the squares you have drawn and write the codes and the areas of the squares in a table. Include the square from the beginning of this chapter. Its code is (2, 0) and its area is 4 square units. Include all the other squares you have already studied in this chapter.

Look at the codes and their areas. See if you can find a rule linking them together.

Code	Area
(2, 0)	4
(2, 1)	5
(3, 2)	13
(2, 2)	8

You will have found that squaring each code number and then adding the squares together gives the area.

$2^2 + 0^2 = 4 + 0 = 4$
$2^2 + 1^2 = 4 + 1 = 5$
$3^2 + 2^2 = 9 + 4 = 13$
$2^2 + 2^2 = 4 + 4 = 8$

The rule linking them together is called **Pythagoras' theorem**.

Squaring the numbers in the code and adding them is the same as squaring the lengths of the sides of the triangle.

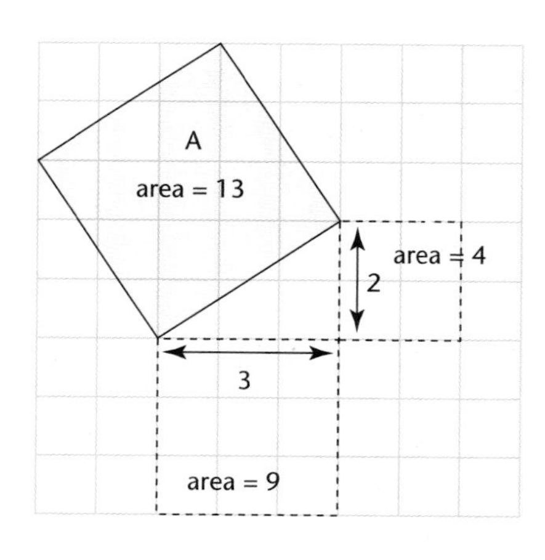

Here is square A again.

Can you see that you can calculate the area of the largest square by adding together the areas of the smaller squares?

The largest square will always be on the longest side of the triangle – this is called the **hypotenuse** of the right-angled triangle.

Pythagoras' theorem can be stated like this.

The area of the square on the hypotenuse = the sum of the areas of the squares on the other two sides.

Foundation Gold/ Higher Bronze

Here is a shorter version.

The square on the hypotenuse = the sum of the squares on the other two sides.

Activity 2

Copy this diagram.

Cut up the two smaller squares so that they fit together to make square A.

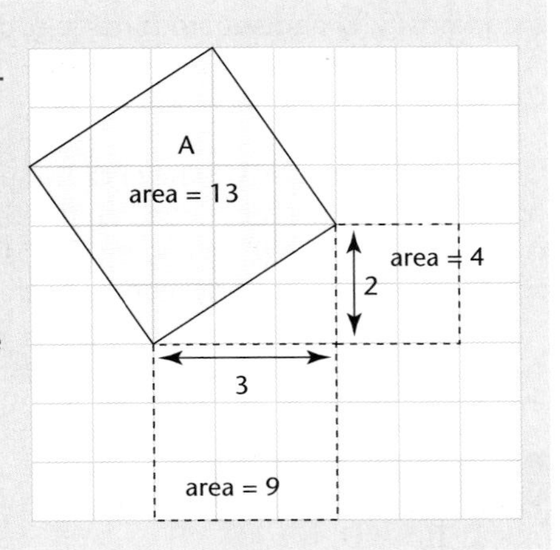

If you know the lengths of two sides of a right-angled triangle

Exercise 9.1

Calculate the missing area in each of these diagrams.

1

10 cm²

5 cm²

?

2

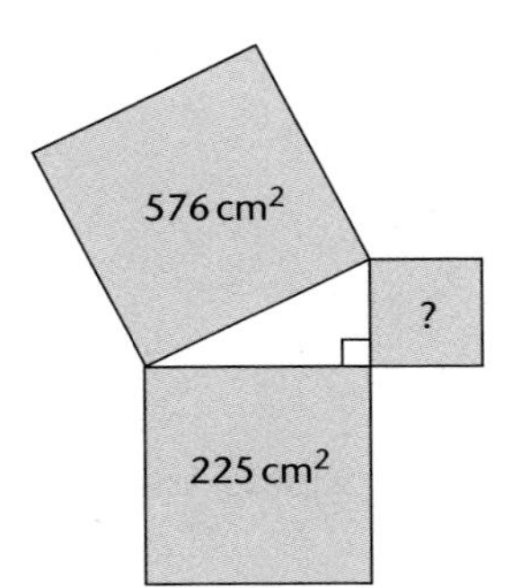

3

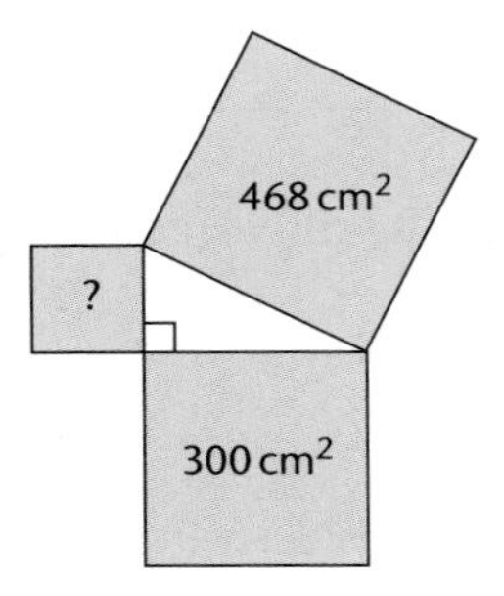

4

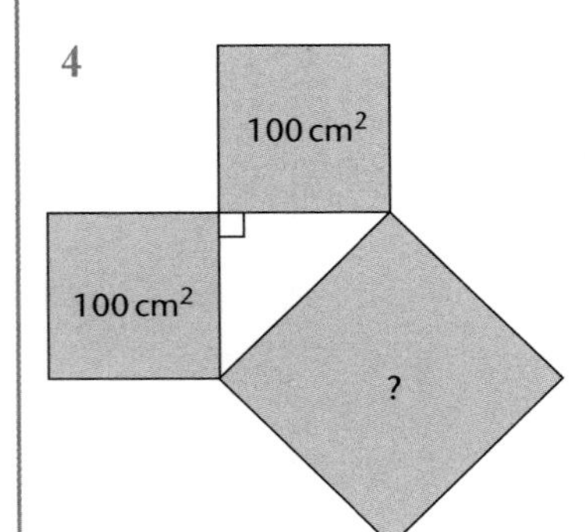

5

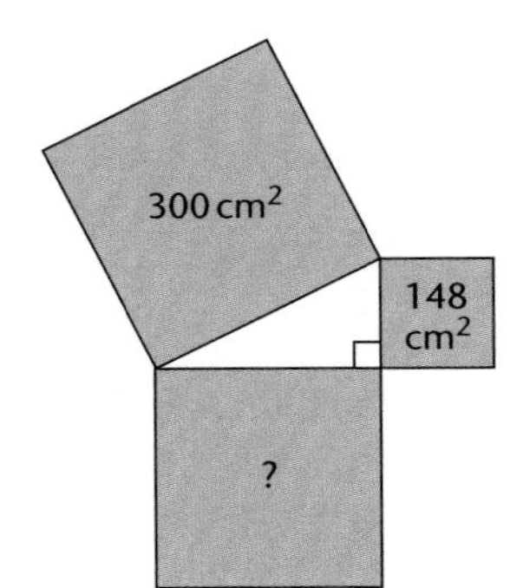

you can use Pythagoras' theorem to find the length of the third side.

The unknown area = 64 + 36 = 100 cm^2

This means that the sides of the unknown square have a length of $\sqrt{100}$ = 10 cm.

When using Pythagoras' theorem, you don't need to draw the squares – you can simply use the rule.

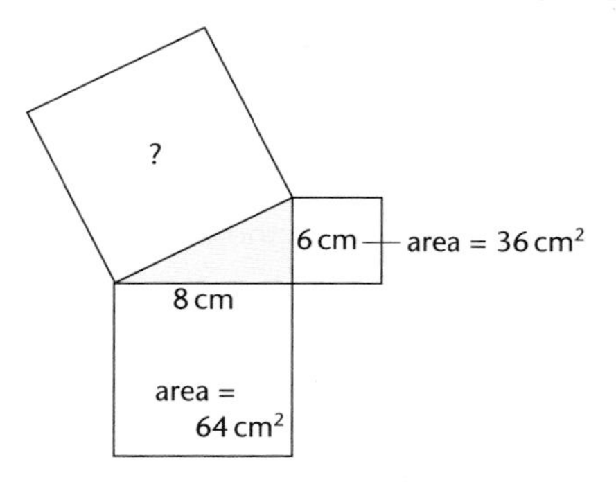

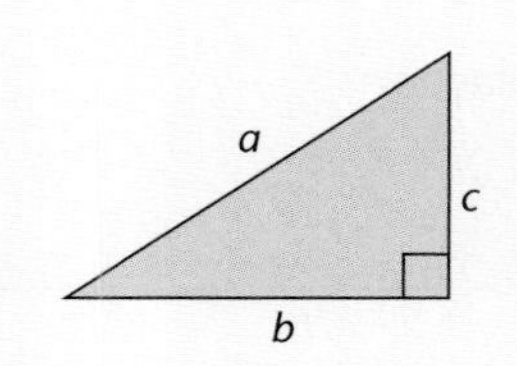

$$a^2 = b^2 + c^2$$

Example 1

Find the length a in the diagram.

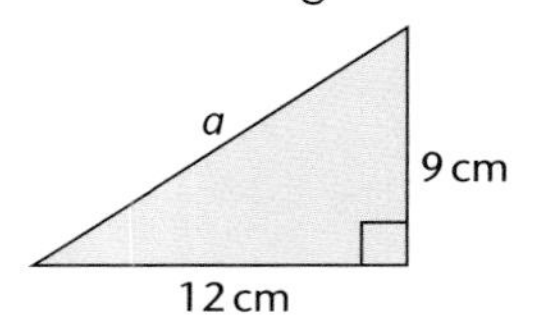

Solution

$a^2 = 9^2 + 12^2$
$= 81 + 144$
$= 225$
$a = \sqrt{225}$
$= 15$ cm

Example 2

Find the length a in the diagram.

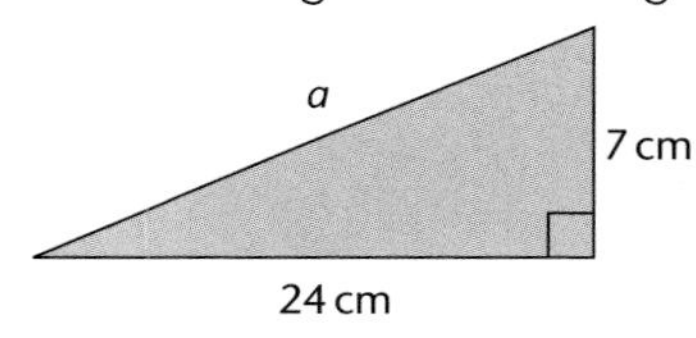

Solution

$a^2 = 7^2 + 24^2$
$= 49 + 576$
$= 625$
$a = \sqrt{625}$
$= 25$ cm

Exercise 9.2

Find the length of the hypotenuse, a, in each of these triangles.

1

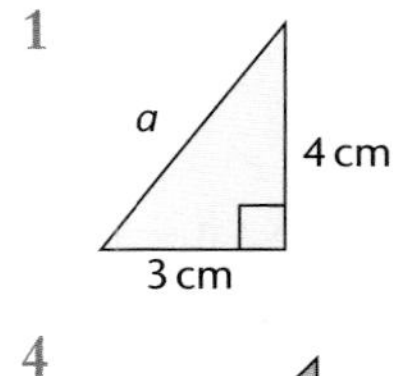

2

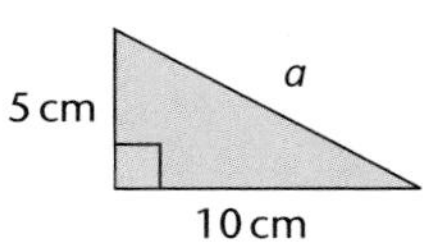

3

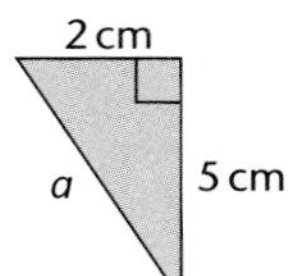

4

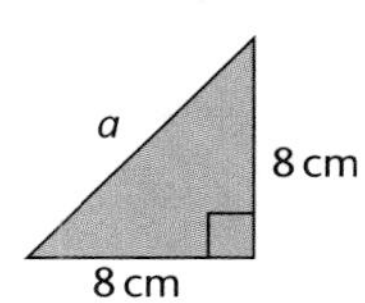

5

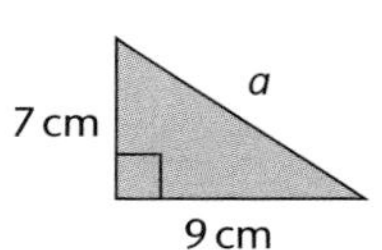

6

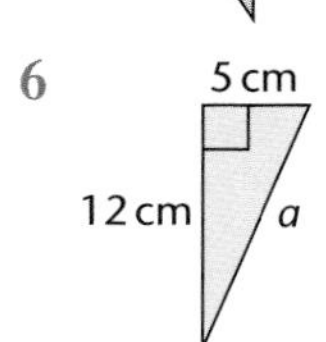

If you know the length of the hypotenuse and the length of one other side, you can find the length of the third side.

Example 3

Find the length c in the diagram.

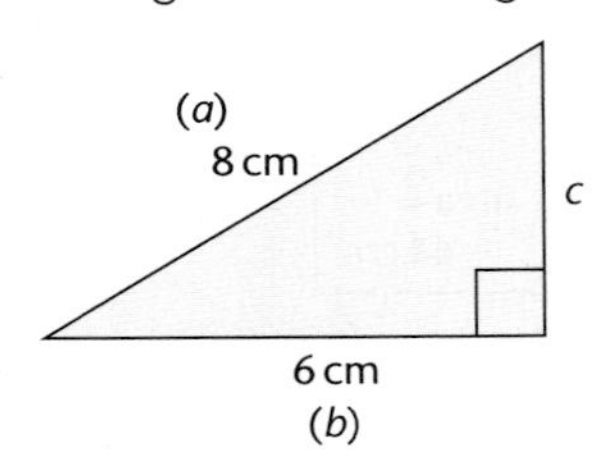

Solution

$a^2 = b^2 + c^2$

$8^2 = 6^2 + c^2$

$64 = 36 + c^2$

$c^2 = 64 - 36 = 28$

$c = \sqrt{28} = 5.29$ cm (to 2 decimal places)

Exercise 9.3

Calculate the length of the third side in each of the triangles in questions **1** to **8**.

Give your answers either exactly or correct to 2 decimal places.

1 17 cm, 15 cm, b

2 7 cm, 9 cm, b

3 20 cm, c, 12 cm

4 30 cm, 8 cm, b

5 5 cm, 169 cm, b

6 b, 3 cm, 5 cm

7 14 cm, c, 20 cm

8 a, 4 cm, 8 cm

For each of the triangles in questions **9** to **17**, find the length marked x.

Give your answers either exactly or correct to 2 decimal places.

9 5 cm, x, 12 cm

10 5 m, x, 3 m

11 5 cm, 8 cm, x

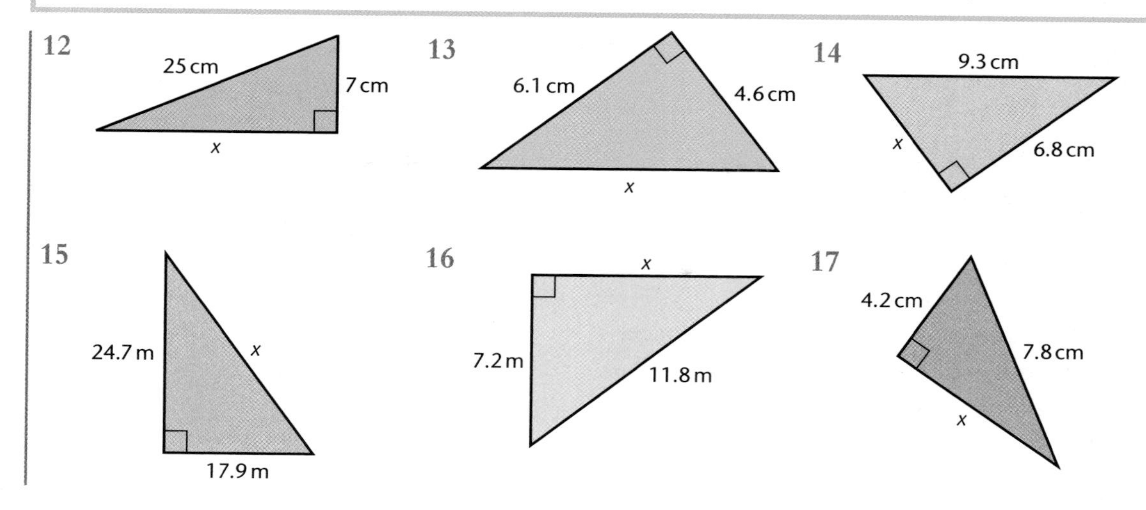

You can use Pythagoras' theorem to solve problems. It is a good idea to draw a sketch if a diagram isn't given. Try to draw it roughly to scale and mark on it any lengths you know.

Example 4	Solution
Tom is standing 115 m from a vertical tower. The tower is 20 m tall. Work out the distance from Tom directly to the top of the tower.	$x^2 = 115^2 + 20^2$ $= 13625$ $x = \sqrt{13625}$ $= 116.7$ m (to 1 decimal place) Tom is 116.7 m (to 1 decimal place) from the top of the tower.

B
x
20 m
T
A
115 m

Exercise 9.4

Solve these problems. In each case draw a diagram first to help you.

1 A rectangular field is 225 m long and 110 m wide.
Find the length of the diagonal path across it.

2 A rectangular field is 25 m long.
A footpath 38.0 m long crosses the field diagonally.
Find the width of the field.

3 A ladder is 7 m long.
It is resting against a wall, with the top of the ladder 5 m above the ground.
How far from the wall is the base of the ladder?

4 Harry is making a kite for his sister. This is his diagram of the kite.

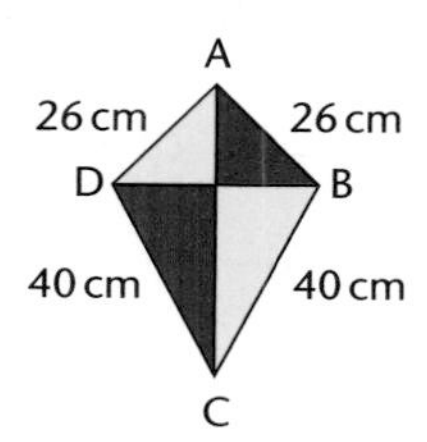

The kite is 30 cm wide.
Harry needs to buy some cane to make the struts AC and DB.
What length of cane does he need to buy?

Foundation Gold/
Higher Bronze

5 This is the side view of a shed.

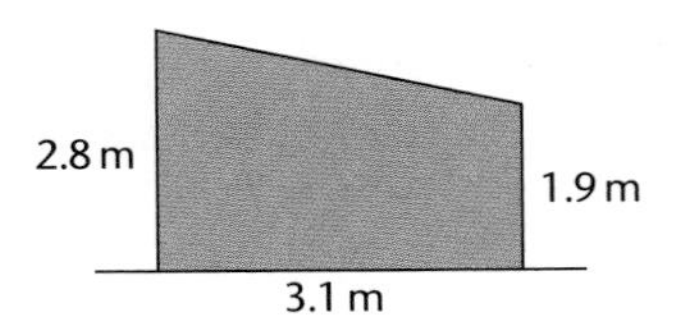

Find the length of the sloping roof.

6 This is the cross-section of a roof space.
The roof timbers AB and BC are each 6.5 m long.
The floor joist AC is 12 m.

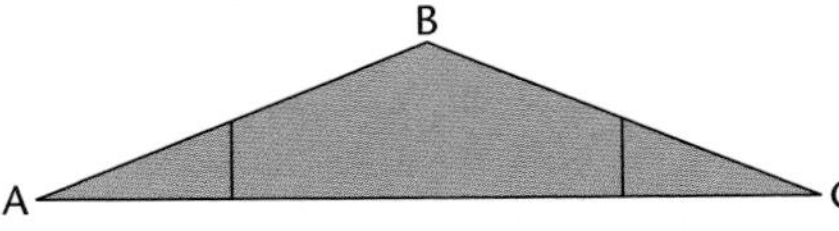

What is the maximum height in the roof space?

7 A tent pole is secured by guy ropes which are 2.4 m long.
They reach the ground, which is horizontal, 1.6 m away from the base of the pole.
How high up the pole are the guy ropes fastened?

Pythagorean triples

Look at these triangles.

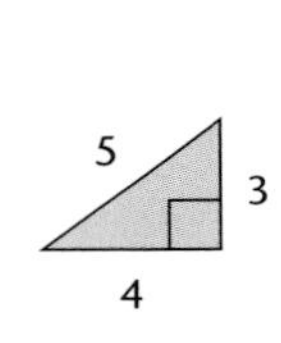

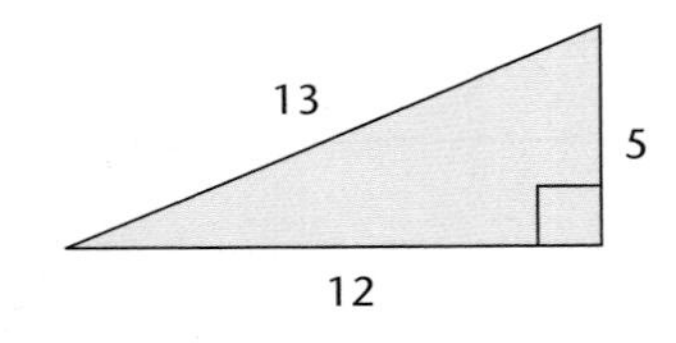

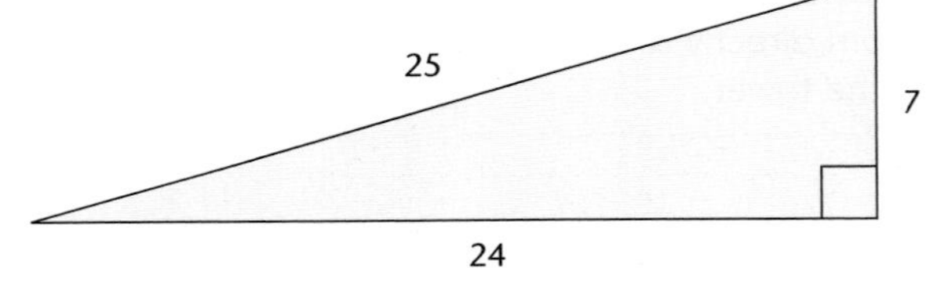

$3^2 + 4^2 = 5^2$ $\quad$ $5^2 + 12^2 = 13^2$ $\quad$ $7^2 + 24^2 = 25^2$

These are examples of **Pythagorean triples**, three numbers that exactly fit Pythagoras' relationship.

3, 4, 5 $\quad$ 5, 12, 13 $\quad$ and $\quad$ 7, 24, 25

are the most well-known whole-number Pythagorean triples.

Activity 3

Check that 2.5, 6, 6.5 is also a Pythagorean triple.

You can also use Pythagoras' theorem in reverse.

If the lengths of the three sides of a triangle form a Pythagorean triple, then the triangle is right-angled.

Exercise 9.5

State whether or not each of these triangles is right-angled. Show your working.

1

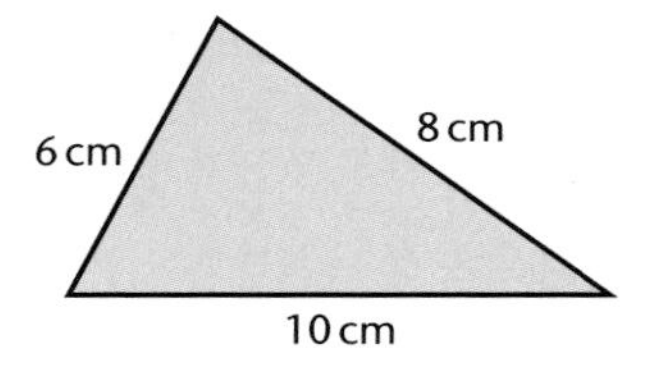

2

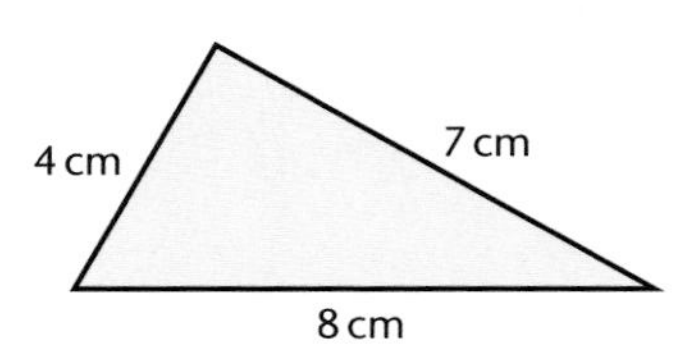

3

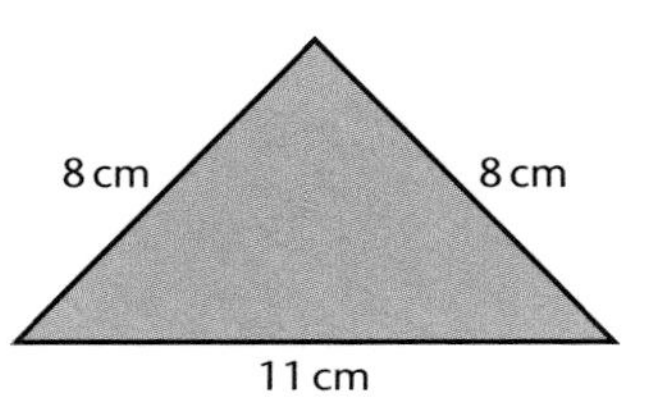

4

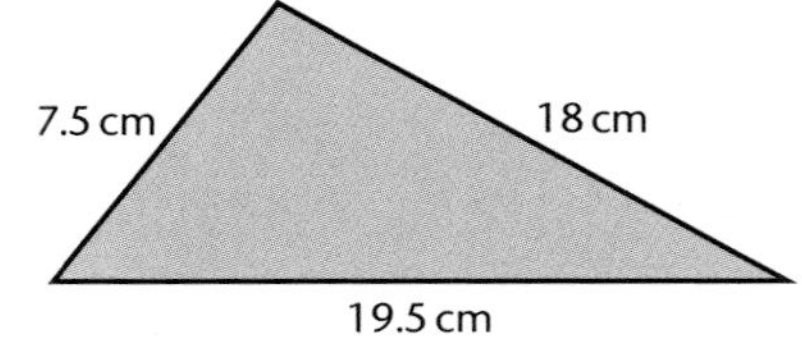

5

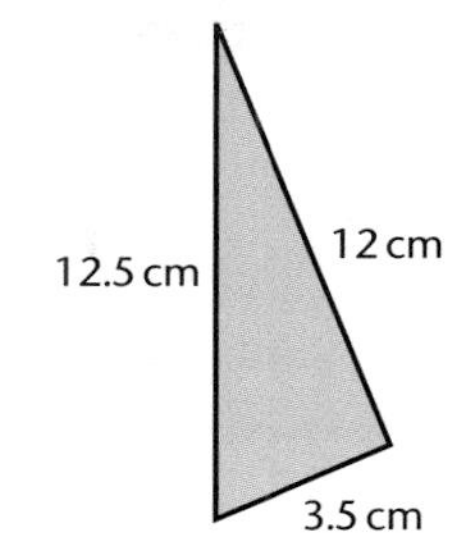

6

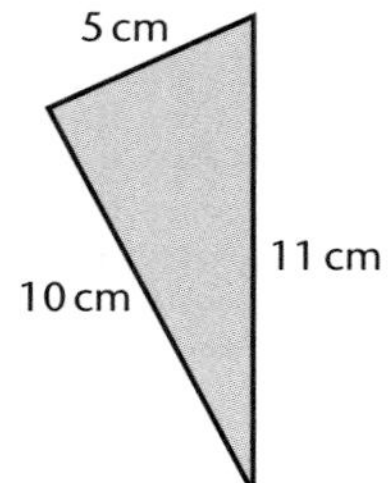

7

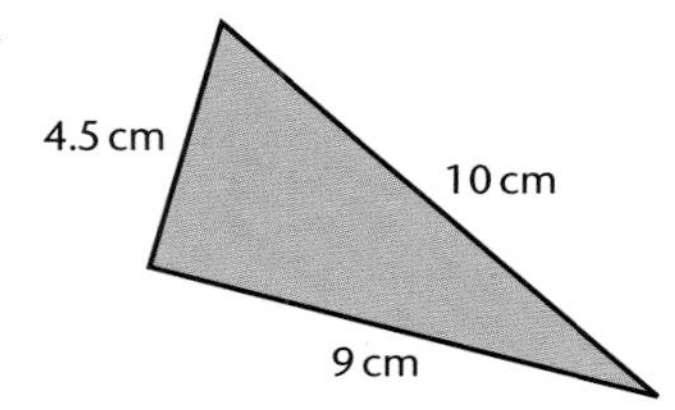

8

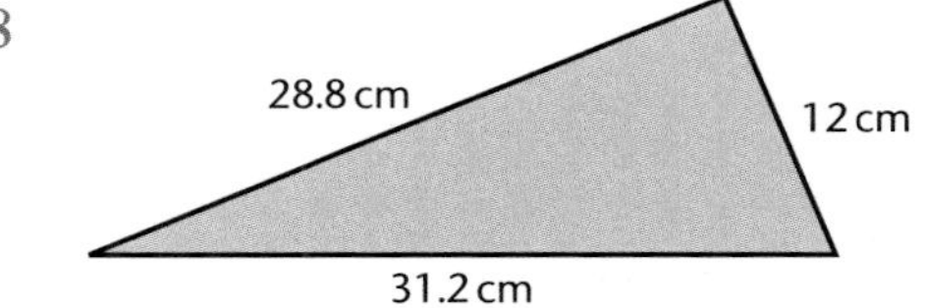

Challenge 1

Plot the points A(3, 1) and B(7, 4) on squared paper.

Complete the right-angled triangle ABC.

Find the length of AB.

$AB^2 = AC^2 + BC^2$
$AC = 7 - 3 = 4$ and $BC = 4 - 1 = 3$
$AB^2 = 4^2 + 3^2 = 25$
$AB = 5$

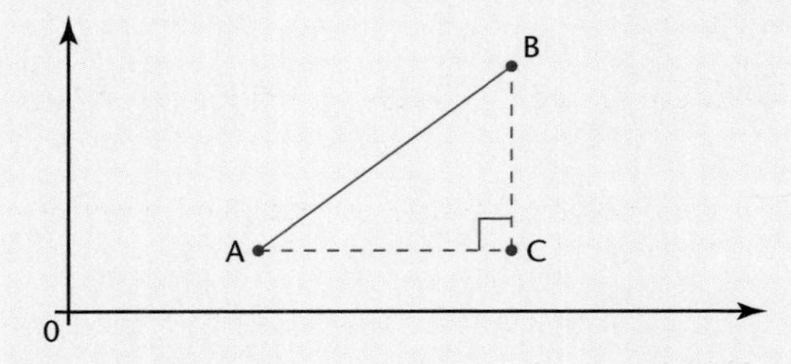

Now find the distance between each of these pairs of points.

a) (1, 1) and (5, 5)
b) (6, 2) and (2, 1)
c) (3, 4) and (0, 0)
d) $(^{-}2, ^{-}1)$ and (4, 1)

Key Ideas

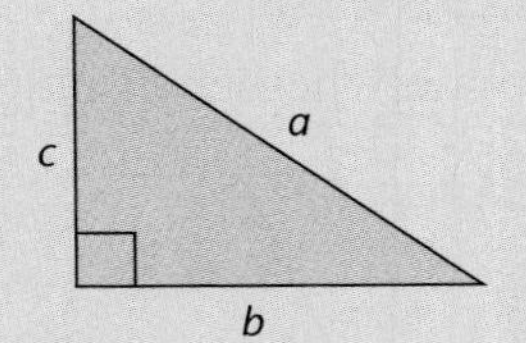

- For a right-angled triangle, Pythagoras' theorem states that $a^2 = b^2 + c^2$.
- A Pythagorean triple is a set of three numbers that exactly fits Pythagoras' relationship.
- If the lengths of the three sides of a triangle form a Pythagorean triple, then the triangle is right-angled.

Chapter 10 Quadratic graphs

YOU WILL LEARN ABOUT	YOU SHOULD ALREADY KNOW
○ The shape of a quadratic graph ○ Drawing a quadratic graph, given its equation ○ Using a quadratic graph to solve equations	○ How to draw the graph of a straight line, given its equation

Drawing quadratic graphs

Quadratic graphs are graphs of equations of the form

$y = ax^2 + bx + c$

where a, b and c are constants, and b and c may be zero. Examples of quadratic equations are $y = x^2 + 3$ and $y = x^2 - 4x + 3$. When the graph is drawn, it produces a curve called a **parabola**.

Activity 1

State whether or not each of these is quadratic.

a) $y = x^2$

b) $y = x^2 + 5x - 4$

c) $y = \frac{5}{x}$

d) $y = x^2 - 3x$

e) $y = x^2 - 3$

f) $y = x^3 + 5x^2 - 2$

g) $y = x(x - 2)$

Example 1

Draw the graph of $y = x^2 + 4$, for values of x from $^-3$ to 3.

Exam tip

Complete a table of values to work out the points. If you are not given a table then make your own.

Solution

First make a table of values.

x	$^-3$	$^-2$	$^-1$	0	1	2	3
x^2	9	4	1	0	1	4	9
$+ 4$	4	4	4	4	4	4	4
$y = x^2 + 4$	13	8	5	4	5	8	13

Now label the axes. The values of x go from $^-3$ to 3 and the values of y go from 4 to 13. It is better to include 0 in the values of y, so let them go from 0 to 15.

On this graph, the scales used are 1 cm to 1 unit on the x-axis and 1 cm to 5 units on the y-axis.

Plot the points and join them with a smooth curve. Label the curve.

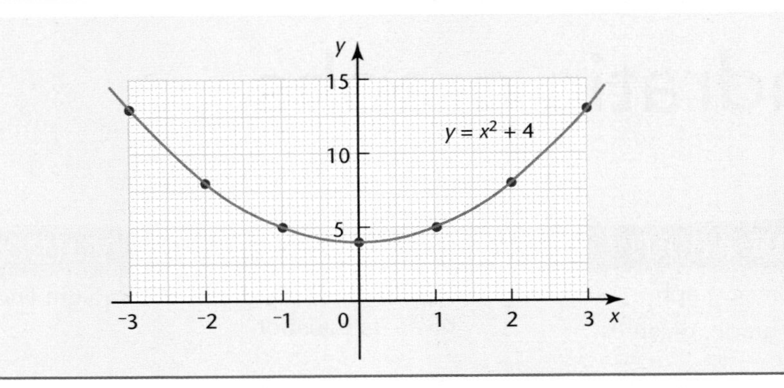

Exam tip

Practise drawing smooth curves. Keep your pencil on the paper. Make the curve a thin single line. Look ahead to the next point as you draw the line.

Example 2

Draw the graph of $y = x^2 + 3x$, for values of x from $^-5$ to 2.

Solution

First make a table of values.

x	$^-5$	$^-4$	$^-3$	$^-2$	$^-1$	0	1	2
x^2	25	16	9	4	1	0	1	4
$+ 3x$	$^-15$	$^-12$	$^-9$	$^-6$	$^-3$	0	3	6
$y = x^2 + 3x$	10	4	0	$^-2$	$^-2$	0	4	10

Add the numbers in rows 2 and 3 to give each value of y.

Label the axes from $^-5$ to 2 for x and from $^-5$ to 10 for y.

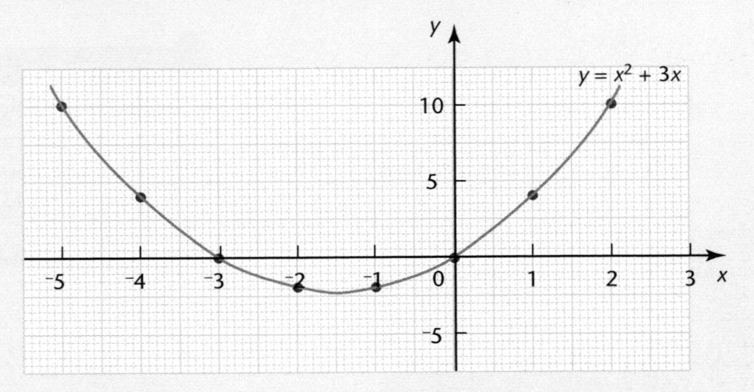

Here you can see that $x = {}^-1$ and $x = {}^-2$ both give $y = {}^-2$, and the lowest value of y (called the **minimum**) will be when x is between $^-1$ and $^-2$.

It is useful to work out the value of y when $x = {}^-1.5$. To do this, add some more values to your table.

When $x = {}^-1.5$, $x^2 = 2.25$ and $3x = {}^-4.5$, so $y = {}^-2.25$.

Plot ($^-1.5$, $^-2.25$), which is the lowest point of the graph.

Exam tip

A common error is to include the value in the x-row when adding to find y so separate this row off clearly.

Exam tip

If you find that you have two equal lowest (or highest) values for y, the curve will go below (or above) that value. To find the y-value between the equal values, find the value of x halfway between the two points. Substitute to find the corresponding value of y.

Example 3

a) Draw the graph of $y = x^2 - 2x - 3$, for values of x from $^-2$ to 4.
Label the axes from $^-2$ to 4 for x, and from $^-5$ to 5 for y.

b) Find the values of x for which $y = 0$.

Solution

a)

x	$^-2$	$^-1$	0	1	2	3	4
x^2	4	1	0	1	4	9	16
$-2x$	4	2	0	$^-2$	$^-4$	$^-6$	$^-8$
-3	$^-3$	$^-3$	$^-3$	$^-3$	$^-3$	$^-3$	$^-3$
$y = x^2 - 2x - 3$	5	0	$^-3$	$^-4$	$^-3$	0	5

To find the value of y, add together the values in rows 2, 3 and 4.

Use a scale of 1 cm to 1 unit on the x-axis and 2 cm to 5 units on the y-axis.

Now plot the points ($^-2$, 5), ($^-1$, 0), and so on. Join them with a smooth curve.

Label the curve.

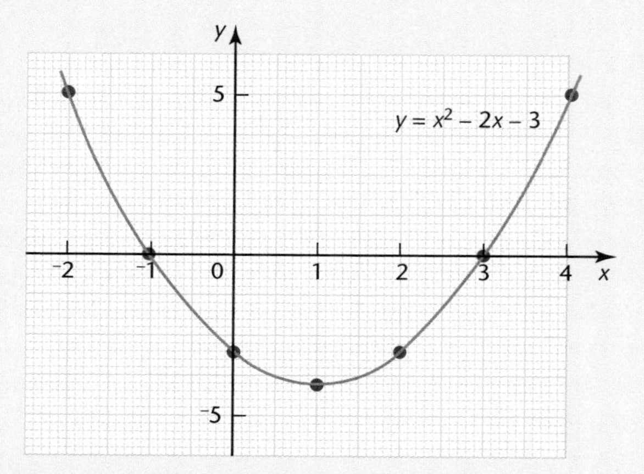

b) $y = 0$ when the curve crosses the x-axis.
This is when $x = {}^-1$ or $x = 3$.

Sometimes questions will be put in context. They will be about a real-life situation, rather than just being about a graph in terms of x and y.

Exam tip

Sometimes a table will be given with only the two rows for the x- and y-values. You may find it useful to include all the rows and then just add the correct rows to get the values of y.

Exam tip

x^2 is always positive so $^-x^2$ is always negative.
If the equation involves $^-x^2$, the parabola will be the opposite way up.
If one point does not fit the pattern, check that point again.

Example 4

The cost C, in pounds, of circular plates is given by the formula

$C = \frac{x^2}{10} + 2$, where x is the radius of the plate in centimetres.

a) Draw up a table of values and complete it.

b) (i) Draw the graph of C against x, for values of x from 5 to 20.

(ii) From your graph find the size of plate that would cost £16.40.

Solution

a)

x	5	8	10	15	20
$\frac{x^2}{10}$	2.5	6.4	10	22.5	40
$+ 2$	2	2	2	2	2
$C = \frac{x^2}{10} + 2$	4.5	8.4	12	24.5	42

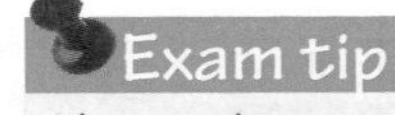

Always draw your curve in pencil so you can rub out any errors.

b) (i)

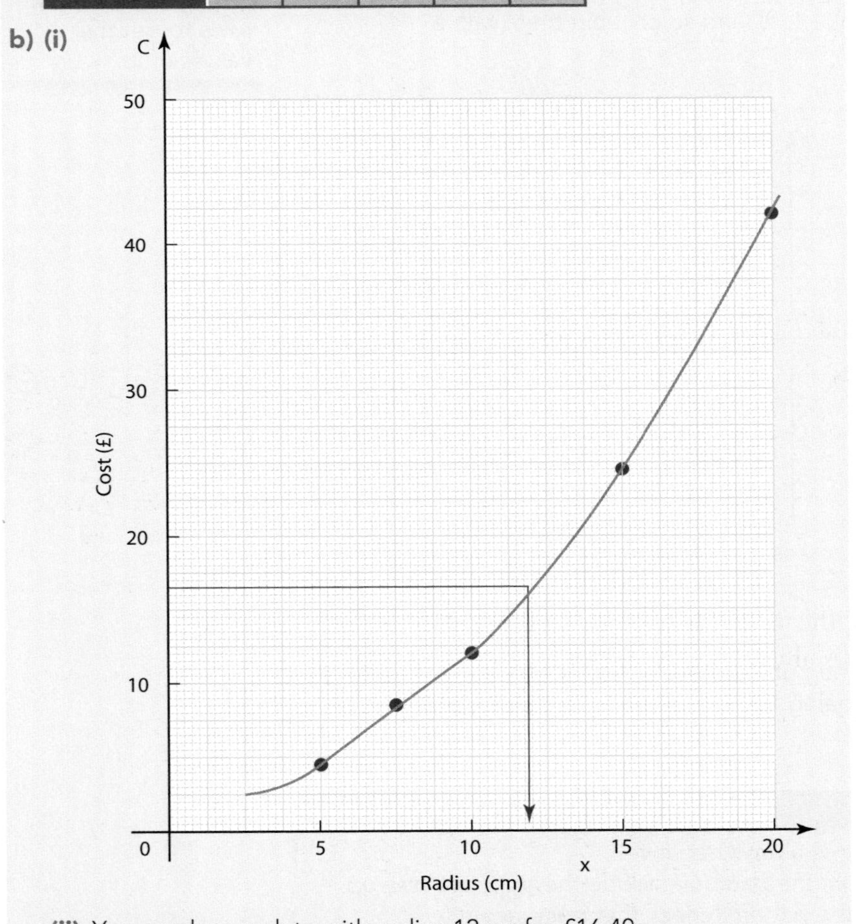

(ii) You can buy a plate with radius 12 cm for £16.40.

Exercise 10.1

1 Copy and complete this table for $y = x^2 + 5$. Do not draw the graph.

x	$^{-}3$	$^{-}2$	$^{-}1$	0	1	2	3
x^2	9						
$+ 5$	5						
$y = x^2 + 5$	14						

2 Copy and complete this table for $y = x^2 + 6$. Do not draw the graph.

x	$^{-}3$	$^{-}2$	$^{-}1$	0	1	2	3
x^2		4					
$+ 6$		6					
$y = x^2 + 6$		10					

3 Copy and complete this table for $y = x^2 + 3x - 7$. Do not draw the graph.
Notice the extra value at $x = {}^{-}1.5$.

x	$^{-}4$	$^{-}3$	$^{-}2$	$^{-}1$	0	1	2	$^{-}1.5$
x^2			4					
$+ 3x$			$^{-}6$					
$- 7$			$^{-}7$					
$y = x^2 + 3x - 7$			$^{-}9$					

4 Copy and complete this table for $y = 2x^2 - 8$. Do not draw the graph.

x	$^{-}3$	$^{-}2$	$^{-}1$	0	1	2	3
x^2	9						
$2x^2$	18						
$- 8$	$^{-}8$						
$y = 2x^2 - 8$	10						

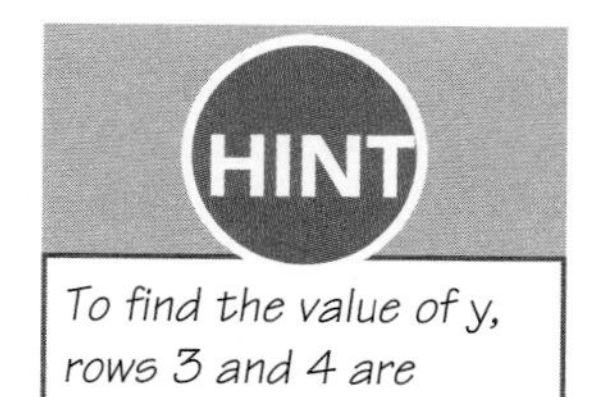

To find the value of y, rows 3 and 4 are added.

5 Copy and complete this table for $y = {}^{-}x^2 - 5x + 6$. Do not draw the graph.
Remember that $^{-}x^2$ is always negative. Notice the extra value at $^{-}2.5$.

x	$^{-}6$	$^{-}5$	$^{-}4$	$^{-}3$	$^{-}2$	$^{-}1$	0	1	2	$^{-}2.5$
$^{-}x^2$			$^{-}16$						$^{-}4$	
$- 5x$			20						$^{-}10$	
$+ 6$			6						6	
$y = {}^{-}x^2 - 5x + 6$			10						$^{-}8$	

6 Make a table of values for $y = x^2 - 3x + 1$, for values of x from $^-2$ to 4. Do not draw the graph.

7 Make a table of values for $y = x^2 - 5x + 8$, for values of x from $^-2$ to 4. Do not draw the graph.

Use 2 mm graph paper for questions **8** to **20**.

8 a) Copy and complete this table for $y = x^2 - 2$.

x	$^-3$	$^-2$	$^-1$	0	1	2	3
x^2							
-2							
$y = x^2 - 2$							

b) Draw the graph of $y = x^2 - 2$, for values of x from $^-3$ to 3. Label the x-axis from $^-3$ to 3 and the y-axis from $^-5$ to 10. Use a scale of 1 cm to 1 unit on the x-axis and 2 cm to 5 units on the y-axis.

9 a) Copy and complete this table for $y = x^2 - 3x$.

x	$^-1$	0	1	2	3	4	5	1.5
x^2								
$-3x$								
$y = x^2 - 3x$								

b) Draw the graph of $y = x^2 - 3x$, for values of x from $^-1$ to 5. Label the x-axis from $^-1$ to 5 and the y-axis from $^-5$ to 10. Use a scale of 1 cm to 1 unit on the x-axis and 2 cm to 5 units on the y-axis.

10 Draw the graph of $y = {}^-x^2 + 4$, for values of x from $^-3$ to 3. Label the x-axis from $^-3$ to 3 and the y-axis from $^-5$ to 5. Use a scale of 1 cm to 1 unit on the x-axis and 2 cm to 5 units on the y-axis.

11 Draw the graph of $y = x^2 + 4x$, for values of x from $^-6$ to 2. Label the x-axis from $^-6$ to 2 and the y-axis from $^-5$ to 15. Use a scale of 1 cm to 1 unit on the x-axis and 1 cm to 5 units on the y-axis.

12 Draw the graph of $y = {}^-x^2 + 2x + 6$, for values of x from $^-2$ to 4.
Label the x-axis from $^-2$ to 4 and the y-axis from $^-5$ to 10.
Use a scale of 1 cm to 1 unit on the x-axis and 1 cm to 5 units on the y-axis.

13 Draw the graph of $y = x^2 - 4x + 3$, for values of x from $^-1$ to 5.

14 Draw the graph of $y = x^2 - 6x + 5$, for values of x from $^-1$ to 6.

15 Draw the graph of $y = {}^-x^2 + 4x - 3$, for values of x from $^-1$ to 5.

16 **a)** Draw the graph of $y = x^2 - 5x + 2$, for values of x from $^-1$ to 6.
b) Find the values of x on your graph when $y = 0$.
Give your answers correct to 1 decimal place.

An extra point at $x = 2.5$ might be useful for question 16.

17 **a)** Draw the graph of $y = 2x^2 - 5x + 1$, for values of x from $^-2$ to 4.
b) Find the value of x on your graph where $y = 0$.
Give your answers correct to 1 decimal place.

Note that for question 17 the values of y are not symmetrical.

18 **a)** Draw the graph of $y = 2x^2 - 12x$, for values of x from $^-1$ to 7.
b) Write down the values of x where the curve crosses $y = 5$.

19 When a stone is dropped from the edge of a cliff, the distance, d metres, it falls is given by the formula $d = 5t^2$, where t is the time in seconds.
a) Work out the values of d for values of t from 0 to 5.
b) Draw the graph for $t = 0$ to 5.
c) The cliff is 65 metres high.
How long does it take the stone to reach the bottom of the cliff?
Give your answer correct to 1 decimal place.

20 The surface area of a cube (S) is given by the formula $S = 6x^2$, where x is the length of an edge of the cube.
a) Draw the graph of $S = 6x^2$ for values of x from 0 to 6.
b) From your graph, find the length of the edge of a cube with surface area 140 cm^2.

Graphical methods of solving equations

One way of finding solutions to quadratic equations is to draw and use a graph.

Example 5

a) Draw the graph of $y = x^2 - 2x - 8$, for values of x from $^-3$ to 5.

b) Solve the equation $x^2 - 2x - 8 = 0$.

c) Solve the equation $x^2 - 2x - 8 = 5$.

Solution

a)

x	$^-3$	$^-2$	$^-1$	0	1	2	3	4	5
x^2	9	4	1	0	1	4	9	16	25
$-2x$	6	4	2	0	$^-2$	$^-4$	$^-6$	$^-8$	$^-10$
-8	$^-8$	$^-8$	$^-8$	$^-8$	$^-8$	$^-8$	$^-8$	$^-8$	$^-8$
$y = x^2 - 2x - 8$	7	0	$^-5$	$^-8$	$^-9$	$^-8$	$^-5$	0	7

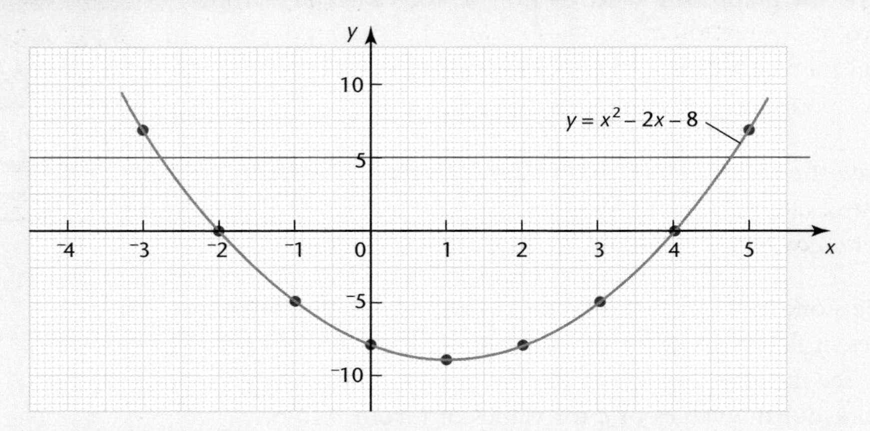

b) The solution of $x^2 - 2x - 8 = 0$ is when $y = 0$, where the curve cuts the x-axis.
The solution is $x = {}^-2$ or $x = 4$.

c) The solution of $x^2 - 2x - 8 = 5$ is when $y = 5$.
Draw the line $y = 5$ on your graph and read off the values of x where the curve cuts the line.
The solution is $x = {}^-2.7$ or $x = 4.7$, to 1 decimal place.

Exercise 10.2

1 a) Draw the graph of $y = x^2 - 7x + 10$, for values of x from 0 to 7.
b) Solve the equation $x^2 - 7x + 10 = 0$.

2 a) Draw the graph of $y = x^2 - x - 2$, for values of x from $^-2$ to 3.
b) Solve the equation $x^2 - x - 2 = 0$.

3 a) Draw the graph of $y = x^2 - 8$, for values of x from $^-4$ to 4.
b) Solve the equation $x^2 - 8 = 0$.
c) Solve the equation $x^2 - 8 = 3$.

4 a) Draw the graph of $y = x^2 + x - 3$, for values of x from $^-3$ to 2.
b) Solve the equation $x^2 + x - 3 = 0$.
c) Solve the equation $x^2 + x - 3 = {}^-2$.

5 a) Draw the graph of $y = x^2 - 4x + 3$, for values of x from $^-1$ to 5.
b) Solve the equation $x^2 - 4x + 3 = 0$.

6 a) Draw the graph of $y = x^2 - 3x$, for values of x from $^-2$ to 5.
b) Solve the equation $x^2 - 3x = 0$.

7 a) Draw the graph of $y = x^2 - 5$, for values of x from $^-3$ to 3.
b) Solve the equation $x^2 - 5 = 0$.
c) Solve the equation $x^2 = 7$.

8 a) Draw the graph of $y = x^2 - 3x - 2$, for values of x from $^-2$ to 5.
b) Solve the equation $x^2 - 3x - 2 = 0$.
c) Solve the equation $x^2 - 3x - 2 = 8$.

Foundation Gold/ Higher Bronze

Key Ideas

- Quadratic equations have the form $y = ax^2 + bx + c$, where a, b and c are constants, and b and c may be zero.
- The graph of a quadratic equation is a parabola. It has this shape.

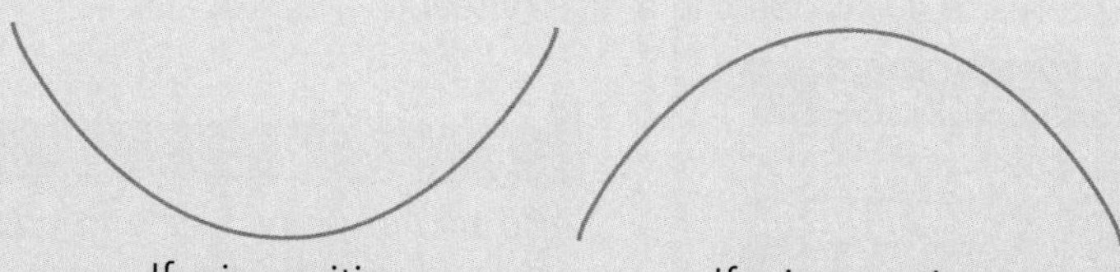

If a is positive　　　　If a is negative

- To solve the quadratic equation $ax^2 + bx + c = k$ graphically, find the values of x where the graph of $y = ax^2 + bx + c$ crosses the line $y = k$.

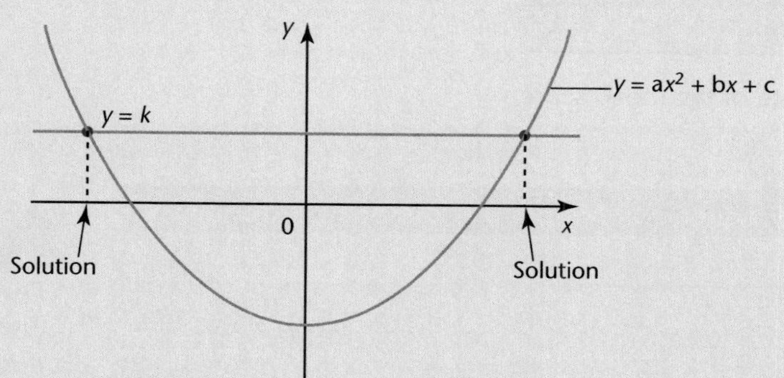

- The solutions are the values of x where the graphs cross.

CHAPTER

Finding the mean of grouped data

YOU WILL LEARN ABOUT	YOU SHOULD ALREADY KNOW
o Finding the mean of grouped discrete data o Finding the mean of grouped continuous data	o How to find the mean of non-grouped data o How to construct and interpret frequency polygons and bar graphs o How to find the mean of data presented in a discrete frequency table

Finding the mean of grouped discrete data

When working out the mean of grouped data in a table, you do not know the exact value for each item of data. The midpoint value is therefore chosen to represent each group, and this is used to calculate an **estimate** of the mean. The midpoint value is multiplied by the frequency of the group.

One method for finding the midpoint of a group is to add the end values of the group and divide by 2. For example, if the group is $60 \leqslant s < 80$, then the midpoint is $\frac{60 + 80}{2} = 70$.

Example 1

The table shows the scores of 40 students in a history test.

Score (s)	Frequency
$0 \leqslant s < 20$	2
$20 \leqslant s < 40$	4
$40 \leqslant s < 60$	14
$60 \leqslant s < 80$	16
$80 \leqslant s < 100$	4
Total	40

Exam tip

Add two columns to the frequency table to help you work out the mean: one column for the midpoint of each group and one for the midpoint multiplied by the frequency of the group.

Calculate an estimate of the mean score.

Solution

Score (s)	Frequency	Midpoint	Midpoint × frequency
$0 \leqslant s < 20$	2	10	20
$20 \leqslant s < 40$	4	30	120
$40 \leqslant s < 60$	14	50	700
$60 \leqslant s < 80$	16	70	1120
$80 \leqslant s < 100$	4	90	360
Total	40		2320

Mean = 2320 ÷ 40 = 58

Exam tip

Don't forget to divide by the *total frequency*, not the number of groups.

Exercise 11.1

1 The table shows a summary of the marks scored by students in a test.
Calculate an estimate of the mean score for these students.

Marks (m)	Frequency
$0 \leqslant m < 20$	1
$20 \leqslant m < 40$	3
$40 \leqslant m < 60$	15
$60 \leqslant m < 80$	9
$80 \leqslant m < 100$	2

2 The table summarises the number of words in each sentence on one page of a book. Calculate the mean length of a sentence.

Number of words	Frequency
1 – 3	0
4 – 6	2
7 – 9	7
10 – 12	6
13 – 15	15

3 The table shows a summary of the attendance at the first 30 games of a football club.
Calculate an estimate of the mean attendance.

Attendance (a)	Frequency
$0 \leqslant a < 4000$	0
$4000 \leqslant a < 8000$	2
$8000 \leqslant a < 12000$	7
$12000 \leqslant a < 16000$	6
$16000 \leqslant a < 20000$	15

4 The table shows a summary of the number of points scored by a school basketball player.
Calculate an estimate of their mean points score.

Points (p)	Frequency
$0 \leqslant p < 5$	2
$5 \leqslant p < 10$	3
$10 \leqslant p < 15$	8
$15 \leqslant p < 20$	9
$20 \leqslant p < 25$	12
$25 \leqslant p < 30$	3

5 The number of road accidents in a small town was recorded each week over two years. The table summarises the information gathered.
Calculate an estimate of the mean number of accidents per week.

Number of accidents	Frequency
0 – 1	1
2 – 3	7
4 – 5	12
6 – 9	36
10 – 14	28
15 – 17	19
18 – 20	1

Finding the mean of grouped continuous data

The previous section dealt with discrete data (numbers that can be counted). In this section, the data is continuous (the result of measurement).

You use the same method to calculate an estimate of the mean of grouped continuous data as for grouped discrete data.

Example 2

The table shows the heights of the students in Year 11 at Sandish School.

Height (h cm)	Frequency
$155 \leqslant h < 160$	2
$160 \leqslant h < 165$	6
$165 \leqslant h < 170$	18
$170 \leqslant h < 175$	25
$175 \leqslant h < 180$	9
$180 \leqslant h < 185$	4
$185 \leqslant h < 190$	1
Total	65

Calculate an estimate of the mean height.

Solution

Height (h cm)	Frequency	Midpoint	Midpoint × frequency
$155 \leqslant h < 160$	2	157.5	315
$160 \leqslant h < 165$	6	162.5	975
$165 \leqslant h < 170$	18	167.5	3015
$170 \leqslant h < 175$	25	172.5	4312.5
$175 \leqslant h < 180$	9	177.5	1597.5
$180 \leqslant h < 185$	4	182.5	730
$185 \leqslant h < 190$	1	187.5	187.5
Total	65		11 132.5

Mean = 11 132.5 ÷ 65 = 171.3 cm, correct to 1 decimal place.

Activity 1

a) Collect some continuous data, for example the masses of school bags for people in your class.

b) Group the data and represent them on a frequency polygon or bar graph. Calculate the mean of the data.

Exercise 11.2

1 Calculate an estimate of the mean of these times.

Time (t seconds)	Frequency
$0 \leqslant t < 2$	4
$2 \leqslant t < 4$	6
$4 \leqslant t < 6$	3
$6 \leqslant t < 8$	2
$8 \leqslant t < 10$	7

2 Calculate an estimate of the mean of these heights.

Height (h cm)	Frequency
$50 \leqslant h < 60$	15
$60 \leqslant h < 70$	23
$70 \leqslant h < 80$	38
$80 \leqslant h < 90$	17
$90 \leqslant h < 100$	7

3 Calculate an estimate of the mean of these lengths.

Length (l metres)	Frequency
$1.0 \leqslant l < 1.2$	2
$1.2 \leqslant l < 1.4$	7
$1.4 \leqslant l < 1.6$	13
$1.6 \leqslant l < 1.8$	5
$1.8 \leqslant l < 2.0$	3

4 Calculate an estimate of the mean length for this distribution.

Length (y cm)	Frequency
$10 \leqslant y < 20$	2
$20 \leqslant y < 30$	6
$30 \leqslant y < 40$	9
$40 \leqslant y < 50$	5
$50 \leqslant y < 60$	3

5 Calculate an estimate of the mean mass of these tomatoes.

Mass of tomato (t g)	Frequency
$35 < t \leqslant 40$	7
$40 < t \leqslant 45$	13
$45 < t \leqslant 50$	20
$50 < t \leqslant 55$	16
$55 < t \leqslant 60$	4

6 Calculate an estimate of the mean of these times.

Time (t seconds)	Frequency
$0 \leqslant t < 20$	4
$20 \leqslant t < 40$	9
$40 \leqslant t < 60$	13
$60 \leqslant t < 80$	8
$80 \leqslant t < 100$	6

7 Calculate an estimate of the mean of these heights.

Height (h metres)	Frequency
$0 < h \leqslant 2$	12
$2 < h \leqslant 4$	26
$4 < h \leqslant 6$	34
$6 < h \leqslant 8$	23
$8 < h \leqslant 10$	5

8 Calculate an estimate of the mean of these lengths.

Length (l cm)	Frequency
$3.0 \leqslant l < 3.2$	3
$3.2 \leqslant l < 3.4$	8
$3.4 \leqslant l < 3.6$	11
$3.6 \leqslant l < 3.8$	5
$3.8 \leqslant l < 4.0$	3

9 Calculate an estimate of the mean of these masses.

Mass (w kg)	Frequency
$30 \leqslant w < 40$	5
$40 \leqslant w < 50$	8
$50 \leqslant w < 60$	2
$60 \leqslant w < 70$	4
$70 \leqslant w < 80$	1

10 Calculate an estimate of the mean of these lengths.

Length (x cm)	Frequency
$0 < x \leqslant 5$	8
$5 < x \leqslant 10$	6
$10 < x \leqslant 15$	2
$15 < x \leqslant 20$	5
$20 < x \leqslant 25$	1

Challenge 1

The bar graph shows the masses of a sample of 50 eggs.

a) Make a frequency table for this information.

b) Calculate an estimate of the mean mass of these eggs.

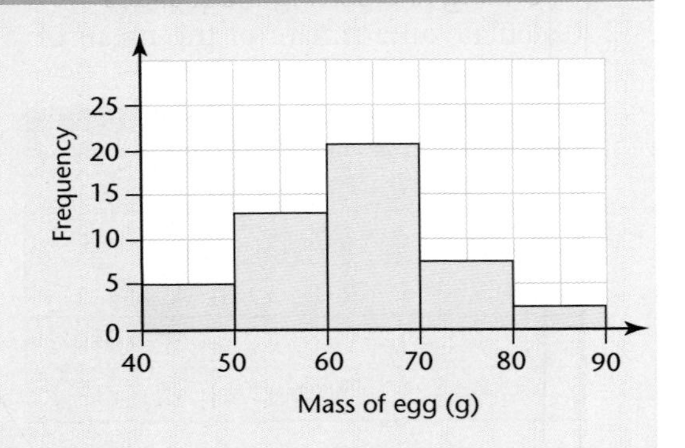

Challenge 2

The bar graph shows the heights of students in Year 9.

a) Make a frequency table for this information.

b) Calculate an estimate of the mean height.

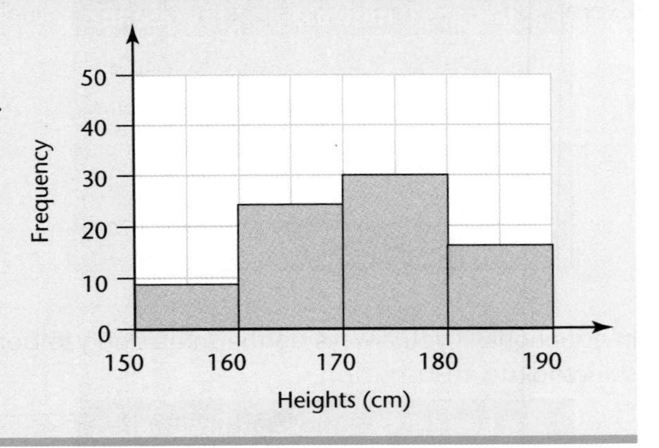

Key Ideas

- To find an estimate of the mean of grouped data
  - use the midpoint of each group as the value for the group,
  - multiply each value by the group frequency,
  - add the results and
  - divide by the sum of their frequencies.

Inequalities

YOU WILL LEARN ABOUT	YOU SHOULD ALREADY KNOW
○ The meaning of the four inequality symbols ○ Solving linear inequalities ○ Showing the solution to an inequality on a number line ○ Forming and solving inequalities	○ How to write a simple formula in letters ○ How to collect together simple algebraic terms ○ How to multiply out expressions such as 3(2x – 5) ○ How to solve linear equations

Foundation Gold/ Higher Bronze

Inequalities

$a < b$ means 'a is less than b'

$a \leqslant b$ means 'a is less than or equal to b'

$a > b$ means 'a is greater than b'

$a \geqslant b$ means 'a is greater than or equal to b'

Expressions involving these signs are called **inequalities**.

Example 1	Solution
Find the integer (whole number) values of x for each of these inequalities. **a)** $^-3 < x \leqslant {^-1}$ **b)** $1 \leqslant x < 4$	**a)** If $^-3 < x \leqslant {^-1}$, then $x = {^-2}$ or $^-1$. Note that $^-3$ is not included but $^-1$ is. **b)** If $1 \leqslant x < 4$, then $x = 1$, 2 or 3. Note that 1 is included but 4 is not.

In equations, if you always do the same thing to both sides the equality is still valid.

The same is usually true for inequalities, but there is one important exception.

Consider the inequality $5 < 7$.

Add 2 to each side:	$7 < 9$	Still true
Subtract 5 from each side:	$2 < 4$	Still true
Multiply each side by 3:	$6 < 12$	Still true
Divide each side by 2:	$3 < 6$	Still true
Multiply each side by $^-2$:	$^-6 < {^-12}$	No longer true
But reverse the inequality sign:	$^-6 > {^-12}$	Now true

If you multiply or divide an inequality by a negative number, you must reverse the inequality sign.

Otherwise inequalities can be treated in the same way as equations.

Example 2

Solve each of these inequalities and show the solution on a number line.

a) $3x + 4 < 10$ **b)** $2x - 5 \leqslant 4 - 3x$ **c)** $x + 4 < 3x - 2$

Solution

a) $3x + 4 < 10$

$3x < 6$ Subtract 4 from each side.

$x < 2$ Divide each side by 3.

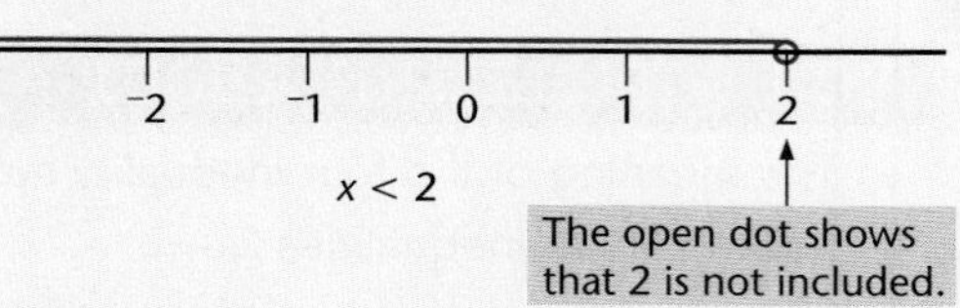

b) $2x - 5 \leqslant 4 - 3x$

$2x \leqslant 9 - 3x$ Add 5 to each side.

$5x \leqslant 9$ Add 3x to each side.

$x \leqslant 1.8$ Divide each side by 5.

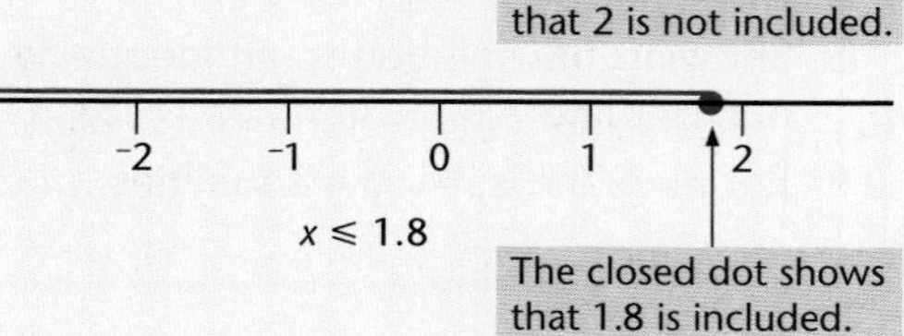

c) $x + 4 < 3x - 2$

$x < 3x - 6$ Subtract 4 from each side.

$^{-}2x < {}^{-}6$ Subtract 3x from each side.

$x > 3$ Divide each side by $^{-}2$ and change $<$ to $>$ (when dividing by $^{-}2$).

You can avoid dealing with negative numbers by using an alternative method.

$x + 4 < 3x - 2$

$x + 6 < 3x$ Add 2 to each side.

$6 < 2x$ Subtract x from each side.

$3 < x$ Divide each side by 2.

$x > 3$ Rewrite the inequality with x as the subject.

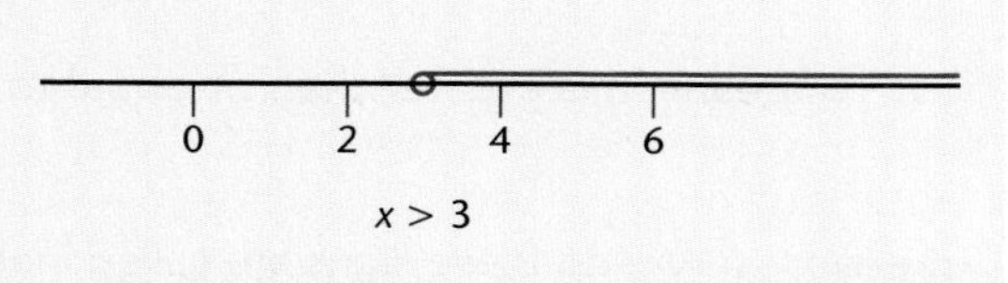

Exercise 12.1

1 Write down two possible values of x for the inequality $x < {}^{-}2$.

2 Write down the integer values of x for each of these inequalities.

a) $^{-}4 \leqslant x < 0$ **b)** $1 < x \leqslant 5$ **c)** $1 < x \leqslant 4$ **d)** $^{-}5 < x \leqslant {}^{-}1$

Solve the inequalities in questions **3** to **24**.
For questions **3** to **10**, represent your solution on a number line.

3 $x - 3 \leqslant 4$ **4** $x + 7 > 9$ **5** $2x - 3 < 5$ **6** $3x + 4 \leqslant 7$

7 $x - 2 < 5$ **8** $2x + 3 > 6$ **9** $3x - 4 \leqslant 8$ **10** $3x \geqslant x - 2$

11 $2x \geqslant x + 5$ **12** $5x > 3 - x$ **13** $2x + 1 < 7$ **14** $4x > 2x + 5$

15 $3x - 6 \geqslant x + 2$ **16** $4x + 2 < 3$ **17** $5a - 3 > 2a$ **18** $2x - 3 < x + 1$

19 $3x + 2 \geqslant x - 1$ **20** $3x + 7 < x + 3$ **21** $8x - 10 > 3x + 25$ **22** $2x - 7 \geqslant 5x + 8$

23 $6x + 11 \leqslant 18 - x$ **24** $2x + 9 > 4x + 5$

Challenge 1

Given that x is an integer, solve each of these inequalities.

a) $3 \leqslant 2x - 1 \leqslant 5$ b) $^{-}4 \leqslant 3x + 2 \leqslant 11$

Forming and solving inequalities

Everyday problems can often be solved by forming inequalities and solving them.

Example 3

Jake is organising a disco but he has only £60 to spend on it.
He has to pay £10 to hire the room and £4 for every person attending.

How many people can he invite to the disco?

Solution

Let the number of people be n.
Write down an inequality involving n and solve it to find the largest number that can attend the disco.

Cost of party $\leqslant$ £60

So $10 + 4n \leqslant 60$

$4n \leqslant 50$

$n \leqslant 12.5$

So the largest number of people that can attend the disco is 12.

Exam tip

Always think whether an answer is sensible. In this example, the final answer must be a whole number!

Foundation Gold/ Higher Bronze

Exercise 12.2

1 A decorator is wallpapering a room. It takes her 30 minutes to prepare the paste and 20 minutes to cut and hang each length of paper. She works for 4 hours and hangs x lengths.
Write down an inequality in x and solve it to find the largest number of lengths the decorator can hang in the time.

2 To hire a bus, the charge is £60 plus £2 a mile. The bus company will only hire the bus if they take at least £225.

a) Let the number of miles be x. Write down an inequality for x and solve it.

b) What is the smallest distance that the bus can be hired to go?

3 Ameer has 40 metres of fencing, in 1 metre lengths that cannot be split. He wants to use as much of it as he can to enclose a rectangle that is twice as long as it is wide.

a) Call the width of the rectangle x metres and write down an inequality.

b) Solve it to find the length and width of the biggest rectangle that he can make.

4 Sara is x years old, and Marie is ten years older than Sara. The sum of their ages is less than 50.

a) Write down an inequality and solve it.

b) What is the oldest that Sara can be?

5 Paul goes to a shop and buys two chocolate bars at 30p each and x cans of cola at 45p each. He has £2 and wants to buy as many cans of cola as possible.

a) Write down an inequality in x and solve it.

b) What is the largest number of cans of cola he can buy?

Solving harder inequalities

The inequalities in this section are more complicated than the ones you met earlier in the chapter. For example, you may need to work with brackets.

Example 4

Solve the inequality $x \geqslant 6x - 9$ and show the solution on a number line.

Solution

$x \geqslant 6x - 9$

$x + 9 \geqslant 6x$ Add 9 to each side.

$9 \geqslant 5x$ Subtract x from each side.

$1.8 \geqslant x$ Divide each side by 5.

$x \leqslant 1.8$ Rewrite with x as the subject.

Notice that the inequality has changed direction too.

With inequalities you can't just swap sides.

-2 -1 0 1 2

Example 5

Solve the inequality $2(3x - 1) > 4x + 7$.

Solution

$2(3x - 1) > 4x + 7$

$6x - 2 > 4x + 7$ Multiply out the brackets.

$6x > 4x + 9$ Add 2 to each side.

$2x > 9$ Subtract 4x from each side.

$x > 4\frac{1}{2}$ Divide each side by 2.

Exercise 12.3

Solve these inequalities.

1 $2x + 3 < 5$

2 $5x - 4 > 10 - 2x$

3 $3(2x - 1) > 15$

4 $4(x - 4) \geqslant x - 1$

5 $4n - 2 > 6$

6 $2n + 6 < n + 3$

7 $4n - 9 \geqslant 2n + 1$

8 $3(x - 1) \geqslant 6$

9 $2(3x - 1) > 4x + 6$

10 $2(x + 3) < 1 - 3x$

11 $3(2x - 1) \geqslant 11 - x$

12 $x + 4 > 2x$

13 $2x - 5 < 4x + 1$

14 $3(x - 4) > 5(x + 1)$

15 $x - 2 < 2x + 4$

16 $2x - 1 > x - 4$

17 $3(x + 3) \geqslant 2x - 1$

18 $3(2x - 4) < 5(x - 6)$

19 Fred and Bob are cousins. Both have sisters, Frances and Beryl.
Bob is twice as old as Fred. Frances is 2 years older than Fred. Beryl is 3 years younger than Bob. Frances is younger than Beryl.
Write down an inequality and solve it to find the youngest age that Fred could be.

20 Solve $x^2 > 9$. Show your answer on a number line.

Key Ideas

- $a < b$ means 'a is less than b'.
  $a \leqslant b$ means 'a is less than or equal to b'.
  $a > b$ means 'a is greater than b'.
  $a \geqslant b$ means 'a is greater than or equal to b'.
- Inequalities can be treated like equations, except when multiplying or dividing by a negative number you must reverse the inequality sign.
- You can use algebra to solve a problem by setting up an equation or inequality for the unknown quantity and then solving it.

CHAPTER 13

Compound measures

YOU WILL LEARN ABOUT	YOU SHOULD ALREADY KNOW
○ Compound measures: speed, density, population density and rate of flow ○ Solving problems involving compound measures	○ How to convert one metric unit to another

Speed

Speed is a **compound measure** because it is calculated from two other measurements: distance and time.

$$\text{Average speed} = \frac{\text{total distance travelled}}{\text{total time taken}}$$

The units of your answer will depend on the units you begin with. Speed has units of 'distance per time', for example km/h.

The formula for speed can be rearranged to find the distance travelled or the time taken for a journey.

$$\text{Distance} = \text{speed} \times \text{time} \qquad \text{Time} = \frac{\text{distance}}{\text{speed}}$$

Exam tip

You may find the d.s.t. triangle helpful. Cover up the quantity you are trying to find.

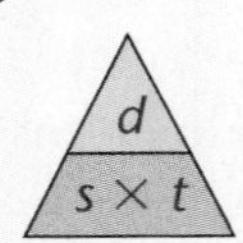

Example 1

Find the average speed of an athlete who runs 100 metres in 20 seconds.

Solution

$$\text{Average speed} = \frac{100\,\text{m}}{20\,\text{s}} = 5\,\text{m/s}$$

Example 2

How many minutes does it take to walk 2 km at a speed of 5 km/h?

Solution

$$\text{Time} = \frac{\text{distance}}{\text{speed}} = \frac{2}{5}\text{ hour} = \frac{2}{5} \times 60 \text{ minutes} = 24 \text{ minutes}$$

Example 3

Find the average speed of a delivery driver who travels 45 km in 30 minutes.

Solution

Average speed $= \frac{45\,\text{km}}{30\text{ minutes}} = 1.5$ km/minute

However, the speed here is more likely to be needed in kilometres per hour.

To find this, first change the time into hours.

30 minutes = 0.5 hour

Average speed $= \frac{45\,\text{km}}{0.5\,\text{h}} = 90\,\text{km/h}$

You may also be able to see other ways of obtaining the results in Examples 2 and 3.

Activity 1

Discuss some everyday situations in which you use speeds. Use appropriate numbers and units.

Rate of flow

Another compound measure is rate of flow, which links volume or mass with time.

Average rate of flow $= \frac{\text{volume or mass}}{\text{total time taken}}$

and is measured in units such as kilograms per hour (kg/h).

Exam tip

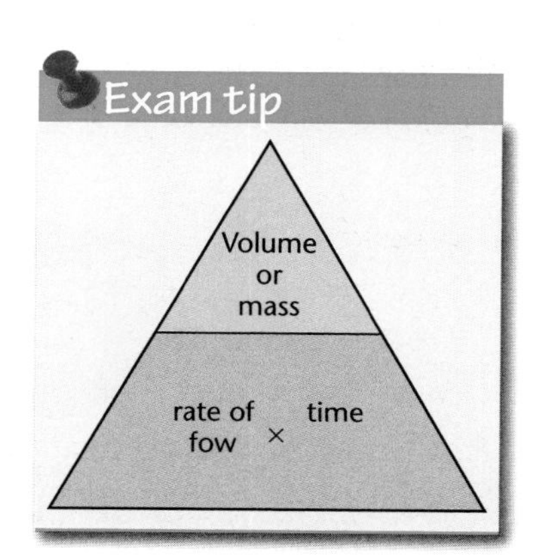

Example 4

It takes half an hour to fill a 1000 litre tank. What is the average rate of flow?

Solution

Rate of flow $= \frac{1000\,\text{l}}{0.5\,\text{h}} = 2000\,\text{l/h}$

Exercise 13.1

1 Water from a tap flows at 25 litres/minute.
How long will it take for this tap to fill a bath holding 500 litres?

2 Flour is loaded into a lorry which holds 20 tonnes. The lorry is filled in 40 minutes.
What is the rate of flow in kilograms per minute?

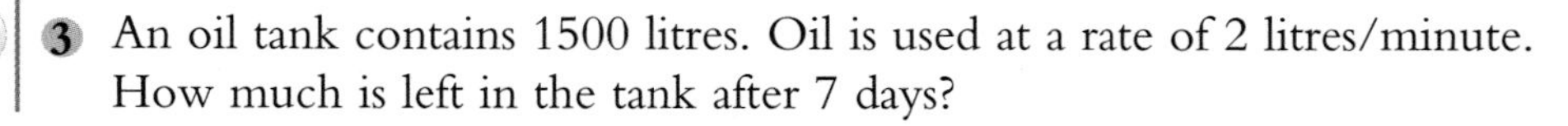

3 An oil tank contains 1500 litres. Oil is used at a rate of 2 litres/minute.
How much is left in the tank after 7 days?

Density

Another example of a compound measure is density, which links mass and volume.

$$\text{Density of a substance} = \frac{\text{mass}}{\text{volume}}$$

It is measured in units such as grams per cubic centimetre (g/cm^3).

Exam tip

As with speed, there is a triangle that can be helpful with these questions.

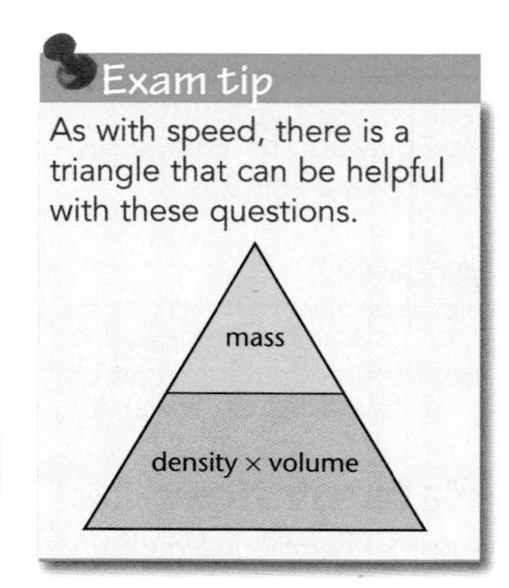

Example 5

The density of gold is 19.3 g/cm^3. Calculate the mass of a gold bar with a volume of 30 cm^3.

Solution

$$\text{Density} = \frac{\text{mass}}{\text{volume}}$$

so mass = density × volume.

The mass of the gold bar = density × volume = $19.3 \times 30 = 579$ g.

Population density

Population density is another example of a compound measure. It gives an idea of how heavily populated an area is. It is measured as the number of people per square kilometre.

$$\text{Population density} = \frac{\text{number of people}}{\text{area}}$$

Exam tip

Once again, a triangle can be helpful with these questions.

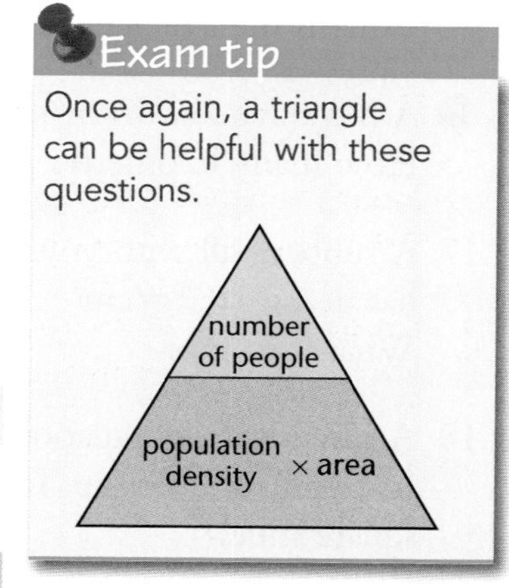

Example 6

In a small town, 300 people live in an area of 2.4 km^2.
Find the population density of the town.

Solution

Population density $= \frac{300}{2.4} = 125$ people/km^2

Exercise 13.2

1 Find the average speed of a car which travels 75 miles in $1\frac{1}{2}$ hours.

2 Find the average speed of a runner who covers 180 m in 40 seconds.

3 Calculate the density of a stone of mass 350 g and volume 40 cm^3.

4 Waring has a population of 60 000 in an area of 8 square kilometres. Calculate its population density.

5 A motorbike travels 1 mile in 3 minutes. Calculate its average speed in miles per hour.

6 Find the average speed of a car which travels 63 miles in $1\frac{1}{2}$ hours.

7 Find the average speed of a runner who covers 180 m in 48 seconds.

8 Calculate the density of a stone of mass 690 g and volume 74 cm^3.
Give your answer to a suitable degree of accuracy.

9 Trenton has a population of 65 000 in an area of 5.8 square kilometres. Calculate its population density, correct to the nearest thousand.

10 A cyclist rides 0.6 mile in 3 minutes. Calculate her average speed in miles per hour.

11 A bus travels at 5 m/s on average. How many kilometres per hour is this?

12 A foam plastic ball with volume 20 cm^3 has density 0.3 g/cm^3.
What is its mass?

13 A town has a population of 200 000. Its population density is 10 000 people per square mile.
What is the area of the town?

14 A runner's average speed in a 80 m race is 7 m/s.
Find the time he takes for the race, to the nearest 0.1 second.

15 A car travels 15 km in 12 minutes.
What is the average speed in km/h?

16 A bus travels at 6.1 m/s on average.
How many kilometres per hour is this?

17 A rubber ball with volume $28.3\,cm^3$ has density $0.7\,g/cm^3$.
What is its mass?

18 A town has a population of 276 300.
Its population density is 9800 people per square mile.
What is the area of the town?

19 A runner's average speed in a 200 m race is 5.3 m/s.
Find the time she takes for the race, to the nearest 0.1 second.

20 A car travels 15 km in 14 minutes.
What is the average speed in km/h?

Key Ideas

- Average speed (measured in units such as m/s), rate of flow (measured in units such as l/h), density (measured in units such as g/cm^3) and population density (measured in units such as $people/km^2$) are compound measures.
- $\text{Average speed} = \dfrac{\text{total distance travelled}}{\text{total time taken}}$
- $\text{Average rate of flow} = \dfrac{\text{volume or mass}}{\text{total time taken}}$
- $\text{Density} = \dfrac{\text{mass}}{\text{volume}}$
- $\text{Population density} = \dfrac{\text{number of people}}{\text{area}}$

14 Reciprocals, factors and multiples

YOU WILL LEARN ABOUT	YOU SHOULD ALREADY KNOW
○ Finding the reciprocal of a number ○ Writing a number as a product of its prime factors ○ Finding the highest common factor (HCF) and lowest common multiple (LCM) of two numbers	○ The meaning of the terms reciprocal, factor and multiple ○ How to convert fractions to decimals

Reciprocals

The reciprocal of a number is $\frac{1}{\text{the number}}$.

Using the rules of fractions gives these results.

The reciprocal of n is $\frac{1}{n}$.

The reciprocal of $\frac{1}{n}$ is n.

The reciprocal of $\frac{a}{b}$ is $\frac{b}{a}$.

To find reciprocals on a calculator, use the reciprocal button. On many calculators it looks like this x^{-1}. Find the reciprocal button on your calculator.

Example 1

Find the reciprocal of each of these.

a) 3 **b)** $\frac{1}{4}$ **c)** $\frac{2}{3}$ **d)** 2.5

Solution

a) The reciprocal of 3 is $\frac{1}{3}$

b) The reciprocal of $\frac{1}{4}$ is 4

c) The reciprocal of $\frac{2}{3}$ is $\frac{3}{2}$ or $1\frac{1}{2}$ or 1.5

d) The reciprocal of 2.5 or $\frac{5}{2}$ is $\frac{1}{2.5} = \frac{2}{5}$ or 0.4

Foundation Gold/ Higher Bronze

Activity 1

What are the reciprocals of these numbers?

a) 2 **b)** 5 **c)** 10 **d)** 0.01 **e)** $\frac{3}{5}$ **f)** $1\frac{1}{2}$

Multiply each number above by its reciprocal.
What do you get?

Now try these products on your calculator and see what you get.
Don't forget to press [=] after each sum.

55×2 (press [=]) $\times \frac{1}{2}$ (press [=])

$55 \times 4 \times \frac{1}{4}$

$7 \times 8 \times \frac{1}{8}$

Try some more until you are sure what is happening.

Example 2

Find the reciprocal of each of these.

a) $1\frac{3}{4}$, giving your answer as a fraction.

b) 1.25, giving your answer as a decimal.

Solution

a) $1\frac{3}{4} = \frac{7}{4}$ so its reciprocal is $\frac{4}{7}$.

b) The reciprocal of $1.25 = \frac{1}{1.25} = 0.8$.

You can use your calculator to do this.

You can press either [1] [÷] [1] [·] [2] [5] [=] or

[1] [·] [2] [5] [x^{-1}] [=].

Exercise 14.1

Do not use your calculator for questions 1 to 4.

1 Write down the reciprocal of each of these numbers.

a) 3 b) 6 c) 49 d) 100 e) 640

f) 4 g) 9 h) 65 i) 10 j) 4.5

2 Write down the numbers of which these are the reciprocals.

a) $\frac{1}{2}$ b) $\frac{1}{10}$ c) $\frac{1}{50}$ d) $\frac{1}{16}$ e) $\frac{1}{9}$

f) $\frac{1}{1000}$ g) $\frac{1}{3}$ h) $\frac{1}{71}$ i) $\frac{1}{100}$ j) $\frac{1}{999}$

3 Calculate the reciprocal of each of these numbers, giving your answer as a fraction or a mixed number.

a) $\frac{3}{5}$ b) $\frac{7}{8}$ c) $1\frac{2}{3}$ d) $8\frac{1}{3}$ e) $\frac{4}{25}$

4 Find the numbers of which these are the reciprocals, giving your answers as fractions or mixed numbers.

a) $\frac{4}{5}$ b) $\frac{3}{8}$ c) $1\frac{3}{5}$ d) $3\frac{1}{3}$

You may use your calculator for question 5.

5 Calculate the reciprocal of each of these numbers, giving your answer as a decimal.

a) 32 b) 1.6 c) 0.5 d) 1.25 e) 62.5

Prime factors

A prime number has exactly two factors, 1 and the number itself.
The first prime numbers are 2, 3, 5, 7, 11, 13, 17, 19, …
Note that 1 is not a prime number. It has only one factor.

To express a number in prime factor form, first divide by 2 until you get an odd number, then try to divide by 3, then by 5, and so on.
The divisions are always by prime numbers.

Example 3

Write 36 as a product of its prime factors.

Solution

2) 36
2) 18
3) 9
3) 3
1

So, written as a product of its prime factors, 36 is $2 \times 2 \times 3 \times 3$, which can be written as $2^2 \times 3^2$.

Activity 2

Copy and complete this table up to $n = 36$, writing the prime factors in index form.
Use the notation F(36) to mean the number of factors of 36.

Number n	Prime factors in index form
2	2^1
3	3^1
4	2^2
5	5^1
6	$2^1 \times 3^1$
:	:
36	$2^2 \times 3^2$

The complete list of factors for 36 is
1, 2, 3, 4, 6, 9, 12, 18, 36.
We can write F(36) = 9, because 36 has nine factors.
But 36 is also equal to $2^2 \times 3^2$. Adding 1 to each of the indices, or powers, gives $(2 + 1)$ and $(2 + 1)$.
Multiplying these numbers gives
$(2 + 1) \times (2 + 1) = 3 \times 3 = 9$, which is the same value as F(36).

1 Investigate prime factors and the number of factors for numbers up to 36.
You might want to add a column to the table you have just completed.

Number n	Prime factors in index form	Number of factors
2	2^1	2
3	3^1	2
4	2^2	3
5	5^1	
6	$2^1 \times 3^1$	
:	:	:
36	$2^2 \times 3^2$	

Check to see if there is a link between the indices of the prime factors and the number of factors.

2 What do you notice about the numbers that have only two factors?

3 Which numbers have an odd number of factors?

Factor tree method for finding prime factors

Rather than systematically dividing by prime numbers, you may prefer to use the factor tree method. In this method, you split the number into any two factors you spot, and then split those factors, continuing until all the factors are prime numbers.

For example, here is a factor tree for the number 126.

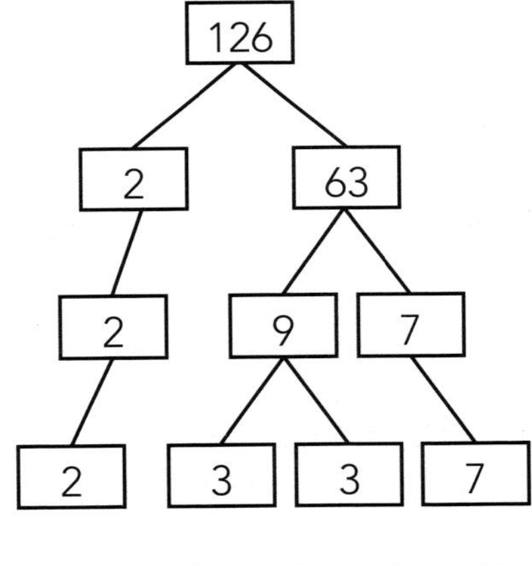

$$126 = 2 \times 3 \times 3 \times 7$$
$$= 2 \times 3^2 \times 7$$

Activity 3

Use the factor tree method to find the prime factors of 64, 100, 144, 350, 540 and 72.

Highest common factor and lowest common multiple

When two or more numbers have been decomposed (split up) into their prime factors, these prime factors can be used to find the **highest common factor** and the **lowest common multiple**.

The highest common factor (HCF) of two numbers is the highest number that is a factor of both numbers.

The lowest common multiple (LCM) of two numbers is the lowest number that is a multiple of both numbers.

Example 4

Find the highest common factor (HCF) of 96 and 180.

Solution

Finding the prime factors of each of the numbers, we obtain

$96 = 2 \times 2 \times 2 \times 2 \times 2 \times 3$ $\quad 96 = 2^5 \times 3$

$180 = 2 \times 2 \qquad\qquad \times 3 \times 3 \times 5$ $\quad 180 = 2^2 \times 3^2 \times 5$

To find the HCF, select the prime numbers that appear in *both* lists and use the *lower* power for each prime.

So the HCF is $2 \times 2 \times 3$.

The HCF of 96 and 180 is $2^2 \times 3 = 12$.

This means that 12 is the highest number that is a factor of both 96 and 180.

Example 5

Find the lowest common multiple (LCM) of 96 and 180.

Solution

$96 = 2 \times 2 \times 2 \times 2 \times 2 \times 3$

$180 = 2 \times 2 \qquad\qquad \times 3 \times 3 \times 5$

To find the LCM, select *all* the prime numbers that appear in the lists and use the *higher* power for each prime.

So the LCM is $2 \times 2 \times 2 \times 2 \times 2 \times 3 \times 3 \times 5$.

The LCM of 96 and 180 is $2^5 \times 3^2 \times 5 = 1440$.

This means that 1440 is the lowest number that has both 96 and 180 as factors. It is the lowest number that is a multiple of both 96 and 180.

Exam tip

Alternatively you can make a list of multiples of 96 and 180 until you reach the same value. This is the LCM of 96 and 180.

Exercise 14.2

1 a) Express each of these numbers as a product of its prime factors.

(i) 48 (ii) 72
(iii) 210 (iv) 350
(v) 75 (vi) 275
(vii) 120 (viii) 198

b) Use your results from part a) to find each of these.

(i) the HCF of 48 and 72
(ii) the LCM of 210 and 350
(iii) the HCF of 75 and 275
(iv) the LCM of 120 and 198

2 a) Decompose each of these numbers into its prime factors.

(i) 495 (ii) 260
(iii) 2700 (iv) 1078
(v) 420 (vi) 1125
(vii) 112 (viii) 1960

b) Use your results from part a) to find each of these.

(i) the HCF of 495 and 2700
(ii) the LCM of 495 and 2700
(iii) the HCF of 420 and 112
(iv) the LCM of 420 and 1960

3 Find the prime factors of these numbers and use this to write down the HCF and LCM of each pair.

a) 64 and 100
b) 18 and 24
c) 50 and 350
d) 72 and 126

4 Find the HCF and the LCM of each pair of numbers.

a) 27 and 63
b) 20 and 50
c) 48 and 84
d) 50 and 64
e) 42 and 49

5 Find the HCF and the LCM of each pair of numbers.

a) 5544 and 2268
b) 2016 and 10 584

Challenge 1

There are two lighthouses on a stretch of the coastline. The light in the first lighthouse flashes every 22 seconds. The light in the second lighthouse flashes every 16 seconds.

At 10 p.m. one evening, both lights are switched on.

What is the next time that the lights flash at the same time?

Challenge 2

Buses to Shenley leave the bus station every 40 minutes. Buses to Winley leave every 15 minutes.

At 8.15 a.m. buses to both Shenley and Winley leave the bus station.

When is the next time that buses to both places leave at the same time?

Challenge 3

Two numbers are **co-prime** if the only integer which 'goes into', or is a factor of, or is a divisor of, both of them is 1.

For example

- 3 and 7 are co-prime because the only factor they have in common is 1
- 4 and 6 are not co-prime because they share a common factor of 2
- 14 and 21 are not co-prime because they share a common factor of 7
- 5 and 23 are co-prime because they have no common factor except 1

1 Try to find four pairs of numbers that are co-prime and four pairs of numbers that are not co-prime.

Look at all the positive integers less than 10.

1 2 3 4 5 6 7 8 9

Of these, 2, 4, 6 and 8 share a common factor of 2 with 10, and 5 divides into 10 and so is also a factor.

Therefore there are four numbers

1 3 7 9

that are less than 10 and are also co-prime with 10.

2 Copy this table and complete it for all the integers up to 24.

Integer n	Integers less than n and co-prime with it	Number of these integers
2	1	1
3	1, 2	2
4	1, 3	2
5	1, 2, 3, 4	4
6	1, 5	2
7	1, 2, 3, 4, 5, 6	6
↓	↓	↓
24	1, 5, 7, 11, 13, 17, 19, 23	8

What do you notice about the numbers in the right-hand column?

We can denote the numbers in the right hand column by $C(n)$.

So $C(6) = 2$.

3 a) Does $C(3) \times C(4) = C(12)$?

b) Does $C(2) \times C(6) = C(12)$?

c) Investigate whether $C(m) \times C(n) = C(mn)$.

Key Ideas

- The reciprocal of n is $\frac{1}{n}$.
- The reciprocal of $\frac{1}{n}$ is n.
- The reciprocal of $\frac{a}{b}$ is $\frac{b}{a}$.
- All positive integers can be decomposed into their prime factors.
- The highest common factor of two numbers is the highest number that is a factor of both numbers.
- The lowest common multiple of two numbers is the lowest number that is a multiple of both numbers.

Changing the subject of a formula

YOU WILL LEARN ABOUT	YOU SHOULD ALREADY KNOW
○ Rearranging a formula ○ Solving a problem by rearranging a formula	○ How to simplify and solve linear equations

Rearranging formulae

Formulae can be treated in the same way as equations. This means they can be rearranged to change the subject.

Example 1

Rearrange each of these formulae to make the letter in brackets the subject.

a) $a = b + c$ (b) **b)** $a = bx + c$ (b)

c) $n = m - 3s$ (s) **d)** $p = \dfrac{q + r}{s}$ (r)

Solution

a)
$a = b + c$
$a - c = b$ Subtract c from both sides.
$b = a - c$ Reverse to get b on the left.

b)
$a = bx + c$
$a - c = bx$ Subtract c from both sides.
$\dfrac{a - c}{x} = b$ Divide both sides by x.
$b = \dfrac{a - c}{x}$ Reverse to get b on the left.

c)
$n = m - 3s$
$n + 3s = m$ Add $3s$ to both sides.
$3s = m - n$ Subtract n from both sides.
$s = \dfrac{m - n}{3}$ Divide both sides by 3.

d)
$p = \dfrac{q + r}{s}$
$sp = q + r$ Multiply both sides by s.
$sp - q = r$ Subtract q from both sides.
$r = sp - q$ Reverse to get r on the left.

Exam tip

If you have difficulty in rearranging a formula, practise by replacing some of the letters with numbers.

Example 2

The formula for the total cost, T, of entry to the cinema for three adults and three children is given by the formula $T = 3(a + c)$, where a is the price of an adult ticket and c is the price of a child ticket.

a) Make c the subject of the formula.

b) Find the cost for a child when the cost for an adult is £5 and the total cost is £24.

Solution

a)

$T = 3(a + c)$

$T = 3a + 3c$ — Multiply out the brackets.

$T - 3a = 3c$ — Subtract $3a$ from both sides.

$\frac{T - 3a}{3} = c$ — Divide both sides by 3.

$c = \frac{T - 3a}{3}$ — Reverse to get c on the left.

b)

$c = \frac{24 - 3 \times 5}{3}$

$= \frac{24 - 15}{3}$

$= \frac{9}{3} = 3$

The cost for a child is £3.

Exercise 15.1

1 Rearrange each formula to make the letter in the brackets the subject.

a) $a = b - c$ (b)

b) $3a = wx + y$ (x)

c) $v = u + at$ (t)

d) $A = \frac{T}{H}$ (T)

e) $C = P - 3T$ (T)

f) $P = \frac{u + v}{2}$ (u)

g) $C = 2\pi r$ (r)

h) $A = p(q + r)$ (q)

i) $p = q + 2r$ (q)

j) $B = s + 5r$ (r)

k) $s = 2u - t$ (t)

l) $m = \frac{pqr}{s}$ (q)

m) $L = 2G - 2F$ (G)

n) $F = \frac{m + 4n}{t}$ (n)

o) $T = \frac{S}{2a}$ (S)

p) $A = t(x - 2y)$ (y)

2 The formula for finding the perimeter P of a rectangle of length l and width w is $P = 2(l + w)$.

a) Rearrange the formula to make l the subject.

b) Find the length of a rectangle of width 8 metres and perimeter 44 metres.

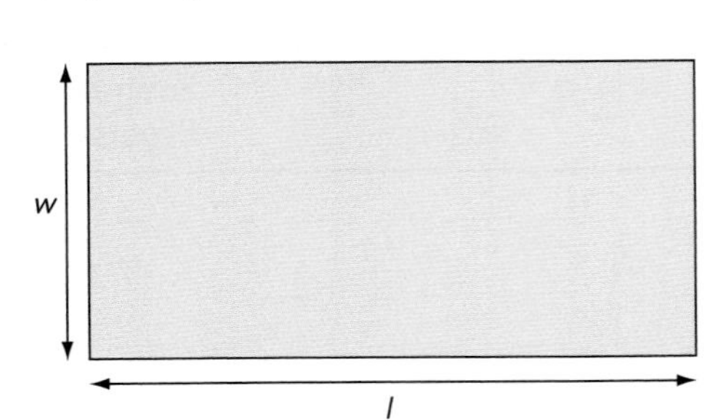

3 The cost (£C) of catering for a conference is given by the formula $C = A + 32n$, where A is the cost of the room and n is the number of delegates.
   a) Rearrange the formula to make n the subject.
   b) Work out the number of delegates when A is £120 and the total cost C is £1912.

4 The cooking time, T minutes, for w kg of meat is given by $T = 45w + 40$.
   a) Make w the subject of this formula.
   b) What is the value of w when the cooking time is 2 hours and 28 minutes?

5 The curved surface area (S cm^2) of a cylinder of radius r cm and height h cm is given by $S = 2\pi rh$.
   a) Rearrange this formula to make r the subject.
   b) Find the radius of a cylinder of height 4 cm, which has a curved surface area of 60.3 cm^2.

6 The formula for the volume V of a cone of height h and base radius r is $V = \frac{1}{3}\pi r^2 h$.

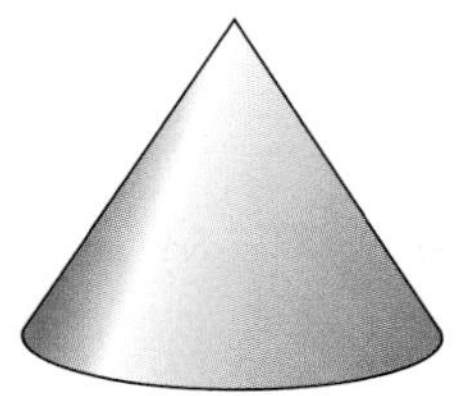

   a) Rearrange the formula to make h the subject.
   b) Find the height of a cone with base radius 9 cm and volume 2290 cm^3.

7 The cost (£C) of a minibus to the airport is given by the formula $C = 20 + \frac{d}{2}$, where d is the distance in miles.
   a) Rearrange the formula to make d the subject.
   b) Work out the distance when it costs £65 to go to the airport.

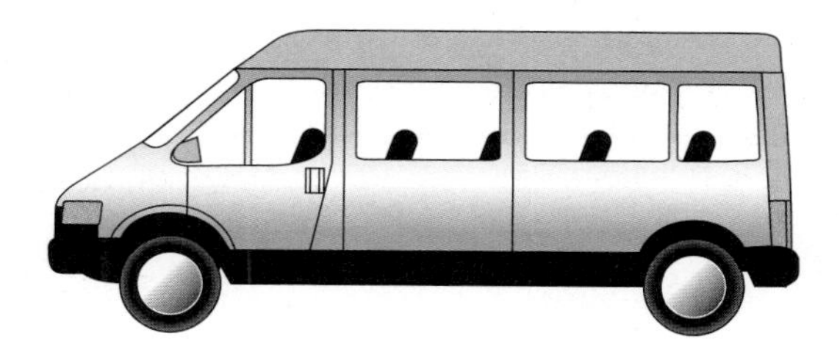

8 The cost (£C) of booking a coach for a party of n people is $C = 40 + 5n$.
   a) Make n the subject of this formula.
   b) Find the number of people when the cost is £235.

9 The total surface area (S cm^2) of a cylinder of radius r cm and height h cm is given by $S = 2\pi rh + 2\pi r^2$.
   a) Rearrange this formula to make h the subject.
   b) Find the height of a cylinder of radius 6 cm, which has a total surface area of 500 cm^2.

10 Using a suitable graph and the formula $S = 2\pi rh + 2\pi r^2$ where S cm^2 is the total surface area of a cylinder of radius r cm and height h cm, find the radius of a cylinder with height 3 cm and surface area 60 cm^2.
Give your answer correct to the nearest millimetre.

Challenge 1

Rearrange each formula to make the letter in brackets the subject.

a) $A = 3r^2$ (r)

b) $a^2 = b^2 + c^2$ (c)

c) $T = 5\sqrt{g}$ (g)

Key Ideas

- You can rearrange a formula using the same steps as you would if it were an equation with numbers instead of letters.

Loci

YOU WILL LEARN ABOUT

- What a locus is
- Constructing the locus of points that are a fixed distance from a given point or line, or equidistant from two fixed points or two intersecting lines, or the shortest distance from a line
- Solving problems that involve one or more loci

YOU SHOULD ALREADY KNOW

- How to use a protractor and compasses
- How to construct a triangle, given three sides
- How to construct a triangle, given two sides and an angle
- How to construct a triangle, given two angles and a side
- How to make scale drawings

Constructing loci

Activity 1

A farmer wants to plant crops but he must keep at least 1 m from the borders of his field.

Trace this diagram.

Sketch the region where he can plant crops.

The local walkers' group have agreed to get a footpath constructed starting at gate G_1 which is equidistant from the two hedges h_1 and h_2.

Construct this path.

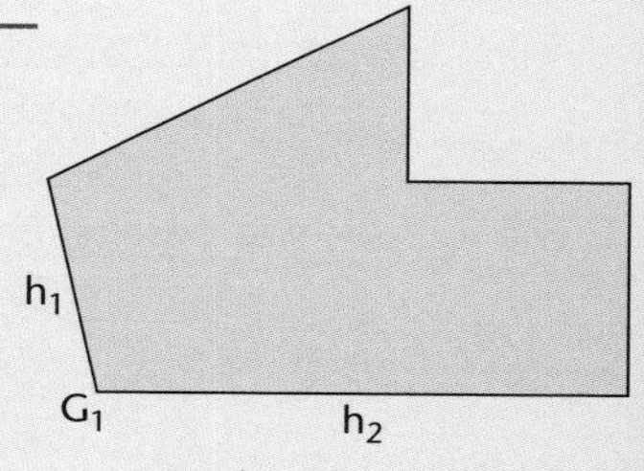

Scale: 1 cm to 20 m

The **locus** of a point is the path or the region that the point covers as it moves according to a particular rule. The plural of locus is **loci**.

Simple loci

The locus of points that are a fixed distance from a given point is a circle, centred at the given point.

This means that the locus of a point 3 cm from A is a circle, centre A and radius 3 cm.

3 cm
A

The locus of a point less than (<) 3 cm from A is the region inside a circle, centre A and radius 3 cm.

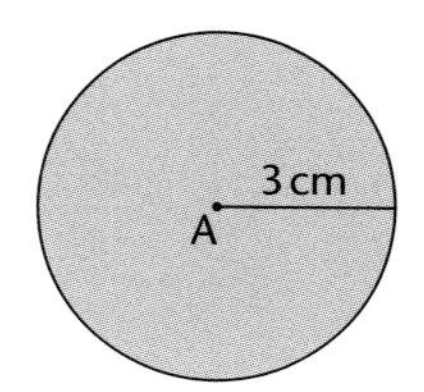

The locus of a point greater than (>) 3 cm from A is the region outside a circle, centre A and radius 3 cm.

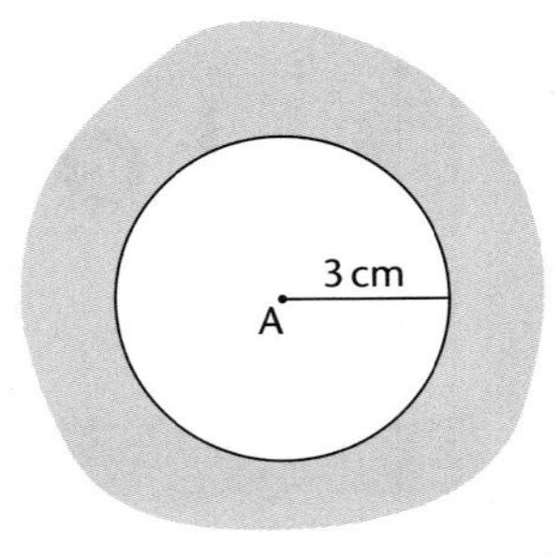

The locus of points that are a fixed distance from a given line is a pair of parallel lines.

For example, the locus of a point 2 cm from a straight line is a pair of lines parallel to that line, 2 cm away from it on either side.

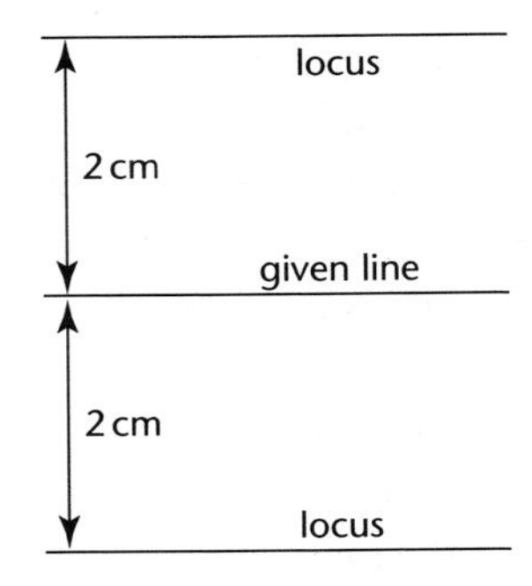

Example 1

A line is 6 cm long. Construct the locus of all the points that are 2 cm from the line.

Solution

The locus is two parallel lines with a semicircle joining them at each end.

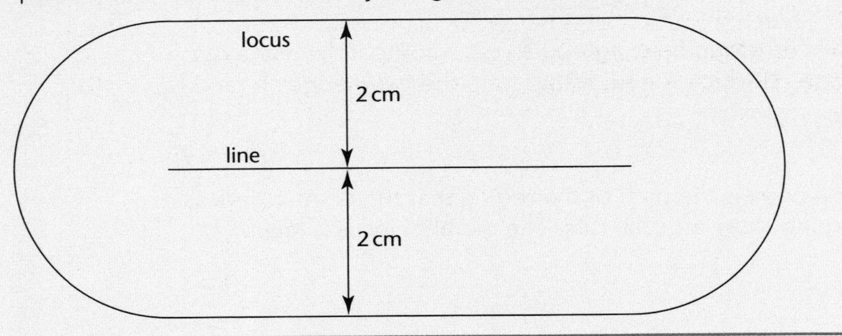

The perpendicular bisector of a line

The locus of a point that stays an equal distance from two points is the perpendicular bisector of the line joining the two points.

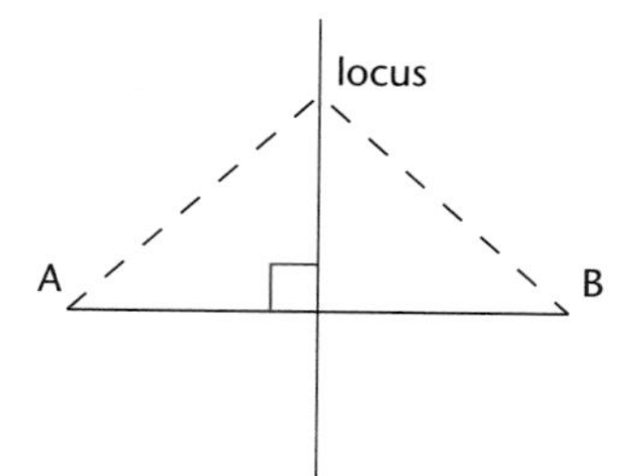

Construction

Draw the line AB.

Open the compasses to a radius that is more than half the length of AB.

Put your compass point at A and draw two arcs, above and below the line.

Keep the compasses set to the same radius.

Put the compass point at B and draw two arcs, above and below the line.

Join the two points where the pairs of arcs cross.

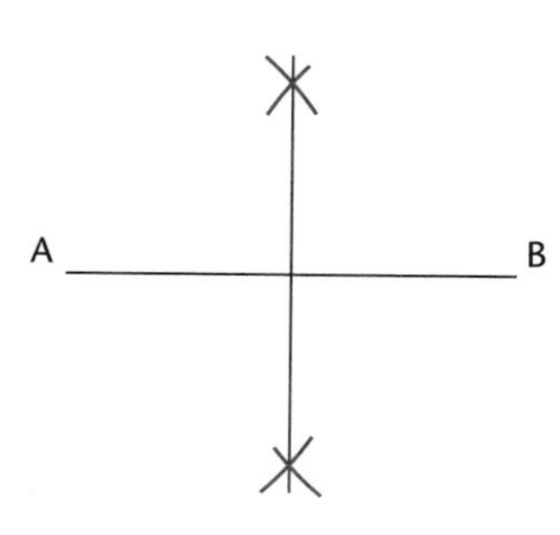

Example 2

Two towns P and Q are 5 kilometres apart.
Draw a diagram and shade the region that is nearer to P than Q. Use a scale of 1 cm to 1 km.

Solution

Mark P and Q 5 cm apart. Draw the perpendicular bisector of the line PQ and shade the region on P's side of the line. The shading could go beyond P and further up or down the page.

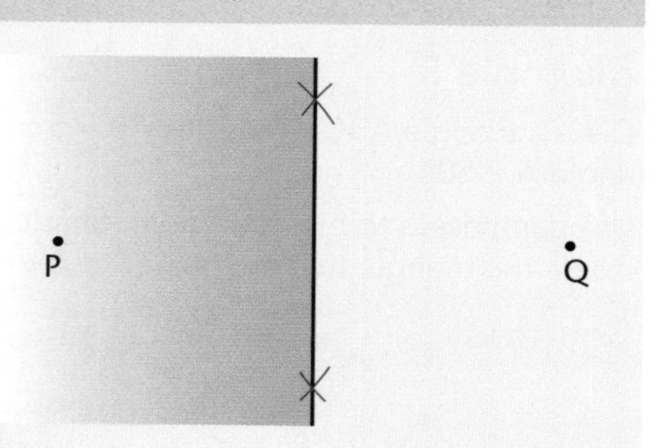

Activity 2

a) **(i)** Draw a triangle. Make it big enough to fill about half of your page.

(ii) Construct the perpendicular bisector of each of the three sides.

(iii) If you have drawn them accurately, the three bisectors should all meet at one point.

Put your compass point on this point, and the pencil on one of the corners of the triangle. Draw a circle.

b) You have drawn the **circumcircle** of the triangle.

What do you notice about this circle?

Foundation Gold/ Higher Bronze

The bisector of an angle

The locus of a point that stays an equal distance from two intersecting lines is the pair of lines that bisect the angles between the lines.

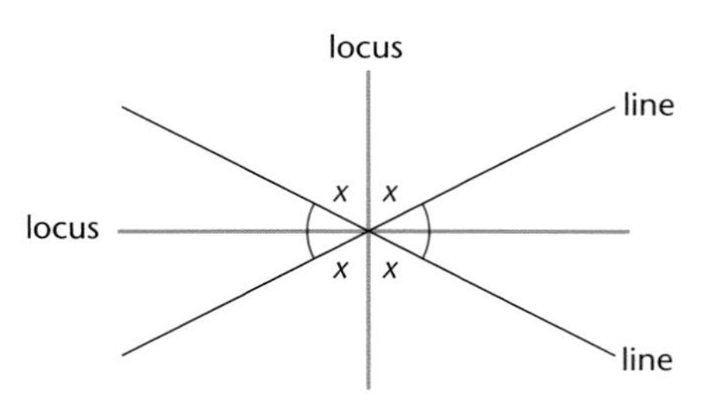

Can you see why this is so?

Drawing the perpendiculars to the lines from a point on the locus creates two congruent triangles.

Construction

Draw an angle and mark the vertex A.

Put your compass point at A and draw an arc to cut the lines forming the angle at B and C.

Put the point at B and draw an arc in the angle.

Keep the compasses set to the same radius.

Put the point at C and draw an arc in the angle to cut the arc just drawn.

Draw a straight line through A and the point where the arcs cross.

The bisector could be continued to the left of A. If the lines are extended, another bisector could be drawn, perpendicular to the first one.

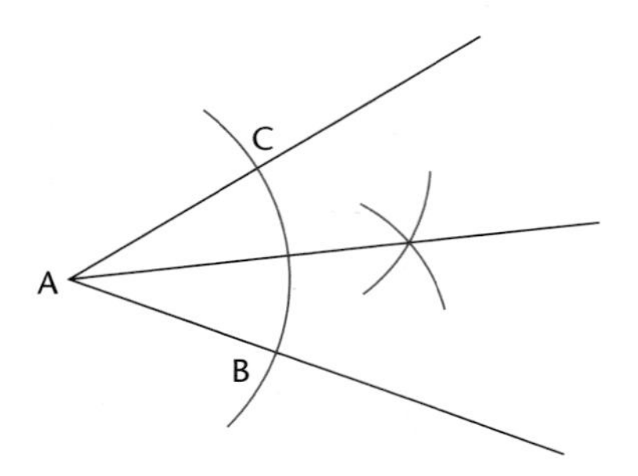

Example 3

Draw a triangle ABC with sides AB = 5 cm and AC = 4 cm, and angle A = 50°.

Use compasses to bisect angle A. Shade in the locus of the points inside the triangle that are nearer to AB than AC.

Solution

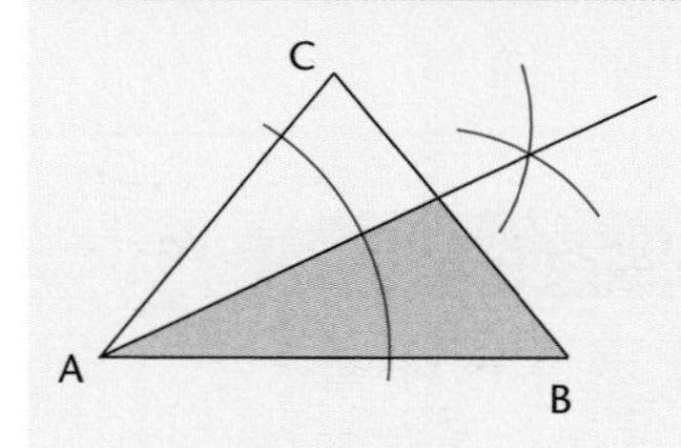

Exam tip

The locus of the points equidistant from two intersecting lines is a pair of lines, but usually you only require one.

The perpendicular from a point to a line

Construction

Use the following method to construct the perpendicular from point P to the given line.

1 Open your compasses to any radius. Put your compass point on P. Draw two arcs, cutting the line at Q and R.

2 Keep the compasses set to the same radius. Put the compass point on Q. Draw an arc below the line. Now put the compass point on R and draw another arc, cutting the first arc at X.

3 Line up your ruler with points P and X. Draw the line PM.

This line is at right angles to the original line.

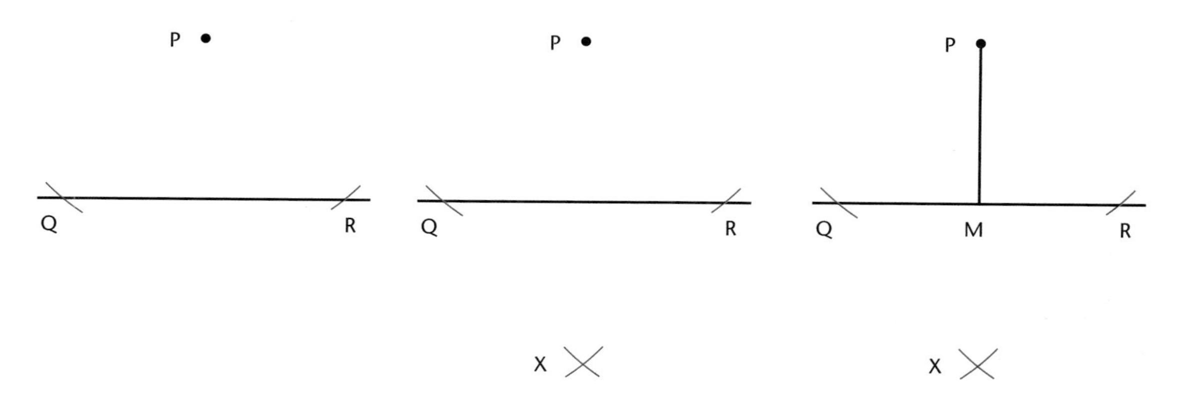

Activity 3

a) (i) Draw a line across your page. Put a cross on one side of the line and label it P.

(ii) Construct the perpendicular from P to the line. Make sure you keep your compasses set to the same radius all the time.

This time, join P to X; don't stop at the original line.

b) (i) Measure PM and XM. What do you notice?

(ii) What can you say about P and X?

Exercise 16.1

1 Draw a circle, centre A and radius 5 cm. Shade the locus of the points that are less than 5 cm from A.

2 Draw a rectangle 4 cm by 5 cm. Sketch the locus of the points outside the rectangle that are 1 cm from the perimeter of the rectangle.

3 Draw a rectangle ABCD with AB = 6 cm and BC = 4 cm. Sketch the locus of the points inside the rectangle that are nearer to A than B.

4 Draw two parallel lines across the page, 4 cm apart. Draw the locus of the points that are 1 cm from the top line and 3 cm from the bottom line.

5 A fox never travels more than 5 miles from its den.
Draw a sketch to show the region where it travels.

6 Draw a line 7 cm long.
Construct the perpendicular bisector of the line.

7 Draw an angle of 70°.
Construct the bisector of the angle.

8 Show, by shading, the locus of the points that are more than 4 cm from a fixed point A.

9 Draw a line 6 cm long.
Show, by shading in a sketch, the locus of the points that are less than 2 cm from the line.

10 Draw an angle of 80°.
Construct the bisector of the angle.

11 Draw a line AB 6 cm long.
Construct the perpendicular bisector of AB.

12 Draw a line 10 cm long. Put a point anywhere above the line.
Construct the perpendicular from your point to the line.

13 Construct a triangle ABC with AB = 8 cm, AC = 7 cm and BC = 5 cm.
Use compasses and a ruler to bisect angle A.

Shade the locus of the points inside the triangle that are nearer to AB than AC.

14 Draw a square ABCD with sides of 6 cm.
Construct the locus of the points that are equidistant from A and C.
What do you notice about the locus?

15 Draw a triangle ABC with AB = 8 cm, angle A = 90° and angle B = 40°.
Do a construction to find the locus of the points inside the triangle that are nearer to AC than BC.

16 Draw a square with sides of 4 cm. Label one corner A.
Show the locus of the points inside the square that are less than 3 cm from A.

17 Draw a rectangle ABCD with sides AB = 7 cm and BC = 5 cm.
Use compasses to construct the line equidistant from AB and AC.

18 Construct the triangle ABC with angle A = 30°, angle B = 50° and AB = 10 cm.
Construct the locus of the points equidistant from A and B.

Exam tip

In most cases you will be asked to construct a locus either to size or to scale. Draw it as accurately as you can.

Exam tip

Do not stop the line of a construction at the intersection of the arcs: draw it through the intersection.

19 Two towns Bimouth and Tritown are 10 miles apart. Phoebe wants to live nearer to Bimouth than Tritown.
Using a scale of 1 cm : 2 miles, make a scale drawing and show, by shading, the region where she can live.

20 Draw a triangle ABC with AB = 7 cm, angle A = 50° and angle B = 40°.
Show, by shading, the locus of the points within the triangle that are nearer to AC than BC.

21 Sonia has a 20 metre cable on her lawnmower and the socket is in the middle of the back wall of her house. The back of the house is 12 m wide and her garden is a rectangle the same width as the house, stretching 24 m from the house.
Using a scale of 1 cm : 4 m, make a drawing of her garden and show, by shading, the region she can reach with the mower.

When you do a construction, leave in your construction lines.

Foundation Gold/
Higher Bronze

Challenge 1

Draw a circle of radius 5 cm. Draw a chord of the circle which is 6 cm long.
Construct the perpendicular from the centre of the circle on to the chord.
Measure the length of each of the two parts of the chord. What do you notice?

Problems involving intersection of loci

Combining all you know about loci, you can answer more complicated questions involving more than one locus.

Example 4

Construct triangle ABC with AB = 7 cm, AC = 6 cm and BC = 4 cm.
By using constructions, find the point that is equidistant from all three vertices. Label this point D.

Solution

First you need the line equidistant from two vertices. If you choose vertices A and B, you need to construct the perpendicular bisector of AB.

Then you need to construct the perpendicular bisector of another side. Where they cross is the required point.

This diagram is not to scale.

You could also bisect the third side, and that line would also pass through the same point.

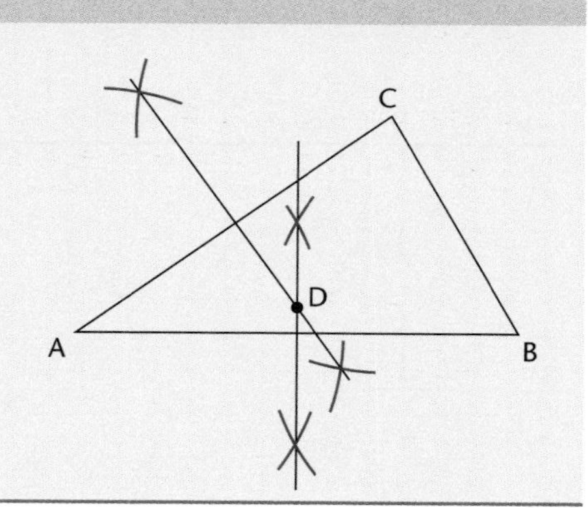

Example 5

Two points A and B are 4 cm apart.

Show, by shading, the locus of the points that are less than 2.5 cm from A, and nearer to B than A.

Solution

You need to draw a circle, radius 2.5 cm and centre A. You also need to draw the bisector of the line AB. The region you require is inside the circle and on the B side of the bisector.

The required region is shaded on the diagram.

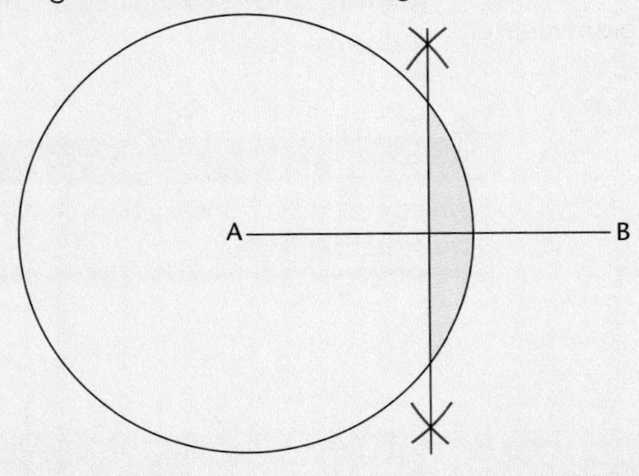

Example 6

Erica wants to put a rocking chair in her room.
She wants the chair more than 0.5 m from a wall and less than 2 m from corner A. This is a sketch of her room.

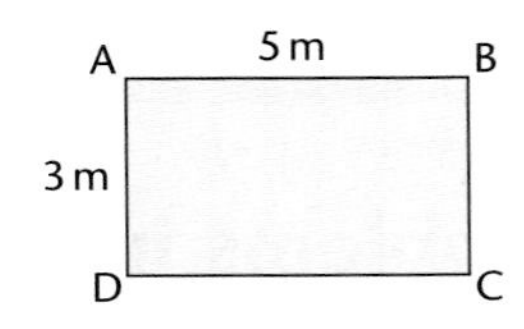

Using a scale of 1 cm : 1 m, make a scale drawing of the room and show, by shading, the region where the chair can be placed.

Solution

Draw the rectangle and then add lines 0.5 cm from each side.
Draw a circle, centre A and radius 2 cm.

In this diagram the regions not required are shaded, leaving the white region where the chair can be placed.

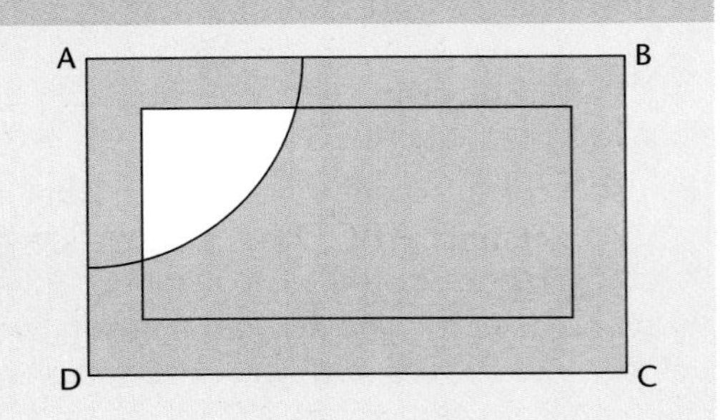

Example 7

Find the centre of the rotation that maps triangle ABC on to triangle A′B′C′.

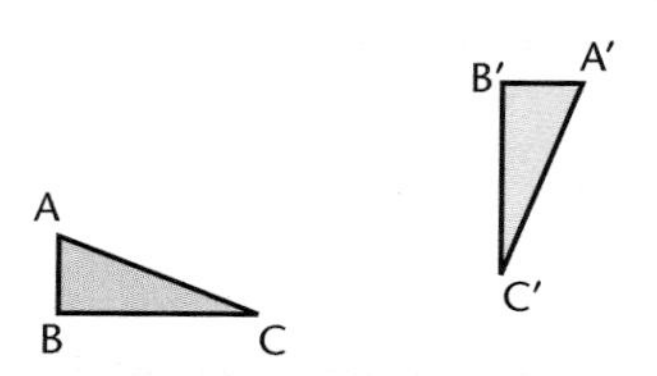

Solution

The centre of rotation must be equidistant from A and A′.
It will be on the perpendicular bisector of AA′.
Arcs have been omitted to make the diagram clearer.

The centre must also be equidistant from C and C′.
The centre of rotation will be the point where the two perpendicular bisectors cross.

The centre must also be equidistant from B and B′.
Construct the perpendicular bisector of BB′ to check.

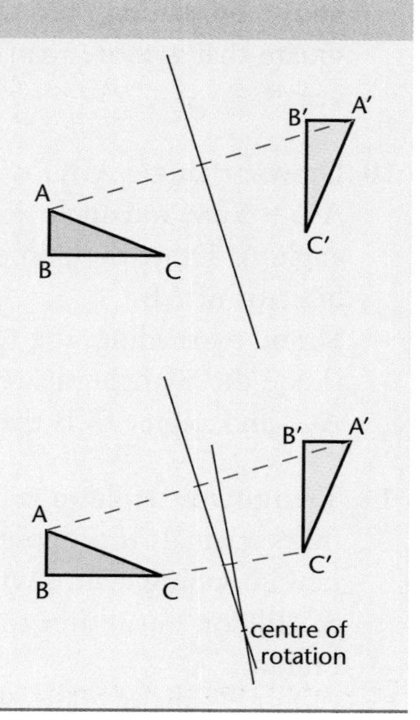

Exercise 16.2

Draw all of these accurately.

1 Two points A and B are 5 cm apart.
Show, by shading, the region that is less than 3 cm from A, and more than 4 cm from B.

2 A rectangle ABCD has sides AB = 5 cm and BC = 4 cm.
Draw the rectangle and show, by shading, the region inside the rectangle that is nearer to AB than CD, and less than 3.5cm from B.

3 Draw a triangle ABC with AB = 6 cm, angle A = 60° and angle B = 55°.
Use constructions to find the point D that is equidistant from all three sides.

4 Draw a rectangle ABCD with sides AB = 4 cm and BC = 3 cm.
Show the points that are equidistant from AB and BC, and 3.5 cm from A.

5 Draw a triangle ABC with sides AB = 9 cm, BC = 6 cm and AC = 5 cm.
Show, by shading, the region inside the triangle that is nearer to AB than BC, and more than 3 cm from C.

6 Show, by shading, the locus of the points that are more than 2 cm from a point A and less than 3 cm from point A.

7 Two points A and B are 4 cm apart.
Show, by shading, the locus of all the points that are less than 2.5 cm from A, and more than 3 cm from B.

8 Draw a triangle ABC with AB = 6 cm, angle A = 40° and angle B = 35°. Use constructions to find the point D that is equidistant from A and B, and 4 cm from C.

9 Draw a square with sides of 4 cm. Show, by shading, the region within the square that is more than 2 cm from every vertex.

10 Draw a triangle ABC with AB = 6 cm, AC = 5 cm and angle A = 55°. Bisect angle A. Draw the perpendicular bisector of AB.
Show, by shading, the region that is inside the triangle, nearer to AB than AC, and nearer to B than A.

11 Two towns, Hilldon and Baton are 20 miles apart. It is proposed to build a new shopping centre within 15 miles of Hilldon but nearer to Baton than Hilldon.
Using a scale of 1 cm : 5 miles, make a drawing and show the region where the shopping centre can be built.

12 Richard's bedroom is rectangular with sides of 4 m and 6.5 m. He wants to put a desk within 1 metre of a longer wall and within 2.5 m of the centre of the window in the middle of one of the shorter walls.
Using a scale of 1 cm : 1 m, make a scale drawing and show, by shading, the region where the desk can be placed.

13 Kirsty has a triangular patio with sides of 6 m, 4 m and 5 m. She wants to put a plant pot on the patio more than 2 m from any corner.
Using a scale of 1 cm : 1 m, make a drawing and show, by shading, where she can put the plant pot.

Exam tip

You can either shade the required region or the regions that are not required. It is often easier to shade regions not required. Give a key or label the diagram to make it clear which you have done.

14 This is a sketch of a plot of land that Arun wants to use for camping.

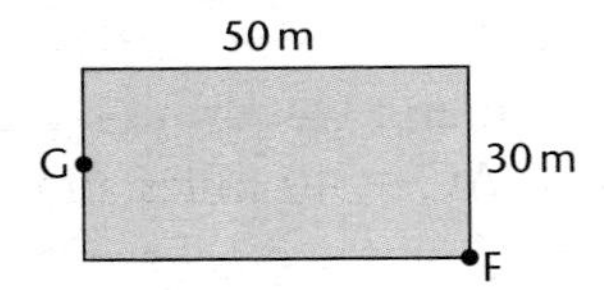

He wants to put a tap in the field within 35 m of the gate, G, which is at the middle of one of the shorter sides. He also wants the tap to be within 25 m of his farm, which is at corner F.
Using a scale of 1 cm : 10 m, make a scale drawing of the land. Show, by shading, the position where the tap can be placed.

15 A field is a quadrilateral ABCD with AB = 25 m, BC = 30 m, angle A = 90°, angle B = 106° and angle C = 65°.
The farmer wants to put a scarecrow within 15 m of corner A and nearer to CD than CB.
Using a scale of 1 cm to 5 m, draw the field and show, by shading, the region where the scarecrow can be placed.

16 Dave and Clare live 7 miles apart. They set out on bikes to meet. They ride directly towards each other. When they meet, Dave has ridden less than 5 miles and Clare has ridden less than 4 miles.
Using a scale of 1 cm : 1 mile, make a scale drawing to show where they could have met.

17 Tariq's garden is a rectangle ABCD with AB = 10 m and BC = 4 m. He wants to put a rotary washing line in the garden. It must be more than 4 m from corner C and more than 1 m from side AB.
Using a scale of 1 cm : 1 m, make a scale drawing of the garden and show where he can put the rotary washing line.

18 The distances between three towns Arbridge, Beaton and Ceborough are AB = 25 miles, AC = 40 miles and BC = 30 miles. A new petrol station is to be built as near as possible to all three towns.
Using a scale of 1 cm : 5 miles, make constructions to find the point D where the petrol station should be placed.

19 Sasha has a rectangular garage measuring 2 m by 5 m. It has a door at one end. She wants to put a hook in the ceiling. It must be midway between the two longer sides, less than 3.5 m from the door end and less than 2.5 m from the other end.
Make a scale drawing of the ceiling using a scale of 1 cm : 1 m. Show, by shading, the region where the hook can be fixed.

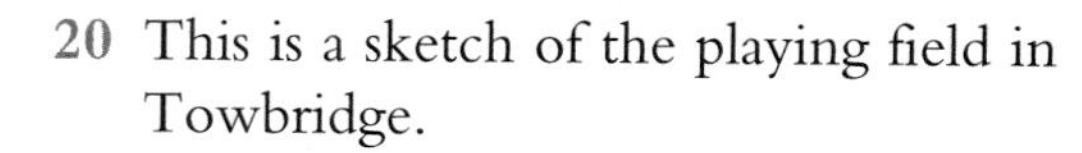

20 This is a sketch of the playing field in Towbridge.

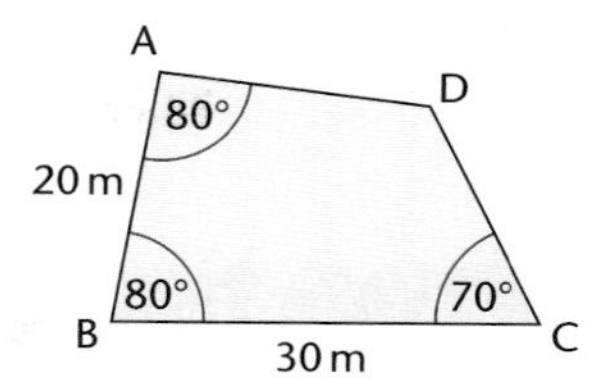

A new swing is to be placed in the field. It must be within 15 m of A and nearer to AB than AD.
Use a scale of 1 cm : 5 m to make a scale drawing and show the region where the swing can be placed.

21 A sailor is shipwrecked at night. She is 140 m from the straight coastline.
She swims straight for the shore.
The coastguard is standing on the beach exactly where the sailor will come ashore. He has a searchlight that can illuminate up to a distance of 50 m.
Make a scale drawing showing the part of the sailor's swim that will be lit up.

Key Ideas

- The locus of a point is the path or the region that the point covers as it moves according to a particular rule.
- The locus of a point x cm from point A is a circle with centre A and radius x cm.
- The locus of a point x cm from a line is a pair of lines parallel to the given line, x cm on either side of it.
- The locus of a point equidistant from two points A and B is the perpendicular bisector of the line AB.
- The locus of a point equidistant from two non-parallel lines is the bisector(s) of the angle(s) between the lines.
- The shortest distance from a point to a line is the perpendicular distance.

CHAPTER 17

Decimals 2

YOU WILL LEARN ABOUT	YOU SHOULD ALREADY KNOW
o Terminating and recurring decimals	o How to turn fractions into decimal equivalents o The decimal equivalents of simple fractions, for example $\frac{1}{2}$, $\frac{1}{4}$ and $\frac{3}{4}$ o How to multiply simple decimals and whole numbers

Terminating and recurring decimals

Activity 1

Write each of these fractions as a decimal.

$\frac{2}{3}$, $\frac{1}{5}$, $\frac{2}{11}$, $\frac{3}{4}$, $\frac{3}{7}$, $\frac{5}{6}$, $\frac{1}{18}$, $\frac{1}{20}$, $\frac{5}{8}$

What do you notice about the decimal values?

Example 1	Solution
Convert $\frac{5}{8}$ to a decimal.	$\frac{5}{8} = 5 \div 8 = 0.625$

Example 2	Solution
Convert $\frac{1}{6}$ to a decimal.	$\frac{1}{6} = 1 \div 6 = 0.16666\ldots$

Look again at Examples 1 and 2.

In Example 1, the decimal equivalent of $\frac{5}{8}$ is exactly 0.625. This is an example of a **terminating decimal**. It is called terminating because it finishes at the digit 5.

In Example 2, the decimal equivalent of $\frac{1}{6}$ goes on forever with the digit 6 repeating itself over and over again. This is an example of a **recurring decimal**.

The dot notation for recurring decimals

To save a lot of writing, there is a special notation for recurring decimals. You put a dot over the figure that recurs.

So, for example, $\frac{1}{3} = 0.333\,333\ldots$ is written as $0.\dot{3}$.

Similarly, $\frac{1}{6} = 0.166\,666\ldots$ is written as $0.1\dot{6}$.

Example 3

Write $\frac{7}{11}$ as a recurring decimal.

Solution

$7 \div 11 = 0.636\,363\ldots$

This time both the 6 and 3 recur. To show this you put a dot over both the 6 and the 3. That is, you write $0.\dot{6}\dot{3}$.

Example 4

Write $\frac{171}{333}$ as a recurring decimal.

Solution

$171 \div 333 = 0.513\,513\,513\ldots$

This time there are three figures that recur. You write this as $0.\dot{5}1\dot{3}$. Put a dot over the first and last figures that recur.

This means that $0.432\,513\,251\,325\,1$, for example, is written as $0.4\dot{3}25\dot{1}$.

Sometimes, because so many figures recur, it is difficult to see the recurring figures on a calculator. For example, $\frac{3}{7}$ will be shown as 0.428 571 428 on many calculators. It is only just noticeable that this means all six digits will recur.

That is, $\frac{3}{7} = 0.428\,571\,428\,571\,428\,57\ldots = 0.\dot{4}2857\dot{1}$.

On calculators which show fewer digits, it may be even less obvious.

Exam tip

When working out a recurring decimal, your calculator may round the last figure using the normal rounding rules. This means that 0.166 666… may appear as 0.166 666 67. Even so, it should be noticed that the decimal recurs. In practical problems, if your answer is a long decimal, it is best to round it to a suitable degree of accuracy.

Activity 2

Use your calculator to work out $\frac{1}{7}, \frac{2}{7}, \frac{3}{7}, \ldots \frac{6}{7}$.

What do you notice about the digits in your answer?

Check the patterns for other fractions which are recurring decimals.

Challenge 1

What can you say about the numbers in the denominators of the fractions that give terminating decimals?

Exercise 17.1

Convert each of the fractions in questions **1** to **32** to a decimal. When the answer is a recurring decimal, use the dot notation.

1 $\frac{3}{8}$ **2** $\frac{5}{6}$ **3** $\frac{5}{16}$ **4** $\frac{11}{40}$

5 $\frac{5}{9}$ **6** $\frac{17}{25}$ **7** $\frac{5}{27}$ **8** $\frac{16}{33}$

9 $\frac{7}{110}$ **10** $\frac{7}{111}$ **11** $\frac{79}{250}$ **12** $\frac{79}{2500}$

13 $\frac{5}{7}$ **14** $\frac{79}{222}$ **15** $\frac{19}{11}$ **16** $\frac{73}{64}$

17 $\frac{7}{8}$ 18 $\frac{7}{9}$ 19 $\frac{9}{11}$ 20 $\frac{7}{32}$

21 $\frac{3}{80}$ 22 $\frac{17}{36}$ 23 $\frac{20}{125}$ 24 $\frac{23}{60}$

25 $\frac{37}{64}$ 26 $\frac{57}{132}$ 27 $\frac{7}{54}$ 28 $\frac{576}{625}$

29 $\frac{457}{1111}$ 30 $\frac{457}{1110}$ 31 $\frac{1}{303}$ 32 $\frac{813}{11\,111}$

33 Given that $\frac{1}{27} = 0.\dot{0}3\dot{7}$, $\frac{1}{3} = 0.\dot{3}$ and $\frac{1}{11} = 0.\dot{0}\dot{9}$, find the decimal equivalent of each of these without using a calculator.

a) $\frac{2}{27}$ b) $\frac{5}{27}$ c) $\frac{2}{3}$ d) $\frac{2}{11}$

e) $\frac{5}{11}$ f) $\frac{6}{11}$

34 Put each set of fractions into ascending order (smallest first).

a) $\frac{1}{3}, \frac{3}{10}, \frac{5}{18}, \frac{4}{11}, \frac{2}{7}, \frac{7}{19}, \frac{2}{5}, \frac{9}{24}$ b) $\frac{3}{5}, \frac{3}{4}, \frac{1}{2}, \frac{5}{9}, \frac{4}{7}, \frac{11}{18}, \frac{8}{15}$

Converting decimals to fractions

All terminating and recurring decimals have fraction equivalents.

Example 5 illustrates the method of changing a terminating decimal to a fraction.

Example 5

Find the equivalent fraction to 0.624.

Solution

From the definitions of the columns in decimals,

$$\begin{aligned}0.624 &= \frac{6}{10} + \frac{2}{100} + \frac{4}{1000}\\ &= \frac{600}{1000} + \frac{20}{1000} + \frac{4}{1000}\\ &= \frac{624}{1000}\\ &= \frac{78}{125} \quad \text{(dividing top and bottom by 8).}\end{aligned}$$

The quick method is simply to write 624 over 1000 straight away, since the 4 represents 'four thousandths'.

Example 6

Convert each of these to a fraction.

a) 0.48 **b)** 0.035

Solution

a) $0.48 = \frac{48}{100} = \frac{12}{25}$

b) $0.035 = \frac{35}{1000} = \frac{7}{200}$

Converting recurring decimals to fractions is more difficult and the general method will be dealt with later in the course. Only simple cases are dealt with here.

These ones are worth remembering.

$0.\dot{3} = \frac{1}{3}$ and $0.\dot{6} = \frac{2}{3}$

so $0.0\dot{3} = \frac{1}{30}$ and $0.0\dot{6} = \frac{2}{30} = \frac{1}{15}$.

$0.\dot{1} = \frac{1}{9}$, $0.\dot{2} = \frac{2}{9}$, $0.\dot{4} = \frac{4}{9}$, and so on

so $0.0\dot{1} = \frac{1}{90}$, $0.0\dot{2} = \frac{2}{90}$, $0.0\dot{4} = \frac{4}{90}$, and so on.

Exercise 17.2

Convert each of the decimals in questions **1** to **20** to a fraction in its lowest terms.

1 0.7 **2** 0.29 **3** 0.85 **4** 0.07

5 0.312 **6** 0.255 **7** 0.056 **8** 0.008

9 0.8 **10** 0.37 **11** 0.68 **12** 0.02

13 0.545 **14** 0.892 **15** 0.018 **16** 0.1345

17 $0.\dot{7}$ **18** $0.0\dot{7}$ **19** $0.\dot{8}$ **20** $0.00\dot{8}$

21 Given that $0.\dot{1} = \frac{1}{9}$, write each of these as a fraction.

a) $0.\dot{2}$ b) $0.\dot{3}$ c) $0.\dot{5}$

22 Given that $0.0\dot{1}\dot{8} = \frac{1}{55}$, write each of these as a fraction.

a) $0.0\dot{3}\dot{6}$ b) $0.0\dot{5}\dot{4}$ c) $0.3\dot{0}\dot{9}$

Key Ideas

- Recurring decimals are written using the dot notation. For example $0.333\,333\,333... = 0.\dot{3}$, $0.342\,342\,342... = 0.\dot{3}4\dot{2}$, $0.018\,181\,818... = 0.0\dot{1}\dot{8}$.
- Terminating decimals are changed to fractions using the fraction that the last digit represents. For example, $0.723 = \frac{723}{1000}$ because the digit '3' represents 3 thousandths.

Foundation Gold/ Higher Bronze

CHAPTER 18

Accuracy

YOU WILL LEARN ABOUT	YOU SHOULD ALREADY KNOW
○ The difference between discrete and continuous data ○ The upper and lower bounds of a measurement	○ How to round a number to the nearest whole number or a given number of decimal places ○ How to convert one metric unit to another, for example millimetres to metres

Discrete and continuous measures

Discrete measures can be counted. They can only take particular values.

Continuous measures include length, time and mass. They cannot be measured exactly, only to a degree of accuracy which depends on the measuring instrument.

Look at this table of data for a bicycle.

Number of wheels	2
Number of gears	15
Diameter of wheel	66 cm
Frame size	66 cm
Price	£99.99

In the table, the discrete measures are

- number of wheels
- number of gears
- price.

The continuous measures are

- diameter of wheel
- frame size.

Exercise 18.1

1 Look at these descriptions from catalogues.
Identify whether each data item is discrete or continuous.

a) Prestige 20 cm polyester golf bag, 6-way graphite-friendly top, 2 accessory pockets

b) Black attaché case, 2 folio compartments, 3 pen holders, size 3.15 cm (Height), 44.5 cm (Width), 11.5 cm (Depth)

c) 16 piece dinner set, 4 dinner plates (diameter 24.5 cm), side plates and bowls

d) Food blender, 1.5 litre working capacity, 3 speed settings, 400 watts

2 Read this extract from a newspaper article.
   a) Give two examples of discrete data in the newspaper article.
   b) Give three examples of continuous data in the article.

Andy James has now scored 108 goals in just 167 games, making him the Town's most prolific scorer ever. In Saturday's game, a penalty brought his first goal after 30 minutes, with Pete Jeffreys having been fouled. Six minutes later, James volleyed into the net again, after a flick on from Neil Matty, five yards outside the penalty box.

3 Read this extract from a newspaper article.
   a) Give two examples of discrete data in the newspaper article.
   b) Give three examples of continuous data in the article.

Lightning killed two people in Hyde Park yesterday as storms swept the south-east. 1.75 inches of rain fell in 48 hours. In Pagham winds of up to 120 mph damaged more than 50 houses and bungalows and several boats. One catamaran was flung 100 feet into the air and landed in a tree.

4 Write a description which includes three discrete measurements and two continuous measurements.

Bounds of measurement

Activity 1

Write down three numbers which round to each of these.

a) 26 **b)** 26.5 **c)** 43 **d)** 43.0 **e)** 50

Suppose a measurement is given as '26 cm to the nearest centimetre'. This means that the next possible measurements on either side are 25 cm and 27 cm. Where does the boundary between these measurements lie?

Any measurement that is nearer to 26 cm than to 25 cm or 27 cm will be counted as 26 cm. This is the marked interval on the number line.

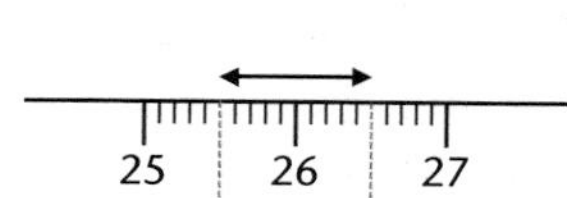

The boundaries of this interval are 25.5 cm and 26.5 cm. These values are exactly halfway between one measurement and the next. Usually when rounding to a given number of decimal places or significant figures, you would round 25.5 up to 26 and 26.5 up to 27.

So this gives

- The interval for 26 cm to the nearest centimetre is m cm where $25.5 \leq m < 26.5$.
- 25.5 cm is called the **lower bound** of the interval.
- 26.5 cm is called the **upper bound** of the interval (although it is not actually included in the interval).

Example 1

Simon won the 200 m race in a time of 24.2 seconds to the nearest tenth of a second. Complete the sentence below.

Simon's time was between … seconds and … seconds.

Solution

As the measurement is stated to the nearest tenth of a second, the next possible times are 24.1 seconds and 24.3 seconds.

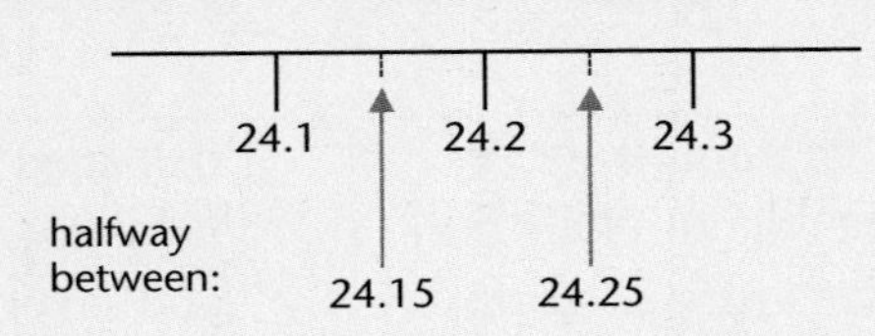

Simon's time was between 24.15 seconds and 24.25 seconds.

Exercise 18.2

1 Give the upper and lower bounds of each of these measurements.

a) Given to the nearest centimetre
(i) 27 cm (ii) 30 cm (iii) 128 cm

b) Given to the nearest 10 cm
(i) 10 cm (ii) 30 cm (iii) 150 cm

c) Given to the nearest millimetre
(i) 5.6 cm (ii) 0.8 cm (iii) 12.0 cm

d) Given to the nearest centimetre
(i) 1.23 m (ii) 0.45 m (iii) 9.08 m

e) Given to the nearest hundredth of a second
(i) 10.62 seconds (ii) 9.81 seconds (iii) 48.10 seconds

f) Given to the nearest centimetre
(i) 34 cm (ii) 92 cm (iii) 210 cm

g) Given to the nearest 10 cm
(i) 20 cm (ii) 60 cm (iii) 210 cm

h) Given to the nearest millimetre
(i) 2.7 cm (ii) 0.2 cm (iii) 18.0 cm

i) Given to the nearest centimetre
(i) 8.17 m (ii) 0.36 m (iii) 2.04 m

j) Given to the nearest hundredth of a second
(i) 15.61 seconds (ii) 12.10 seconds (iii) 54.07 seconds

2 Complete each of these sentences.

a) A mass given as 57kg to the nearest kilogram is between … kg and … kg.

b) A height given as 4.7m to 1 decimal place is between … m and … m.

c) A volume given as 468ml to the nearest millilitre is between … ml and … ml.

d) A winning time given as 34.91 seconds to the nearest hundredth of a second is between … seconds and … seconds.

e) A mass given as 0.634kg to the nearest gram is between … kg and … kg.

f) A mass given as 64kg to the nearest kilogram is between … kg and … kg.

g) A height given as 8.3m to 1 decimal place is between … m and … m.

h) A volume given as 234ml to the nearest millilitre is between … ml and … ml.

i) A winning time given as 27.94 seconds to the nearest hundredth of a second is between … seconds and … seconds.

j) A mass given as 0.256kg to the nearest gram is between … kg and … kg.

Challenge 1

A chimneysweep uses a pole made up of ten identical flexible pieces. Each piece is 1 metre long, measured to the nearest centimetre.

What height of chimney can you be sure that he could reach?

Challenge 2

Emma measures her pencil to be 18 cm and her pencil case states a length of 18.5 cm.

Can Emma be sure that her pencil will fit inside the pencil case? Explain your answer.

Challenge 3

A rectangle measures 12 cm by 5 cm, with both measurements correct to the nearest centimetre.

a) Work out the greatest possible perimeter of the rectangle.

b) Work out the smallest possible area of the rectangle.

Key Ideas

- Discrete measures can be counted. They can only take particular values.
- Continuous measures include length, time, mass, and so on. They cannot be measured exactly.
- A time of 5.7 seconds to the nearest tenth of a second lies between 5.65 seconds and 5.75 seconds.

CHAPTER 19

Indices

YOU WILL LEARN ABOUT	YOU SHOULD ALREADY KNOW
o Simplifying expressions using powers o The three basic rules for combining indices o Calculating proficiently with squares, cubes, square roots and cube roots	o The meaning of the terms square number and cube number o The meaning of a power such as 2^2 and 2^3

Indices (or **powers**) are a form of mathematical shorthand.

$3 \times 3 \times 3 \times 3$ is written as 3^4 and

$2 \times 2 \times 2 \times 2 \times 2 \times 2 \times 2 \times 2$ is written as 2^8.

Activity 1

$2^2 \times 2^5 = (2 \times 2) \times (2 \times 2 \times 2 \times 2 \times 2) = 2^7$.

By writing the powers out fully as above, find each missing index indicated with a question mark.

a) $2^3 \times 2^2 = 2^?$ **b)** $2^4 \times 2^5 = 2^?$ **c)** $3^6 \div 3^4 = 3^?$

d) $4^8 \div 4^3 = 4^?$ **e)** $(3^3)^2 = 3^?$

What do you notice?

Multiplying numbers in index form

$$\begin{aligned} 3^4 \times 3^8 &= (3 \times 3 \times 3 \times 3) \times (3 \times 3 \times 3 \times 3 \times 3 \times 3 \times 3 \times 3) \\ &= 3 \times 3 \times 3 \times 3 \times 3 \times 3 \times 3 \times 3 \times 3 \times 3 \times 3 \times 3 \\ &= 3^{12} \end{aligned}$$

The indices are added:

$$\begin{aligned} 3^4 \times 3^8 &= 3^{4+8} \\ &= 3^{12} \end{aligned}$$

The rule is $n^a \times n^b = n^{a+b}$

Example 1	Solution
Write these in index form.	
a) $5 \times 5 \times 6 \times 6 \times 6 \times 6$	**a)** $5 \times 5 \times 6 \times 6 \times 6 \times 6 = 5^2 \times 6^4$
b) $3x^3 \times 4x^5$	**b)** $3x^3 \times 4x^5 = 12x^8$

Dividing numbers in index form

$$2^6 \div 2^4 = \frac{2 \times 2 \times 2 \times 2 \times 2 \times 2}{2 \times 2 \times 2 \times 2}$$

$= 2 \times 2$
$= 2^2$

The indices are subtracted: $2^6 \div 2^4 = 2^{6-4} = 2^2$

The rule is $n^a \div n^b = n^{a-b}$

Example 2	Solution
Write this in index form. $25x^4 \div 5x^2$	$25x^4 \div 5x^2 = \frac{25x^4}{5x^2} = 5x^2$

Exercise 19.1

1 Write these in a simpler form, using indices.

a) $3 \times 3 \times 3 \times 3 \times 3$
b) $7 \times 7 \times 7$
c) $8 \times 8 \times 8 \times 8 \times 8$
d) $4 \times 4 \times 4 \times 4 \times 4$
e) $8 \times 8 \times 8$
f) $2 \times 2 \times 2 \times 2 \times 2$

2 Write these in a simpler form, using indices.

a) $5 \times 4 \times 4 \times 4 \times 5$
b) $3 \times 3 \times 5 \times 5 \times 5$
c) $2 \times 2 \times 2 \times 3 \times 3 \times 4 \times 4 \times 4 \times 4 \times 4$
d) $7 \times 7 \times 7 \times 8 \times 8 \times 9 \times 9 \times 9$

3 Write these in a simpler form, using indices.

a) $5^2 \times 5^3$ b) $6^2 \times 6^7$
c) $10^3 \times 10^4$ d) $3^6 \times 3^5$
e) $8^3 \times 8^2$ f) $4^2 \times 4^3$
g) $9^2 \times 9^7$ h) $6^2 \times 6^6$

4 Write these in a simpler form, using indices.

a) $10^5 \div 10^2$ b) $3^5 \div 3^2$
c) $8^4 \div 8^2$ d) $7^5 \div 7^3$
e) $6^3 \div 6^2$

5 Work out these, giving your answers in index form.

a) $\frac{3^9}{3^5 \times 3^2}$ b) $\frac{2^4 \times 2^3}{2^5}$
c) $\frac{5^4 \times 5^5}{5^2 \times 5^3}$ d) $\frac{4^{12}}{4^5 \times 4^4}$
e) $\frac{2^5 \times 2^6}{2^4}$ f) $\frac{6^5 \times 6^4}{6^2 \times 6}$

6 Write these as a single power of a.

a) $a^2 \times a^3$ b) $a^4 \times a^5$
c) $a^4 \times a^2$ d) $a^3 \times a^6$

7 Write these as a single power of a.

a) $a^6 \div a^4$ b) $a^7 \div a^3$
c) $a^8 \div a^2$ d) $a^5 \div a^2$

8 Simplify these.

a) $2a^2 \times 3a^3$ b) $4a^4 \times 3a^5$
c) $3a^4 \times 4a^2$ d) $5a^3 \times 3a^6$

9 Simplify these.

a) $6a^6 \div 2a^4$ b) $10a^7 \div 5a^3$
c) $6a^8 \div 3a^2$ d) $12a^5 \div 4a^2$

Challenge 1

a) Simplify these as much as possible.

(i) $(x)^2$ (ii) $(3^4)^2$ (iii) $(a^3)^5$ (iv) $(2y^2)^3$ (v) $(3x^3)^4$

b) Use your answers to part a) to complete statements 2 and 3.

1 Powers of numbers and letters can also be raised to powers.

2 If the numbers and letters are in brackets, the indices are …

3 The rule is $(n^a)^b = \ldots$

Powers and roots

Exam tip

You are expected to know the squares up to $15^2 = 225$, so you should learn these.

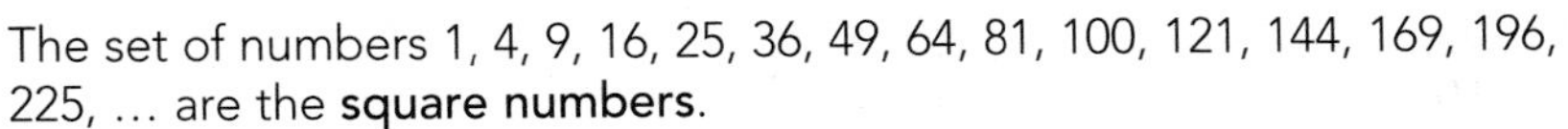

The set of numbers 1, 4, 9, 16, 25, 36, 49, 64, 81, 100, 121, 144, 169, 196, 225, … are the **square numbers**.

Each of them can be written as the square of a counting number.

Because $16 = 4^2$, the 'square root' of 16 is 4. This is written as $\sqrt{16} = 4$.

Similarly, $\sqrt{36} = 6$ and $\sqrt{81} = 9$.

But $(^-4)^2 = 16$ as well as $4^2 = 16$. It follows that the square root of 16 could be 4 or $^-4$. This is often written ±4.

Similarly, the square root of $81 = \pm9$.

Exam tip

In many practical problems where the answers are square roots, the negative answer would not have a meaning and so it should be left out.

The numbers 1, 8, 27, 64, 125, …, 1000, … are **cube numbers** because each of them can be written as the cube of a whole number.

Number	1	8	27	64	125	…	1000	…
Cube	1^3	2^3	3^3	4^3	5^3	…	10^3	…

Exam tip

The cube numbers shown are the ones you are expected to remember and so you should learn them.

Because $27 = 3^3$, the cube root of 27 is 3. This is written as $\sqrt[3]{27} = 3$.

You may get easy square roots and cube roots in the non-calculator section of an examination paper, but make sure you can find and use the [$\sqrt[3]{\ }$] and [$\sqrt{\ }$] buttons on your calculator.

On older calculators, you may have to put the number in first and then press the correct button. On most modern calculators, you press the root button first, then the number, then [=].

So [$\sqrt{\ }$] [5] [2] [=] 7.211…

Exercise 19.2

Do not use your calculator for questions 1 to 7.

1 Write down the square of each number.

a) 7 b) 12
c) 8 d) 11
e) $\sqrt{10}$

2 Write down the square root of each number.

a) 36 b) 81
c) 169 d) 196
e) 23^2

3 Write down the cube of each number.

a) 4 b) 5
c) 3 d) 10
e) $\sqrt[3]{18}$

4 Write down the cube root of each number.

a) 8 b) 1
c) 64 d) 1000
e) 20^3

5 Find the area of a square of side 4 mm.

6 Solve the equation $x^2 = 81$.

7 a) Which number is $3^2 \times 4^2$?
b) Write 36 as a product of prime factors.

You may use your calculator for questions 8 to 16.

8 Write down the square of each number.

a) 25 b) 40
c) 35 d) 50
e) 73

9 Write down the square root of each number.

a) 400 b) 289
c) 361 d) 10 000
e) 7921

10 Write down the cube of each number.

a) 7 b) 9
c) 20 d) 25
e) 1.5 f) 2.7
g) 5.4

11 Write down the cube root of each number.

a) 343 b) 729
c) 1331 d) 1 000 000
e) 216 f) 1728
g) 512

12 Work these out.
Give your answers to 2 decimal places.

a) $\sqrt{56}$ b) $\sqrt{27}$
c) $\sqrt{60}$ d) $\sqrt{70}$
e) $\sqrt{39}$ f) $\sqrt{90}$
g) $\sqrt{280}$ h) $\sqrt{678}$
i) $\sqrt{380}$ j) $\sqrt{456}$

13 Find the length of a square of area 14 cm^2.
Give your answer to 2 decimal places.

14 Find the length of a cube whose volume is 45 cm^3.
Give your answer to 2 decimal places.

15 Put these numbers in order, smallest first.

2^3, 3^2, 4^2, $\sqrt{25}$, $\sqrt[3]{343}$, 5^2

16 Find $(\sqrt[3]{16})^3$. You should be able to do it without a calculator. If you could not, try it with a calculator. What do you notice?

Foundation Gold/
Higher Bronze

Challenge 2

I think of a number, I find its cube root and then square that. The answer is 9.

What is the number I thought of?

Challenge 3

Find two numbers less than 200 which are both square numbers and cube numbers.

Key Ideas

- To multiply using powers, use the rule $n^a \times n^b = n^{a+b}$.
- To divide using powers, use the rule $n^a \div n^b = n^{a-b}$.
- You should know the squares from 1^2 to 15^2.
- You should know the values of 1^3, 2^3, 3^3, 4^3, 5^3 and 10^3.
- You should know how to find and use the square root and cube root buttons on your calculator.

Sequences

YOU WILL LEARN ABOUT	YOU SHOULD ALREADY KNOW
○ Recognising a linear sequence ○ Using the formula for the *n*th term of a sequence ○ Finding the formula for the *n*th term of a linear sequence	○ The sets of square and cube numbers

Foundation Gold/ Higher Bronze

Linear sequences

Look at this sequence.

2, 5, 8, 11, 14, ...

The terms of the sequence increase by 3 every time. When the increase is constant like this, the sequence is called a linear sequence.

In the same way, the sequence 11, 9, 7, 5, 3, 1, $^-1$, ... is also linear, since the terms decrease by the same amount (2) every time. Another way of thinking of this is that you are adding $^-2$ each time. This idea will help with finding the *n*th term.

Challenge 1

Work in pairs.

Each of you find as many different sequences as you can where the first two terms are 1 and 3. For each sequence, write down the first four terms on a separate piece of paper. On the back of the paper, write the rule you have used.

Swap a sequence with your partner.

Try to find each other's rule.

Exercise 20.1

Which of these sequences are linear and which are not? For each sequence, write down the next two terms.

1 3 5 7 9

2 43 40 38 35

3 78 75 72 69

4 1 5 9 13

5 4 9 16 25

6 1 6 15 28

7 2 9 16 23

The nth term

It is often possible to find a formula to give the terms of a sequence.

You usually use n to stand for the number of a term.

Example 1	Solution
The formula is nth term = $2n + 1$. Find each of these terms.	
a) 1st term	**a)** The 1st term $= 2 \times 1 + 1 = 3$
b) 2nd term	**b)** The 2nd term $= 2 \times 2 + 1 = 5$
c) 3rd term	**c)** The 3rd term $= 2 \times 3 + 1 = 7$

Exercise 20.2

Each of these is the formula for the nth term of a sequence.
Find the first four terms of each sequence.

1 $n + 1$	2 $2n$	3 $2n - 1$	4 $n + 5$
5 $3n$	6 $3n + 1$	7 $5n - 3$	8 $10n$
9 $7n - 7$	10 $2 - n$	11 n	12 $n + 3$
13 $4n$	14 $n - 1$	15 $2n + 1$	16 $3n - 1$
17 $6n + 5$	18 $2n - 3$	19 $5 - n$	20 $10 - 2n$

Challenge 2

Each of these is the formula for the nth term of a sequence.
Find the first four terms of each sequence.

a) n^2 b) $n^2 + 2$ c) $n^2 - 5$ d) $3n^2$ e) n^3

Finding the formula for the nth term

Look back at the formulae in Exercise 20.2 and the sequences in your answers, and notice how they are connected. If the formula contains a '$2n$', the terms increase by 2 each time; if it contains a '$5n$', the terms increase by 5 each time. Similarly, if it contains '^-3n' it increases by $^-3$ (or goes down 3) each time. So to find a formula for a given sequence, find how much more (or less) each term is than the one before it.

Exam tip

This will always work if the differences are the same each time.

Example 2

Find the formula for the *n*th term of this sequence.

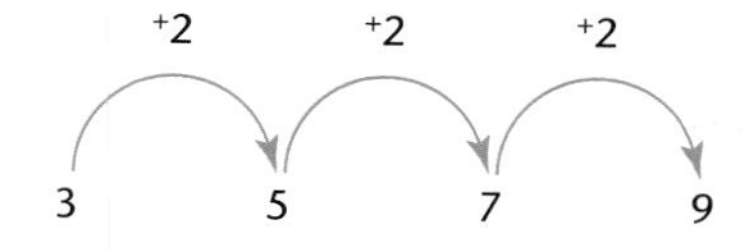

Solution

The differences between all the terms are 2, so the formula will include a '$2n$'.

When $n = 1$, $2n = 2$.

But the first term of the sequence is 3, which is 1 more.

The formula will be *n*th term $= 2n + 1$.

Example 3

Find the formula for the *n*th term of this sequence.

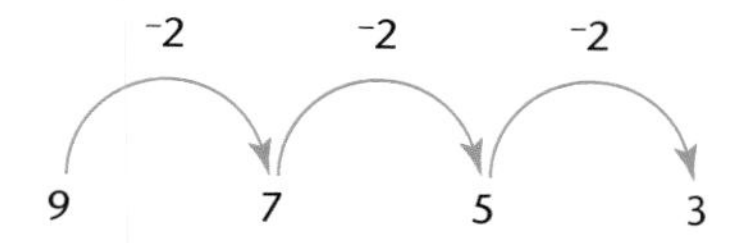

Solution

The differences here are still 2 but they must be $^-2$, as the terms are getting smaller. This time the formula will include a '^-2n', to make the terms get smaller.

When $n = 1$, $^-2n = {}^-2$.

But the first term of the sequence is 9, which is 11 more.

The formula for the *n*th term is $^-2n + 11$, or $11 - 2n$.

Exercise 20.3

Find the formula for the *n*th term for each of these sequences.

1 1 2 3 4

2 4 6 8 10

3 4 8 12 16

4 0 2 4 6

5 7 11 15 19

6 1 7 13 19

7 11 21 31 41

8 5 8 11 14

9 101 201 301 401

10 0 1 2 3

11 2 5 8 11

12 7 9 11 13

13 4 9 14 19

14 15 20 25 30

15 $^{-}1$ 3 7 11

16 5 7 9 11

17 101 102 103 104

18 4 3 2 1

19 7 4 1 $^{-}2$

20 25 23 21 19

21 Snooker balls are arranged in a triangular pattern as shown.

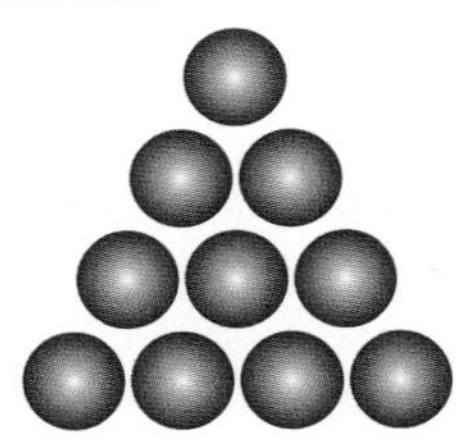

Find a formula for the number of balls in the nth row of the triangle.

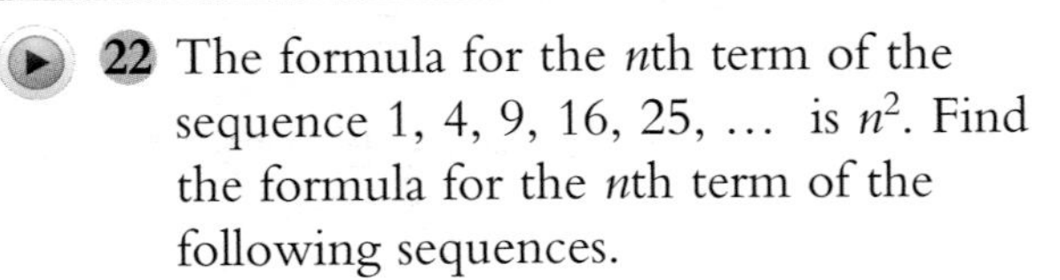

22 The formula for the nth term of the sequence 1, 4, 9, 16, 25, … is n^2. Find the formula for the nth term of the following sequences.

a) 3, 6, 11, 18, 27, …

b) $^{-}4$, $^{-}1$, 4, 11, 20, …

c) 2, 6, 12, 20, 30, …

Exam tip

Check that your formula for the nth term is correct by trying it out for the first few terms. Put $n = 1, 2, 3, \ldots$

Key Ideas

- If the differences between the terms of a sequence are all the same, the sequence is linear.
- If this difference is a, the formula for the nth term of the sequence will be $an + b$. Put in a value of n to find the value of b.

Prisms and units

YOU WILL LEARN ABOUT	YOU SHOULD ALREADY KNOW
○ Finding the volume of a prism, in particular the volume of a cylinder ○ Solving problems involving the volume of a prism ○ Finding the surface area of a prism, in particular the surface area of a cylinder ○ Converting between metric units of area and volume	○ Common metric units for length, area and volume ○ How to find the area of a rectangle, triangle, parallelogram and trapezium ○ How to find the circumference and area of a circle ○ How to find lengths using Pythagoras' theorem ○ How to round answers to a suitable degree of accuracy

The volume of a prism

A **prism** is a three-dimensional shape that has the same cross-section throughout its length.

A **cuboid** is a prism with a rectangular cross-section.

This is how to find the volume of a cuboid.

Volume of a cuboid = length × width × height

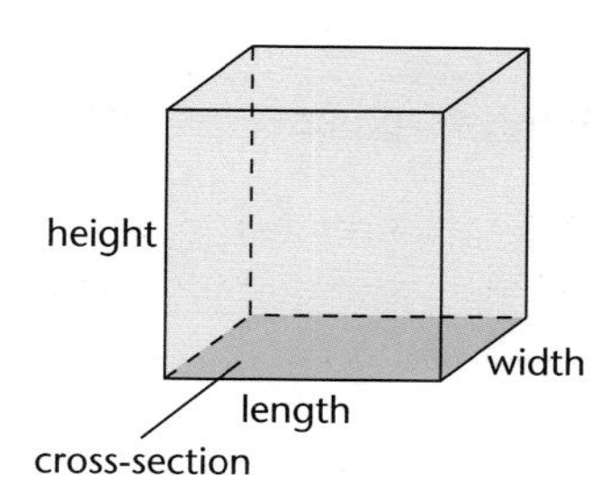

You can also think of this as

Volume of a cuboid = area of cross-section × height

This is an example of a general formula for the volume of a prism. When laid on its side, along its length,

Volume of a prism = area of cross-section × length

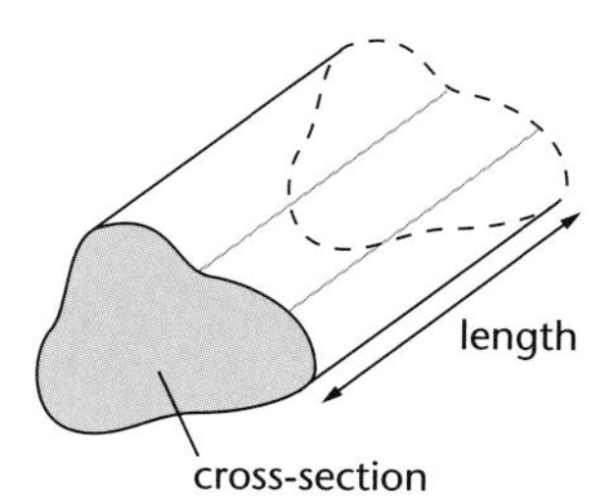

Another important prism is the **cylinder**. Its cross-section is a circle, which has area πr^2.

Volume of a cylinder = $\pi r^2 h$

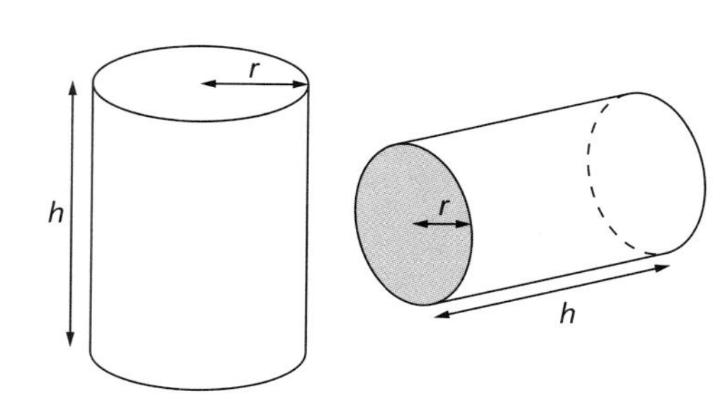

Example 1

Calculate the volume of a cylinder with base diameter 15 cm and height 10 cm.

Solution

Radius of base $\frac{15}{2} = 7.5$ cm

Volume of a cylinder $= \pi r^2 h$

$= \pi \times 7.5^2 \times 10$

$= 1767$ cm^3, to the nearest whole number.

Example 2

A chocolate box is a prism with a trapezium as cross-section, as shown.

Calculate the volume of the prism.

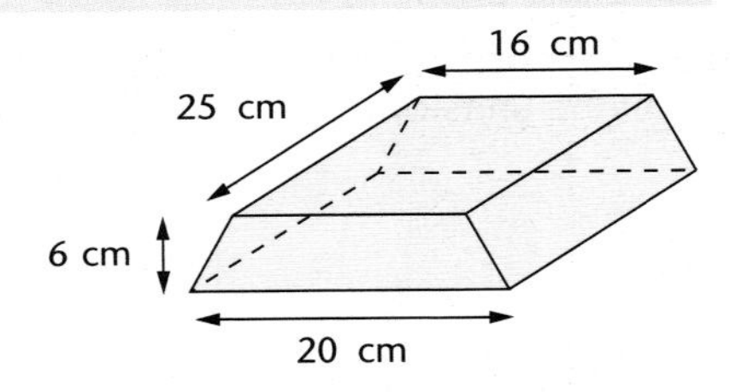

Solution

Area of a trapezium $= \frac{1}{2}(a + b)h$

$= \frac{1}{2}(20 + 16) \times 6$

$= 108$ cm^2

Volume of a prism = area of cross-section × length

$= 108 \times 25$

$= 2700$ cm^3

Example 3

A cylinder has volume 100 cm^3 and is 4.2 cm high.

Find the radius of its base. Give your answer to the nearest millimetre.

Solution

Volume of a cylinder $= \pi r^2 h$

$100 = \pi \times r^2 \times 4.2$

$r^2 = \frac{100}{\pi \times 4.2}$

$= 7.578\ldots$

$r = \sqrt{7.578\ldots}$

$= 2.752\ldots$

$= 2.8$ cm, to the nearest millimetre.

Exercise 21.1

1 Calculate the volume of a cylinder with base radius 5.6 cm and height 8.5 cm.

2 A cylindrical stick of rock is 12 cm long and has radius 2.4 cm.
Find its volume.

3 A cylinder has diameter 8 cm and height 8 cm.
Calculate its volume.

4 Calculate the volume of a prism 15 cm long with each of these cross-sections.

a)

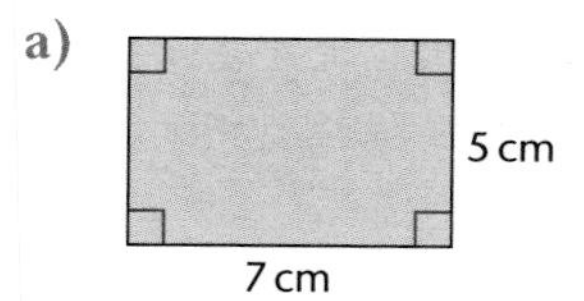

b)

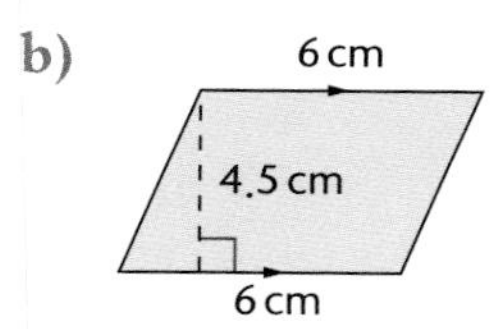

c)

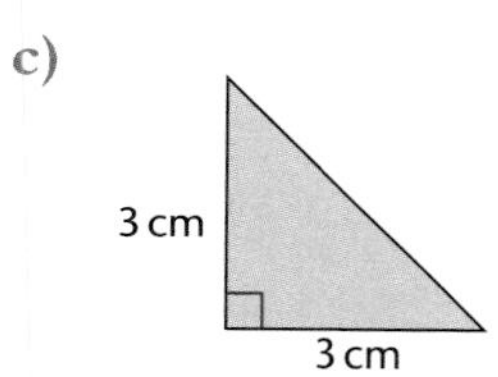

5 A chocolate bar is in the shape of a triangular prism.
Calculate its volume.

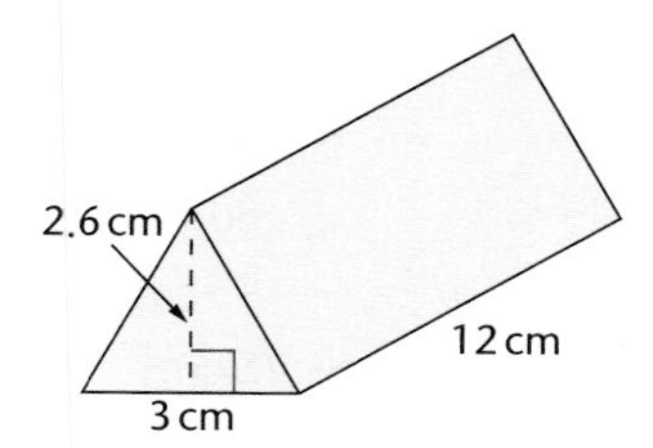

6 Calculate the volume of a cylinder with base radius 4.3 cm and height 9.7 cm.

7 A cylindrical water tank is 4.2 m high and has radius 3.6 m.
Find its volume.

8 A cylinder has diameter 9 cm and height 12 cm.
Calculate its volume.

9 Calculate the volume of a prism 12 cm long with each of these cross-sections.

a)

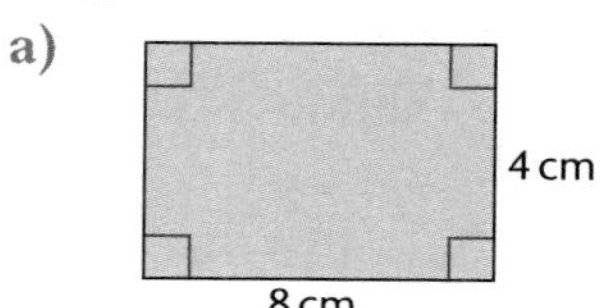

b)

5.6 cm
5 cm

c)

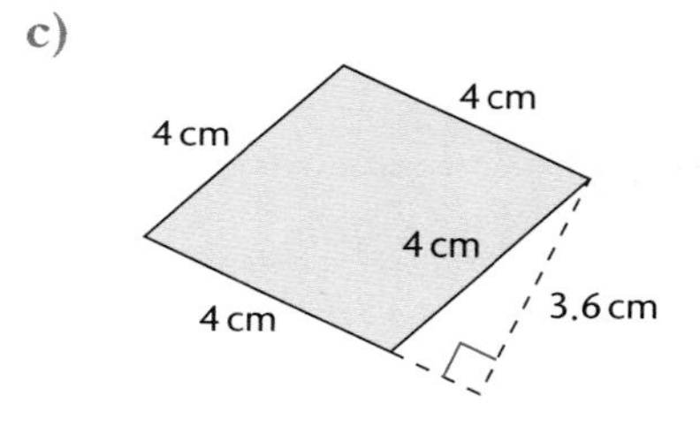

10 A gift box is a prism with a triangular base.
Calculate its volume.

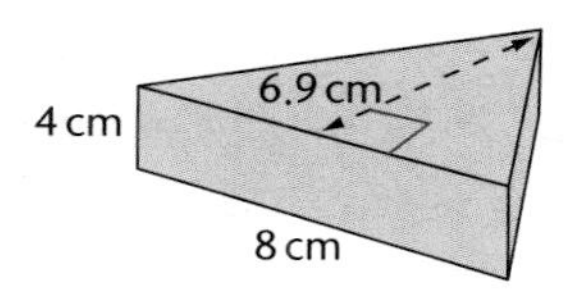

11 A pencil-box is a prism with a trapezium as its cross-section, as shown.
Calculate the volume of the box.

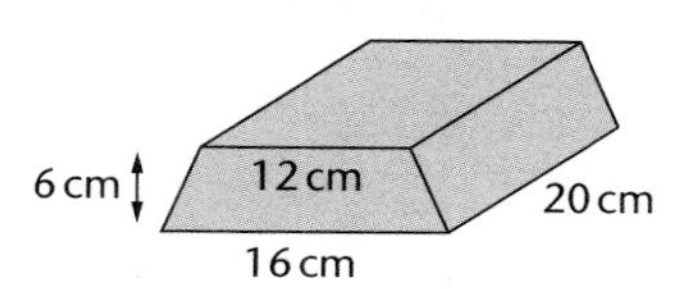

12 The area of cross-section of a prism is 75 cm^2. Its volume is 1200 cm^3.
Calculate its length.

13 The volume of a cylinder is $800\,\text{cm}^3$. Its radius is 5.3 cm.
Calculate its height.

14 A cylinder has volume $570\,\text{cm}^3$ and height 7 cm.
Find its base radius. Give your answer to the nearest millimetre.

15 The volume of a cylindrical tank is $600\,\text{m}^3$. Its height is 4.6 m.
Calculate the radius of its base.

16 A vase is a prism with a trapezium as its base. The internal measurements are as shown.

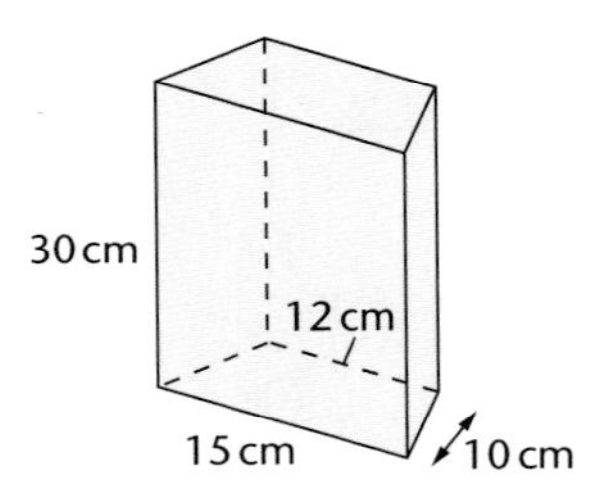

How much water can the vase hold?
Give your answer in litres.
(1 litre = $1000\,\text{cm}^3$)

17 The area of cross-section of a prism is $90\,\text{cm}^2$. Its volume is $1503\,\text{cm}^3$.
Calculate its length.

18 The volume of a cylinder is $1500\,\text{cm}^3$.
Its radius is 7.5 cm.
Calculate its height. Give your answer to the nearest millimetre.

19 A cylinder has volume $620\,\text{cm}^3$ and height 8 cm.
Find its base radius. Give your answer to the nearest millimetre.

20 The volume of a cylinder is $1100\,\text{cm}^3$.
Its height is 10.8 cm.
Calculate its radius. Give your answer to the nearest millimetre.

The surface area of a prism

To find the total surface area of a prism, simply add up the surface area of all the individual surfaces.

Example 4

Find the total surface area of this prism.

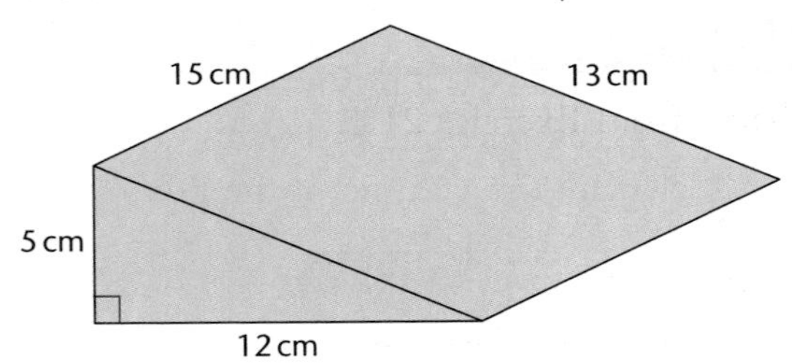

Solution

Area of end $= \frac{1}{2} \times 12 \times 5 = 30\,\text{cm}^2$
Area of other end $= 30\,\text{cm}^2$
Area of base $= 12 \times 15 = 180\,\text{cm}^2$
Area of top $= 13 \times 15 = 195\,\text{cm}^2$
Area of back $= 5 \times 15 = 75\,\text{cm}^2$
Total surface area $= 510\,\text{cm}^2$

For a cylinder there are three surfaces.

The two ends each have area = πr^2.

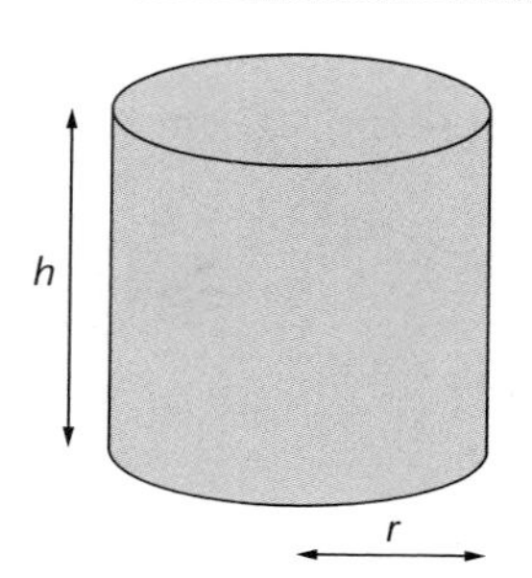

If the cylinder were made out of paper, the curved surface would open out to a rectangle.

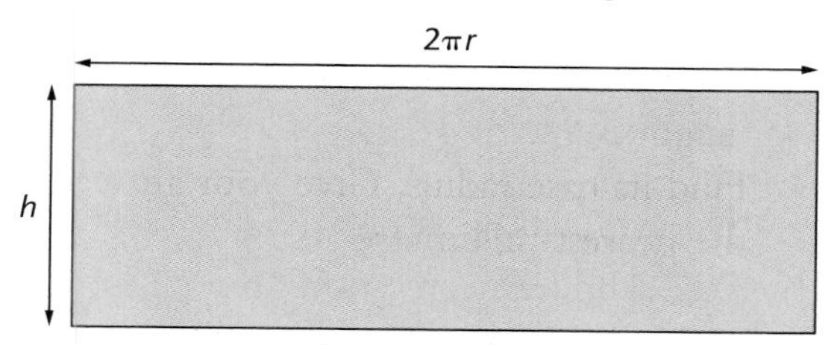

The length of the rectangle = the circumference of the cylinder
= $2\pi r$.

The curved surface area is therefore $2\pi r \times h = 2\pi rh$.

Total surface area of a cylinder = $2\pi rh + 2\pi r^2$

Example 5

Calculate the total surface area of a cylinder with base diameter 15 cm and height 10 cm.

Solution

Radius of base = $15 \div 2 = 7.5$ cm

Area of two ends = $2 \times \pi r^2 = 2 \times \pi \times 7.5^2 = 353.4\,\text{cm}^2$

Curved surface area = $2\pi rh = 2 \times \pi \times 7.5 \times 10 = 471.2\,\text{cm}^2$

Total surface area = $353.4 + 471.2 = 825\,\text{cm}^2$, to the nearest whole number.

Exam tip

Before calculating the surface area of a prism, make a rough sketch of the net of the prism. This should stop you missing out a face.

Exercise 21.2

1 Find the total surface area of each of these shapes from Exercise 21.1.

a) The cylinder in question **1**
b) The cylinder in question **2**
c) The cylinder in question **3**
d) The prism in question **4a)**
e) The cylinder in question **6**
f) The cylinder in question **7**
g) The cylinder in question **8**
h) The prism in question **9a)**

2 Find the total surface area of each of these prisms.
(You may wish to sketch the net of the prism first.)

a)

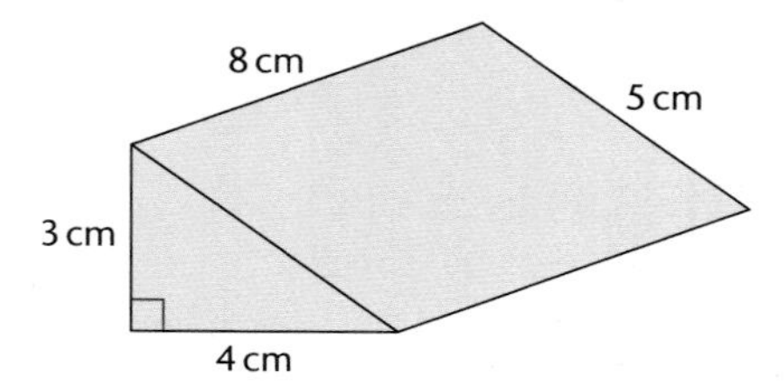

b)

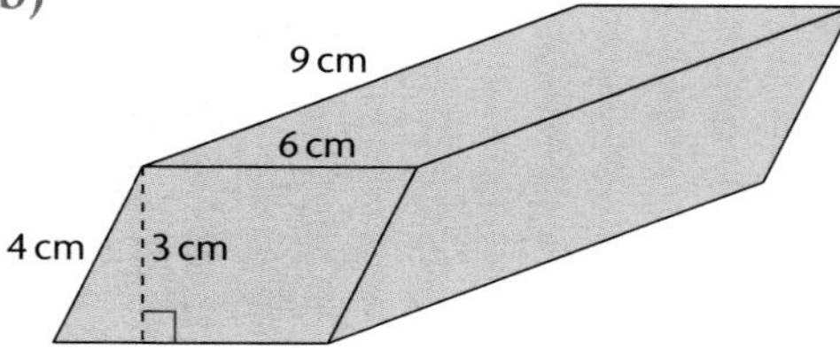

3 Find the total surface area of each of these shapes from Exercise 21.1.
(You will need your answers to Exercise 21.1.)

a) The cylinder in question **13**
b) The cylinder in question **14**
c) The cylinder in question **15**
d) The cylinder in question **18**
e) The cylinder in question **19**
f) The cylinder in question **20**

Challenge 1

A prism has a cross-section which is an equilateral triangle with sides of 6 cm.

The length of the prism is 10 cm.

Find the total surface area of the prism.

HINT

You will need to use Pythagoras' theorem to find the height of the triangle.

Challenge 2

A prism has a cross-section which is an isosceles triangle with sloping sides of 10 cm and base 16 cm.

The length of the prism is 4.5 cm.

Find the total surface area of the prism.

Converting between measures

You already know the basic relationships between **linear** metric measures. 'Linear' means 'to do with length'.

You can use these relationships to work out the relationships between metric units of area and volume.

For example:

$1\text{ cm} = 10\text{ mm}$

$1\text{ cm}^2 = 1\text{ cm} \times 1\text{ cm}$
$= 10\text{ mm} \times 10\text{ mm}$
$= 100\text{ mm}^2$

$1\text{ cm}^3 = 1\text{ cm} \times 1\text{ cm} \times 1\text{ cm}$
$= 10\text{ mm} \times 10\text{ mm} \times 10\text{ mm}$
$= 1000\text{ mm}^3$

$1\text{ m} = 100\text{ cm}$

$1\text{ m}^2 = 1\text{ m} \times 1\text{ m}$
$= 100\text{ cm} \times 100\text{ cm}$
$= 10\,000\text{ cm}^2$

$1\text{ m}^3 = 1\text{ m} \times 1\text{ m} \times 1\text{ m}$
$= 100\text{ cm} \times 100\text{ cm} \times 100\text{ cm}$
$= 1\,000\,000\text{ cm}^3$

Exam tip

1 litre = 1000 cm^3

Example 6

Change these units.

a) $5\,m^3$ to cm^3 b) $5600\,cm^2$ to m^2

Solution

a) $5\,m^3 = 5 \times 1\,000\,000\,cm^3 = 5\,000\,000\,cm^3$

Convert $1\,m^3$ to cm^3 and multiply by 5.

b) $5600\,cm^2 = 5600 \div 10\,000\,m^2 = 0.56\,m^2$

To convert from m^2 to cm^2 you multiply, so to convert from cm^2 to m^2 you divide.

Make sure you have done the right thing by checking that your answer makes sense. If you had multiplied by 10 000, you would have got $56\,000\,000\,m^2$, which is obviously a much larger area than $5600\,cm^2$.

Exercise 21.3

Change the units in questions **1** to **5**.

1 a) 25 m to cm
b) 42 cm to mm
c) 2.36 m to cm
d) 5.1 m to mm

2 a) $3\,m^2$ to cm^2
b) $2.3\,cm^2$ to mm^2
c) $9.52\,m^2$ to cm^2
d) $0.014\,cm^2$ to mm^2

3 a) $90\,000\,mm^2$ to cm^2
b) $8140\,mm^2$ to cm^2
c) $7\,200\,000\,cm^2$ to m^2
d) $94\,000\,cm^2$ to m^2

4 a) $3.2\,m^3$ to cm^3
b) $42\,cm^3$ to m^3
c) $5000\,cm^3$ to m^3
d) $6.42\,m^3$ to cm^3

5 a) 2.61 litres to cm^3
b) 9500 ml to litres
c) 2.4 litres to ml
d) 910 ml to litres

6 What is wrong with this statement?
'The trench I have just dug is 5 m long, 2 m wide and 50 cm deep.
To fill it in, I would need $500\,m^3$ of concrete.'

7 This carton holds 1 litre of juice.
How high is it?

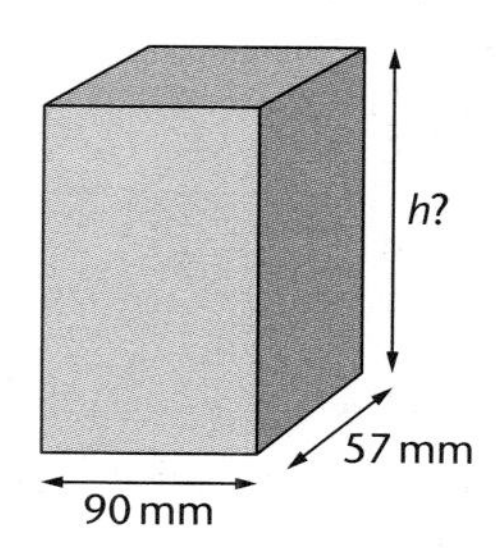

8 How many litres are there in 1 cubic metre?

Foundation Gold/ Higher Bronze

9 The cross-section of this piece of guttering is a trapezium.
The length of the guttering is 2.4 m.
How many litres can it hold?

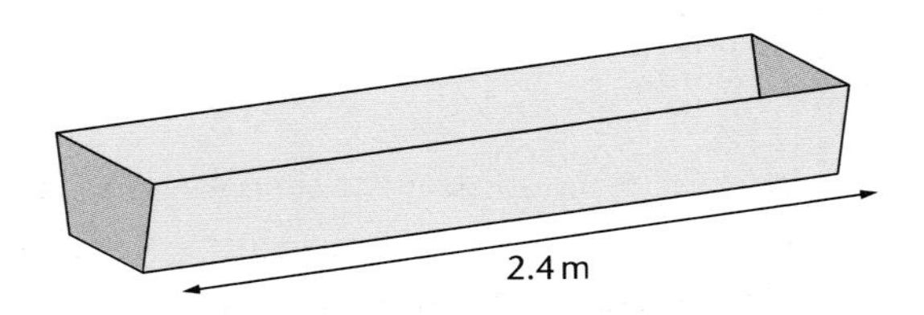

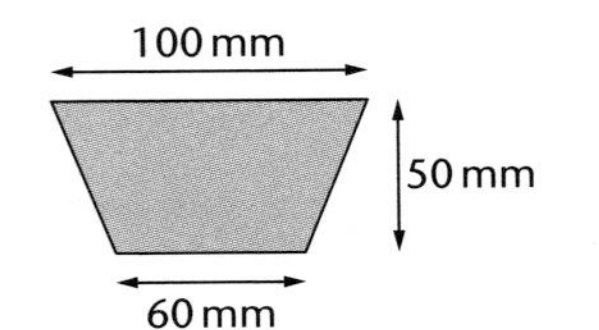

10 This barrel is a cylinder and it holds 500 litres. It is 1.5 m high.
Work out its diameter.

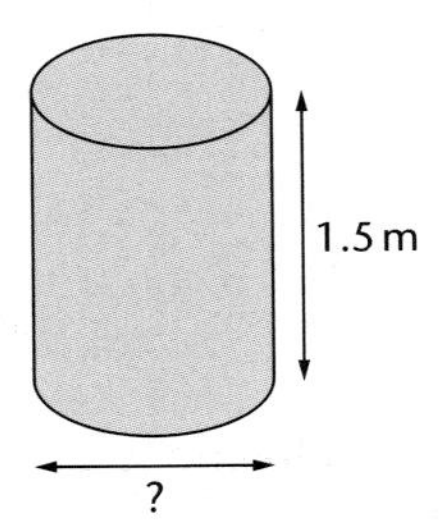

11 1 hectare = 10 000 m^2.
How many hectares are there in 1 km^2?

12 A sugar cube has sides of 15 mm.
Find how many will fit in a box measuring 11 cm by 11 cm by 5 cm.

Challenge 3

Cleopatra is reputed to have had a bath filled with asses' milk.
Today her bath might be filled with cola!
Assuming that a can of drink holds 330 millilitres, approximately how many cans would she need to have a bath in cola?

Key Ideas

- Volume of a prism = area of cross-section × length
- The total surface area of a solid is the total area of all the faces.
- 1 m^2 = 10 000 cm^2 = 1 000 000 mm^2
- 1 litre = 1000 cm^3
- 1 m^3 = 1 000 000 cm^3 = 1000 litres

Fractions

YOU WILL LEARN ABOUT	YOU SHOULD ALREADY KNOW
○ Adding and subtracting fractions including mixed numbers ○ Multiplying and dividing fractions including mixed numbers	○ How to add and subtract simple fractions ○ How to multiply and divide simple fractions ○ How to convert between improper fractions and mixed numbers

Adding and subtracting mixed numbers

To add or subtract mixed numbers, you deal with the whole numbers first.

Example 1

Work out these.

a) $1\frac{1}{4} + 2\frac{1}{2}$

b) $2\frac{3}{5} + 4\frac{2}{3}$

Solution

a) $1\frac{1}{4} + 2\frac{1}{2} = 1 + 2 + \frac{1}{4} + \frac{1}{2}$ Add the whole numbers first.

$= 3 + \frac{1}{4} + \frac{2}{4}$ Common denominator = 4.

$= 3\frac{3}{4}$

b) $2\frac{3}{5} + 4\frac{2}{3} = 6 + \frac{3}{5} + \frac{2}{3}$ Add the whole numbers first.

$= 6 + \frac{9}{15} + \frac{10}{15}$ Common denominator = 15.

$= 6\frac{19}{15}$ $\frac{19}{15}$ is improper ('top heavy').

$= 7\frac{4}{15}$ $\frac{19}{15} = 1\frac{4}{15}$

Example 2

Work out these.

a) $3\frac{3}{4} - 1\frac{1}{3}$

b) $5\frac{3}{10} - 2\frac{3}{4}$

Solution

a) $3\frac{3}{4} - 1\frac{1}{3} = 3 - 1 + \frac{3}{4} - \frac{1}{3}$ Subtract the whole numbers first.

$= 2 + \frac{9}{12} - \frac{4}{12}$ Common denominator = 12.

$= 2\frac{5}{12}$

In part **b)** there is a slight added difficulty.

b) $5\frac{3}{10} - 2\frac{3}{4} = 5 - 2 + \frac{3}{10} - \frac{3}{4}$ Subtract the whole numbers first.

$= 3 + \frac{6}{20} - \frac{15}{20}$ Common denominator = 20.

$= 2 + \frac{20}{20} + \frac{6}{20} - \frac{15}{20}$ $\frac{6}{20} - \frac{15}{20}$ is negative so take one of the

$= 2\frac{11}{20}$ whole numbers and change it to $\frac{20}{20}$.

Exercise 22.1

1 Add each of these. Write your answers as simply as possible.

a) $1\frac{1}{3} + 3\frac{1}{4}$ b) $1\frac{1}{2} + 2\frac{1}{6}$

c) $3\frac{1}{5} + \frac{7}{10}$ d) $1\frac{4}{5} + 2\frac{1}{10}$

e) $1\frac{3}{4} + 4\frac{2}{5}$ f) $6\frac{1}{6} + 1\frac{4}{9}$

g) $2\frac{5}{6} + 7\frac{4}{9}$ h) $2\frac{4}{7} + 1\frac{2}{3}$

i) $\frac{2}{7} + \frac{1}{2} + \frac{5}{14}$ j) $\frac{4}{5} + 1\frac{3}{4} + 2\frac{1}{2}$

k) $1\frac{1}{2} + \frac{3}{4} + 2\frac{3}{8}$ l) $6\frac{1}{3} + 1\frac{4}{9} + 1\frac{2}{9}$

2 Subtract each of these. Write your answers as simply as possible.

a) $2\frac{4}{5} - 1\frac{3}{5}$ b) $2\frac{2}{3} - 1\frac{1}{6}$

c) $5\frac{3}{8} - 2\frac{1}{4}$ d) $3\frac{5}{8} - 1\frac{1}{4}$

e) $3\frac{2}{3} - \frac{1}{2}$ f) $2\frac{4}{5} - \frac{1}{2}$

g) $3\frac{2}{5} - 1\frac{3}{4}$ h) $4\frac{2}{5} - 1\frac{1}{4}$

i) $5\frac{1}{6} - 3\frac{2}{3}$ j) $8\frac{1}{6} - 5\frac{3}{8}$

k) $5\frac{1}{5} - \frac{2}{3}$ l) $1\frac{1}{4} - \frac{5}{8}$

3 There were three books in a pile on Faisal's desk.
The first book was $2\frac{3}{4}$ inches high, the second $\frac{7}{8}$ inch high and the third $1\frac{5}{6}$ inches high.
What was the total height of the pile?

4 The blade of a knife was $5\frac{3}{4}$ inches long.
The handle was $4\frac{2}{5}$ inches long.
What was the total length of the knife?

5 Caroline bought an internet cable 24 inches long. She cut off two pieces, each $5\frac{5}{8}$ inches long.
How long was the piece she had left?

6 Sam had a piece of wood $28\frac{1}{2}$ inches long.
After putting up a shelf, $9\frac{5}{8}$ inches were left.
What length did he use?

Multiplying and dividing mixed numbers

To multiply and divide mixed numbers, the mixed numbers must first be changed into improper fractions.

Example 3

Work out these.

a) $2\frac{1}{2} \times 4\frac{3}{5}$

b) $2\frac{3}{4} \div 1\frac{5}{8}$

Solution

a) $2\frac{1}{2} \times 4\frac{3}{5} = \frac{5}{2} \times \frac{23}{5}$ First change the mixed numbers into improper fractions.

$= \frac{\cancel{5}^1}{2} \times \frac{23}{\cancel{5}_1}$ The arithmetic is much easier if you cancel the 5s.

$= \frac{23}{2}$ Multiply the numerators and multiply the denominators.

$= 11\frac{1}{2}$ Change the result back to a mixed number.

b) $2\frac{3}{4} \div 1\frac{5}{8} = \frac{11}{4} \div \frac{13}{8}$ Change the mixed numbers to improper fractions.

$= \frac{11}{\cancel{4}_1} \times \frac{\cancel{8}^2}{13}$ Invert the **second** fraction and multiply.

The arithmetic is much easier if you cancel by 4.

$= \frac{22}{13}$ Multiply the numerators and multiply the denominators.

$= 1\frac{9}{13}$ Change back to a mixed number.

Exam tip

A common error is to multiply the whole numbers first. Dealing with the whole numbers separately can only be done with addition and subtraction.

Note that if you are multiplying or dividing by a whole number like 6, you can write it as $\frac{6}{1}$.

Exam tip

When cancelling, divide a term in the numerator and a term in the denominator by the same number.
Only cancel a division calculation when it is at the multiplication stage.

Exercise 22.2

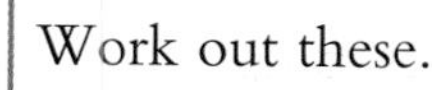

Work out these.

1 a) $4\frac{1}{2} \times 2\frac{1}{6}$ b) $1\frac{1}{2} \times 3\frac{2}{3}$ c) $4\frac{1}{5} \times 1\frac{2}{3}$ d) $3\frac{1}{3} \times 2\frac{2}{5}$
e) $2\frac{2}{5} \times \frac{3}{4}$ f) $3\frac{1}{5} \times 1\frac{2}{3}$

2 a) $2\frac{1}{3} \div 1\frac{1}{3}$ b) $2\frac{2}{5} \div 1\frac{1}{2}$ c) $3\frac{1}{5} \div \frac{4}{15}$ d) $3\frac{1}{8} \div 1\frac{1}{4}$
e) $2\frac{1}{4} \div 3\frac{1}{2}$ f) $1\frac{1}{5} \div \frac{4}{15}$

3 a) $3\frac{1}{2} \times 2\frac{1}{5}$ b) $4\frac{2}{7} \times \frac{1}{2}$ c) $2\frac{3}{4} \div 1\frac{3}{4}$ d) $1\frac{5}{12} \div 3\frac{1}{3}$
e) $3\frac{1}{5} \times 2\frac{5}{8}$ f) $2\frac{7}{8} \div 1\frac{3}{4}$ g) $2\frac{7}{9} \times 3\frac{3}{5}$ h) $5\frac{5}{6} \div 1\frac{3}{4}$
i) $3\frac{5}{7} \times 2\frac{1}{13}$ j) $5\frac{2}{5} \div 2\frac{1}{4}$ k) $5\frac{2}{7} \times 3\frac{1}{2}$ l) $4\frac{1}{12} \div 3\frac{1}{4}$

4 a) $2\frac{1}{2} \times 1\frac{1}{3} \times 1\frac{3}{8}$ b) $1\frac{1}{2} \times 2\frac{2}{3} \div 1\frac{3}{5}$ c) $1\frac{1}{4} \times 3\frac{1}{5} \div 1\frac{1}{2}$ d) $3\frac{1}{2} \times 1\frac{2}{3} \times \frac{5}{7}$
e) $1\frac{1}{4} \times 1\frac{2}{3} \div 1\frac{1}{9}$ f) $3\frac{1}{3} \times 1\frac{1}{4} \div 2\frac{1}{2}$

Challenge 1

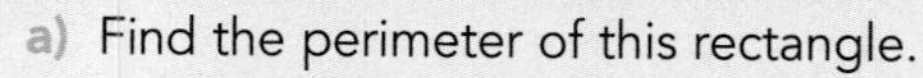

a) Find the perimeter of this rectangle.
b) Find the area of this rectangle.

$5\frac{1}{4}$ cm

$3\frac{2}{3}$ cm

Challenge 2

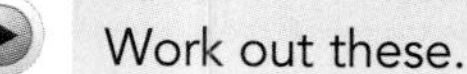

Work out these.

a) $(3\frac{1}{2} + 2\frac{4}{5}) \times 2\frac{1}{2}$ b) $5\frac{1}{3} \div 3\frac{3}{5} + 2\frac{1}{3}$
c) $4\frac{2}{3} \times 2\frac{2}{7} - 4\frac{7}{8}$ d) $(2\frac{4}{5} + 3\frac{1}{4}) \div (3\frac{2}{3} - 2\frac{3}{4})$

Foundation Gold/ Higher Bronze

Activity 1

Fractions on your calculator

You need to be able to calculate with fractions without a calculator. However, when a calculator is allowed you can use the fraction button.

The fraction button looks like this. [a b/c]

To enter a fraction such as $\frac{2}{5}$ into your calculator you press

[2] [a b/c] [5] [=].

Your display will look like this. [2⌋5]

To do a calculation like $\frac{2}{5} + \frac{1}{2}$, the sequence of buttons is

[2] [a b/c] [5] [+] [1] [a b/c] [2] [=].

This is what you should see on your display.

[9⌋10]

To enter a mixed number such as $2\frac{3}{5}$ into your calculator you press

[2] [a b/c] [3] [a b/c] [5] [=].

Your display will look like this. [2⌋3⌋5]

You can also cancel a fraction on your calculator.

When you press [8] [a b/c] [1] [2], you should see [8⌋12].

When you press [=] the display changes to [2⌋3].

When you do calculations with fractions on your calculator, it will automatically give the answer as a fraction in its lowest terms.

Similarly, if you enter an improper fraction into your calculator and press the [=] button, the calculator will automatically change it to a mixed number.

Try working through some of the examples and questions earlier in this chapter using your calculator.

Exam tip

Your calculator may have a different fraction button, e.g. [▭/▭]. Make sure you can calculate with fractions on your own calculator

Key Ideas

- When you add or subtract mixed numbers, you deal with the whole numbers first and then the fraction parts.
- To multiply and divide mixed numbers, you must change the mixed numbers to improper ('top heavy') fractions.

23 Transformations 2

YOU WILL LEARN ABOUT	YOU SHOULD ALREADY KNOW
○ Recognising, describing and carrying out more difficult rotations, reflections, translations and enlargements ○ Using column vectors to describe translations ○ Using negative scale factors ○ Combining transformations and describing the results	○ How to recognise and draw the lines $y = x$ and $y = {}^{-}x$ ○ The terms 'object' and 'image' as they apply to transformations ○ How to recognise and draw reflections ○ How to rotate a shape about a point on the shape or the origin through 90°, 180° or 270° ○ How to recognise and draw translations ○ How to recognise and draw an enlargement of a shape using a centre and scale factor of enlargement

Reflections

When a transformation is a reflection, a tracing of the object can be turned over and will fit on the image.

To describe a reflection, you need to find and give the mirror line.

Example 1

Describe the transformation that maps shape ABC on to shape PQR.

Exam tip

Tracing paper is optional in an examination. Always ask for it when doing transformation questions.

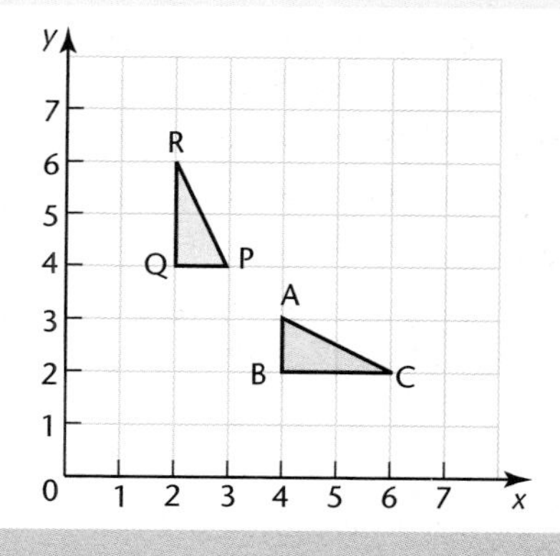

Solution

You can check that the transformation is a reflection by using tracing paper.

To find the mirror line, put a ruler between two corresponding points (B and Q) and mark a point halfway between them, at (3, 3).

Repeat this for two other corresponding points (C and R). The midpoint is (4, 4).

Join the two midpoints to find the mirror line. The mirror line is $y = x$.

The transformation is a reflection in the line $y = x$.
Again, the result can be checked using tracing paper.

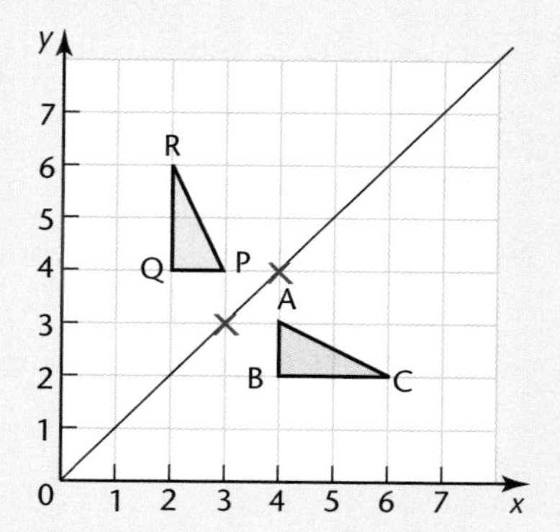

Rotations

When a transformation is a rotation, it should be possible to place a tracing of the object over the image without turning the tracing paper over.

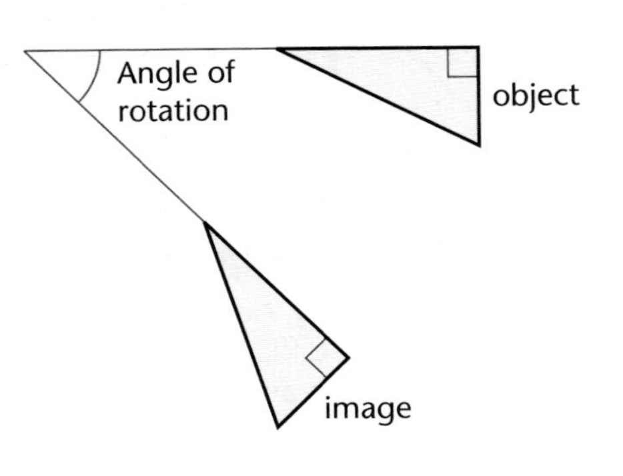

To describe a rotation, you need to give the angle of rotation and the centre of rotation.

To find the angle of rotation, find a pair of sides that correspond in the object and the image. Measure the angle between them. You may need to extend both of these sides to do this.

You can usually find the centre of rotation, either by counting squares or using tracing paper.

Example 2

Describe fully the transformation that maps triangle A on to triangle B.

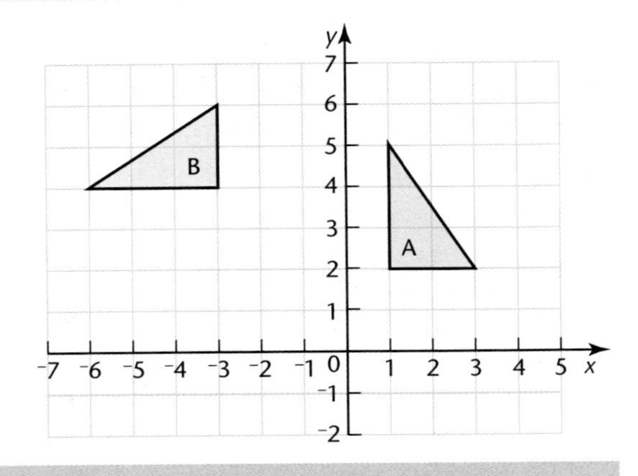

Exam tip

Always remember to state whether the rotation is clockwise or anticlockwise.

Solution

The transformation is a rotation.

The angle is 90° anticlockwise.

You may need to make a few trials, using tracing paper and a compass point centred on different points, to find that the centre of rotation is $(^-2, 1)$.

Exercise 23.1

Label the diagrams you draw in this exercise carefully and keep them as you will need them in a later exercise.

1 a) Plot the points $(1, 0)$, $(1, ^-2)$ and $(2, ^-2)$ and join them to form a triangle.
Label it A.
Reflect triangle A in the line $y = 1$.
Label the image B.

b) Reflect triangle B in the line $y = x$.
Label the image C.

2 a) On a new grid, plot the points $(2, 5)$, $(3, 5)$ and $(1, 3)$ and join them to form a triangle.
Label it D.
Reflect triangle D in the line $x = \frac{1}{2}$.
Label the image E.

b) On the same grid, reflect triangle E in the line $y = {}^-x$.
Label the image F.

3 a) Copy this diagram.
Rotate the flag G through 90° clockwise about the point (1, 2).
Label the image H.

b) On the same grid, rotate the flag H through 180° about the point (2, −1).
Label the image I.

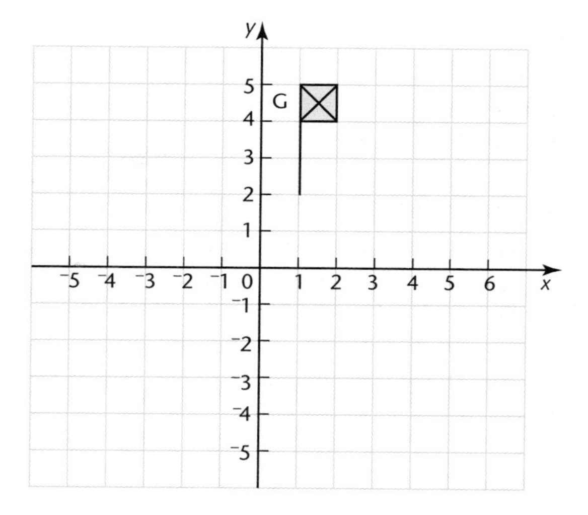

4 a) Plot the points (0, 1), (0, 4) and (2, 3) and join them to form a triangle. Label it J.
Rotate triangle J through 90° anticlockwise about the point (2, 3). Label the image K.

b) On the same grid, rotate triangle K through 90° clockwise about the point (2, −1).
Label the image L.

5 Look at this diagram.
Which of the triangles B, C, D, E, F and G are reflections of triangle A and which are rotations of triangle A?

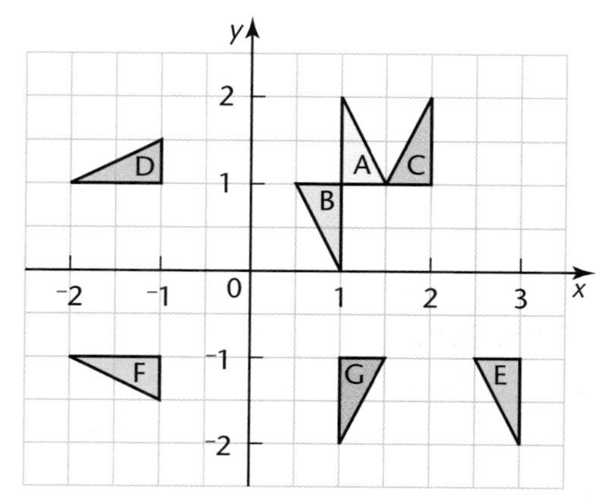

6 Look at this diagram.
Describe fully these single transformations.

a) Triangle A on to triangle B
b) Triangle A on to triangle C
c) Triangle A on to triangle D
d) Triangle D on to triangle E
e) Triangle E on to triangle F
f) Triangle D on to triangle G

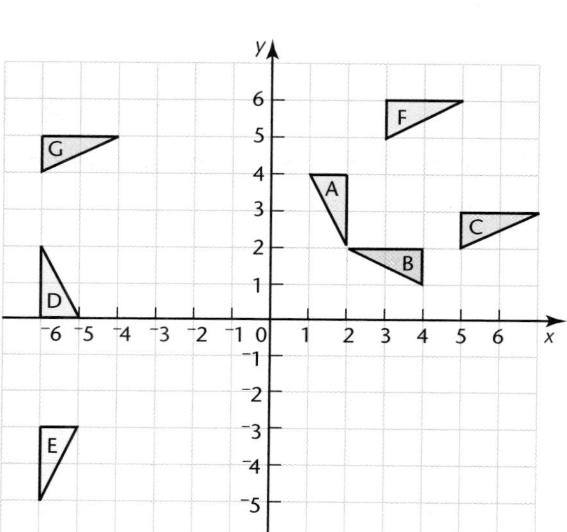

Exam tip

When describing transformations, always state the type of transformation first and then give all the necessary extra information. For reflections this is the mirror line; for rotations it is angle, direction and centre.
You may get no marks unless you name the transformation.

Translations

In a translation, every point of an object moves the same distance in the same direction. The object and the image look identical with no turning or reflection.

The movements can be described as a movement in the *x*-direction and a movement in the *y*-direction. They can be written in the form of a column vector, for example $\begin{pmatrix} 5 \\ -3 \end{pmatrix}$.

In a column vector, the top number represents the *x*-movement and the bottom number represents the *y*-movement.

Example 3

Describe fully the transformation that maps triangle ABC on to triangle PQR.

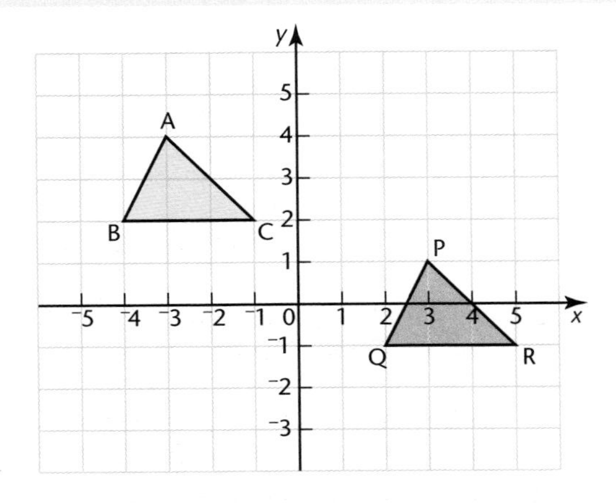

Solution

Point A is translated on to point P. This is a movement of 6 to the right and 3 down.

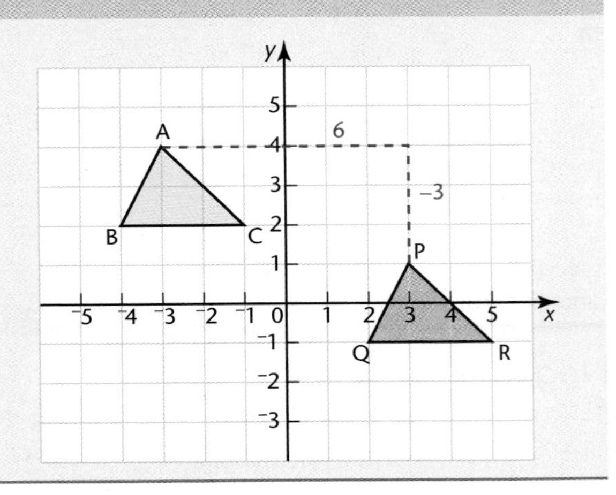

The transformation is a translation by the vector $\begin{pmatrix} 6 \\ -3 \end{pmatrix}$.

Enlargements

An **enlargement** produces an image that is exactly similar in shape to the object, but the image is larger or smaller.

Drawing enlargements

Example 4

Enlarge the triangle ABC with scale factor 2.5 and centre of enlargement O, to form triangle PQR.

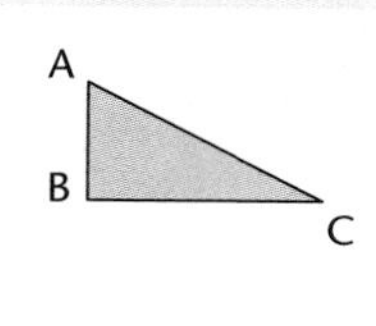

Solution

Draw lines from O to A, O to B and O to C and extend them.

Measure the lengths OA, OB and OC. These are 2.0 cm, 1.5 cm and 2.9 cm respectively.

Multiply these lengths by 2.5 to give OP = 5.0 cm, OQ = 3.75 cm and OR = 7.25 cm.

Measure these distances along the extended lines OA, OB and OC, and mark P, Q and R.

Join P, Q and R to form the triangle.

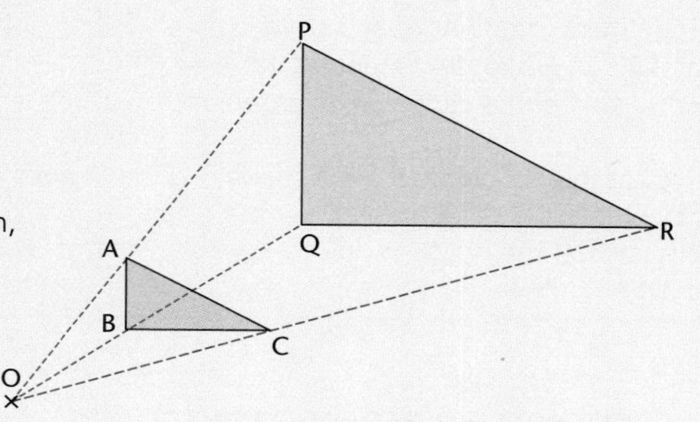

Exam tip

When you have drawn your enlargement, check that the ratio of the sides of the image to the corresponding sides in the object is equal to the scale factor (in this case, 2.5).
If it is not, you have probably measured some or all of your distances from the points of the object and not from O.

Negative scale factors

If the scale factor of an enlargement is negative, the image is on the opposite side of the centre of enlargement from the object, and the image is inverted. This is shown in the next example.

Example 5

Plot the coordinates A(2, 2), B(4, 2) and C(2, 5) and join them to form a triangle.

Enlarge triangle ABC by a scale factor of ⁻3 using the point O(1, 1) as the centre of enlargement.

Solution

Plot the triangle.

Then draw a line from each of the vertices through the centre of enlargement O and extend it on the other side.

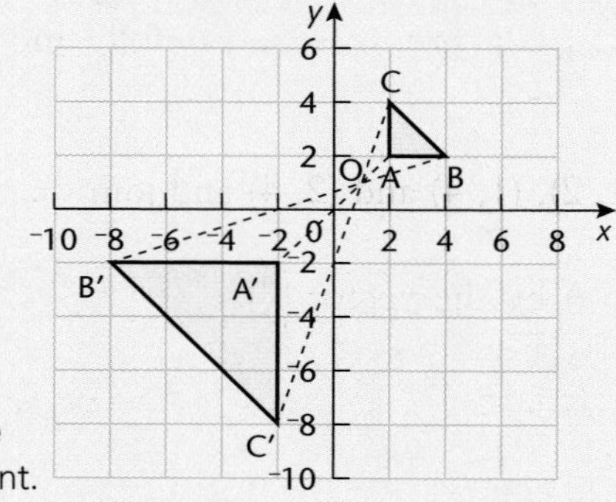

Measure the distance from the vertex A and multiply it by 3.

Then mark the point A′ at a distance of 3 × OA along OA extended on the other side of the centre of enlargement.

Repeat this for points B and C, labelling the images B′ and C′.

Notice that the image is on the opposite side of O.
It is three times the size of the object ABC and is inverted.

Exam tip

When the scale factor is negative, the image will be on the opposite side from the object.

For example, from the centre of enlargement to point B is 3 units in the positive x-direction and 1 unit in the positive y-direction. Multiplying by the scale factor, point B′ is 9 units in the negative x-direction and 3 units in the negative y-direction.

Describing enlargements

To describe an enlargement, you need to give the scale factor of the enlargement and the centre of the enlargement.

Example 6

Describe fully the transformation that maps triangle DEF on to triangle STU.

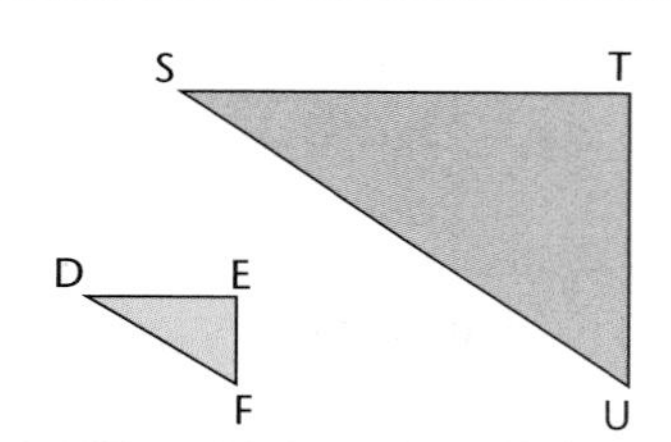

Solution

The shapes are similar, so the transformation is an enlargement.

Since the lengths of the sides of triangle STU are 3 times the lengths of the corresponding sides of triangle DEF, the scale factor is 3.

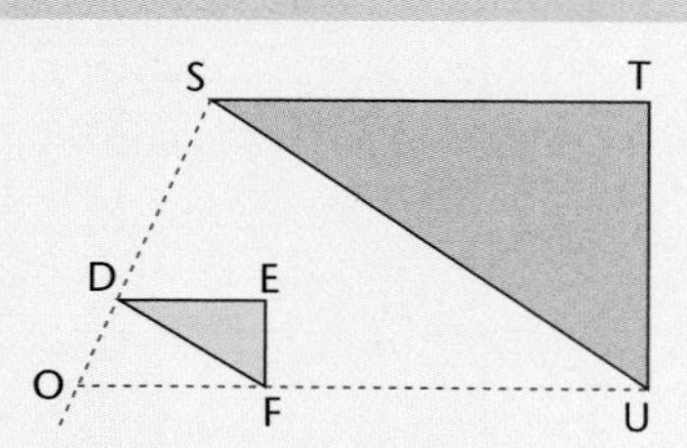

All that remains to be found is the centre of enlargement.

Join SD and extend it. Join UF and extend it to cross the extended line SD. The point where the lines cross, O, is the centre of enlargement.

The transformation is an enlargement, scale factor 3, centre of enlargement O.

If you were working on a grid, you would describe the centre of enlargement by stating the coordinates of the point.

The transformation that maps triangle STU on to triangle DEF in Example 6 is an enlargement with centre O and a scale factor of $\frac{1}{3}$. Note that it is still called an enlargement even though the image is smaller than the object.

Exercise 23.2

Label the diagrams you draw in this exercise carefully and keep them as you will need them in a later exercise.

1 a) Plot the points (1, 2), (1, 4) and (2, 4) and join them to form a triangle.
Label it A.
Translate triangle A by the vector $\begin{pmatrix} 5 \\ 2 \end{pmatrix}$.
Label the image B.

b) On the same grid, translate triangle B by the vector $\begin{pmatrix} 2 \\ ^-4 \end{pmatrix}$.
Label the image C.

2 a) On a new grid, plot the points (0, 2), (1, 4) and (3, 2) and join them to form a triangle.
Label it D.
Translate triangle D by the vector $\begin{pmatrix} ^-4 \\ 2 \end{pmatrix}$.
Label it E.

b) On the same grid, translate triangle E by the vector $\begin{pmatrix} 8 \\ 0 \end{pmatrix}$.
Label the image F.

3 a) Draw a set of axes. Label the x-axis from 0 to 13 and the y-axis from 0 to 15.
Plot the points (1, 2), (2, 4) and (1, 3) and join them to form a triangle.
Label it G.
Enlarge triangle G with scale factor 2 and centre the origin.
Label the image H.

b) On the same grid, enlarge triangle H with scale factor 3 and centre of enlargement (0, 5).
Label the image I.

4 Draw a set of axes, the x-axis from $^-16$ to 6 and the y-axis from $^-8$ to 4.
Plot the points A(2, 2), B(5, 0), C(5, $^-1$), D(2, $^-1$) and join them to form a quadrilateral.
Enlarge this quadrilateral with scale factor $^-3$ and centre of enlargement the origin.

5 Draw a set of axes, the x-axis from $^-16$ to 6 and the y-axis from $^-8$ to 4.
Plot the points P(4, 2), Q(6, $^-1$), R(2, $^-2$) and join them to form a triangle.
Enlarge this triangle with scale factor $^-2$ and centre of enlargement the origin.

6 Draw a set of axes, the x-axis from $^-14$ to 6 and the y-axis from $^-8$ to 4.
Plot the points A(2, $^-2$), B(5, 1), C(2, 2) and join them to form a triangle.
Enlarge this triangle with scale factor $^-2$ and centre of enlargement ($^-1$, $^-1$).
Write down the coordinates of the image.

7 Draw a set of axes, the x-axis from 0 to 10 and the y-axis from 0 to 8. Plot the points A(1, 6), B(1, 7), C(3, 7) and join them to form a triangle. Enlarge this triangle with scale factor $^{-}3$ and centre of enlargement (3, 6). Write down the coordinates of the image.

8 Look at this diagram. Describe fully these single transformations.

a) Triangle A on to triangle B
b) Triangle A on to triangle C
c) Triangle C on to triangle D
d) Triangle A on to triangle E
e) Triangle A on to triangle F
f) Triangle G on to triangle A

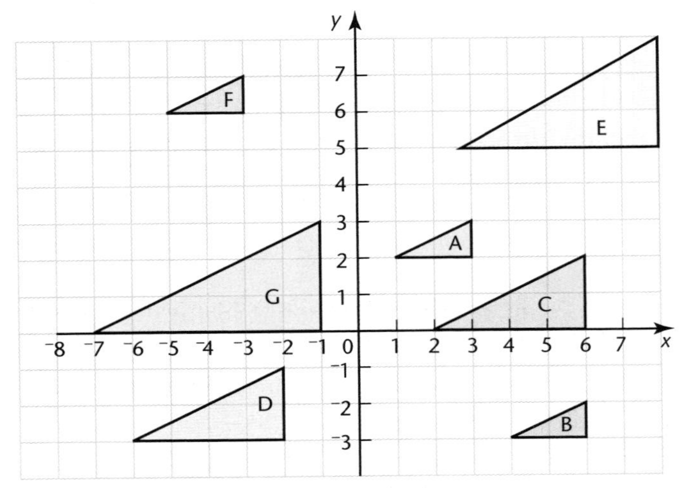

Challenge 1

A triangle ABC has sides AB = 9 cm, AC = 7 cm and BC = 6 cm.

A line XY is drawn parallel to BC through a point X on AB and a point Y on AC.

AX = 5 cm.

a) Draw a sketch of the triangles.

b) (i) Describe fully the transformation that maps ABC on to AXY.

(ii) Work out the length of XY correct to 2 decimal places.

Combining transformations

Sometimes when one transformation is followed by another, the result is equivalent to a single transformation.

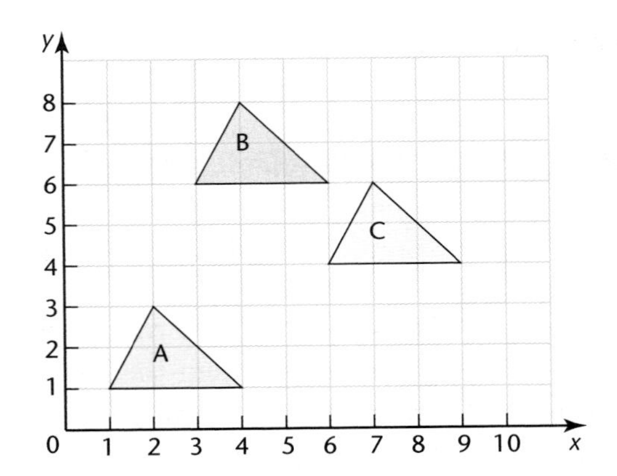

For example, in the diagram, triangle A has been translated by the vector $\begin{pmatrix} 2 \\ 5 \end{pmatrix}$ on to triangle B.

Triangle B has then been translated by the vector $\begin{pmatrix} 3 \\ -2 \end{pmatrix}$ on to triangle C.

Notice that triangle A could have been translated directly on to triangle C by the vector $\begin{pmatrix} 5 \\ 3 \end{pmatrix}$.

So the first transformation followed by the second transformation is equivalent to the single transformation: translation by the vector $\begin{pmatrix} 5 \\ 3 \end{pmatrix}$.

Exam tip

Make sure you do the transformations in the right order, as it usually makes a difference.
If a question asks for a single transformation, do not give a combination of two transformations as this does not answer the question and will usually score no marks.

Example 7

Find the single transformation that is equivalent to a reflection in the line $x = 1$, followed by a reflection in the line $y = {}^-2$.

Solution

Choose a simple shape and draw a diagram. An asymmetrical shape such as a right-angled triangle or a flag is usually best.

In the diagram, reflecting the object flag A in the line $x = 1$ gives flag B.

Reflecting flag B in the line $y = {}^-2$ gives flag C.

The transformation that maps A directly on to C is a rotation through 180°.

The centre of rotation is (1, ⁻2), which is where the mirror lines cross.

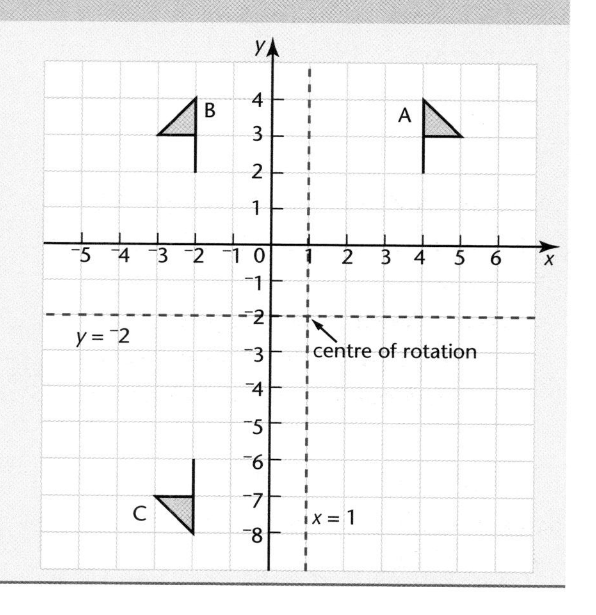

The transformation is a rotation through 180° about the centre of rotation (1, ⁻2).

Use tracing paper to check this.

A rotation of 180° is the only rotation for which you do not need to state the direction, as 180° clockwise is the same as 180° anticlockwise.

Exercise 23.3

In this exercise you will need some of the diagrams you drew in Exercises 23.1 and 23.2.

1 Look back at the diagram you drew for question **1** of Exercise 23.2.
Describe fully the single transformation that is equivalent to a translation by the vector $\begin{pmatrix} 5 \\ 2 \end{pmatrix}$ (A on to B) followed by a translation by the vector $\begin{pmatrix} 2 \\ {}^-4 \end{pmatrix}$ (B on to C).

2 Look back at the diagram you drew for question **2** of Exercise 23.2.
Describe fully the single transformation that is equivalent to a translation by the vector $\begin{pmatrix} {}^-4 \\ 2 \end{pmatrix}$ (D on to E) followed by a translation by the vector $\begin{pmatrix} 8 \\ 0 \end{pmatrix}$ (E on to F).

Foundation Gold/ Higher Bronze

3 Look back at the diagram you drew for question **3** of Exercise 23.2.
Describe fully the single transformation that is equivalent to an enlargement, scale factor 2, centre the origin (G on to H) followed by an enlargement scale factor 3, centre (0, 5) (H on to I).

4 Look back at the diagram you drew for question **1** of Exercise 23.1.
Describe fully the single transformation that is equivalent to a reflection in the line $y = 1$ (A on to B) followed by a reflection in the line $y = x$ (B on to C).

5 Look back at the diagram you drew for question **2** of Exercise 23.1.
Describe fully the single transformation that is equivalent to a reflection in the line $x = \frac{1}{2}$ (D on to E) followed by a reflection in the line $y = {}^{-}x$ (E on to F).

6 Look back at the diagram you drew for question **3** of Exercise 23.1.
Describe fully the single transformation that is equivalent to a rotation through 90° clockwise about the point (1, 2) (G on to H) followed by a rotation through 180° about the point (2, ⁻1) (H on to I).

7 Look back at the diagram you drew for question **4** of Exercise 23.1.
Describe fully the single transformation that is equivalent to a rotation through 90° anticlockwise about the point (2, 3) (J on to K) followed by a rotation through 90° clockwise about the point (2, ⁻1) (K on to L).

8 Copy the diagram.

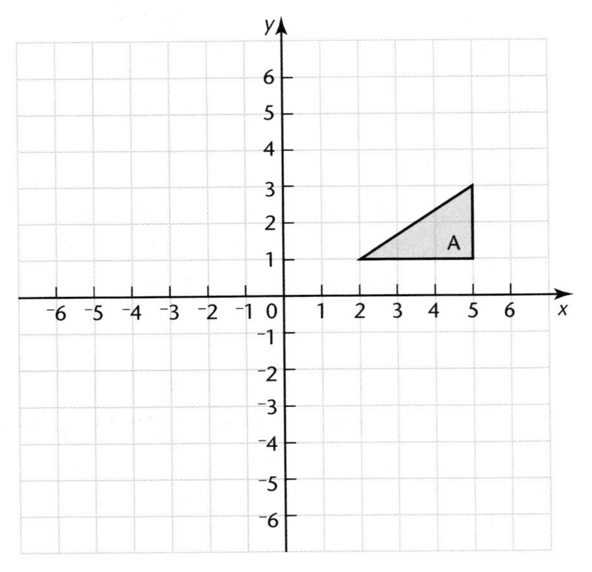

a) Reflect triangle A in the x-axis. Label the image B.
b) Reflect triangle B in the y-axis. Label the image C.
c) Describe fully the single transformation that will map triangle A on to triangle C.

In questions **9** to **15**, carry out the transformations on a simple shape of your choice.

9 Describe fully the single transformation that is equivalent to a reflection in the line $x = 1$ followed by a reflection in the line $x = 5$.

10 Describe fully the single transformation that is equivalent to a reflection in the line $y = 2$ followed by a reflection in the line $y = 6$.

11 Describe fully the single transformation that is equivalent to an enlargement, scale factor 2 and centre the origin, followed by a translation by the vector $\begin{pmatrix} 3 \\ 2 \end{pmatrix}$.

12 Describe fully the single transformation that is equivalent to a rotation through 90° clockwise about the origin, followed by a translation by the vector $\begin{pmatrix}4\\0\end{pmatrix}$.

13 Describe fully the single transformation that is equivalent to a reflection in the x-axis followed by a rotation through 90° anticlockwise about the origin.

14 Describe fully the single transformation that is equivalent to a reflection in the y-axis followed by a rotation through 90° anticlockwise about the origin.

15 Describe fully the single transformation that is equivalent to a reflection in the line $y = x$ followed by a reflection in the line $y = {}^{-}x$.

Challenge 2

Look at your answers to Exercise 23.3 questions 9 and 10.

Try to make a general statement about the result of reflection in a mirror line followed by reflection in a parallel mirror line.

Challenge 3

Look at the answers to Exercise 23.3 questions 4 and 5.

Try to make a general statement about the result of reflection in a mirror line followed by reflection in an intersecting mirror line.

Challenge 4

Draw a pair of axes and label them $^{-}6$ to 6 for x and y.

a) Draw a shape in the positive region near the origin. Label it A.

b) Translate shape A by vector $\begin{pmatrix}2\\1\end{pmatrix}$. Label it B.

c) Translate shape B by vector $\begin{pmatrix}3\\-2\end{pmatrix}$. Label it C.

d) Translate shape C by vector $\begin{pmatrix}-6\\-1\end{pmatrix}$. Label it D.

e) Translate shape D by vector $\begin{pmatrix}1\\2\end{pmatrix}$. Label it E.

f) What do you notice about shapes A and E? Can you suggest why this happens?
Try to find other combinations of translations for which this happens.

Foundation Gold/ Higher Bronze

Challenge 5

Look at your answers to Exercise 23.3, questions 1 and 2.

Try to make a general statement about the result of translation by the vector $\begin{pmatrix} a \\ b \end{pmatrix}$ followed by translation by the vector $\begin{pmatrix} c \\ d \end{pmatrix}$.

Challenge 6

Look again at the answers to Exercise 23.3, question 3.

Try to make a general statement about the result of enlargement with scale factor p followed by enlargement with scale factor q.

Key Ideas

- When describing transformations, always give the name of the transformation and then the extra information required.

Name of transformation	Extra information
Reflection	Mirror line
Rotation	Angle, direction, centre of rotation
Translation	Column vector
Enlargement	Scale factor, centre of enlargement

- When asked to describe a **single** transformation, do not give a combination of transformations.

Index